# Earth Observations for Environmental Sustainability for the Next Decade

# Earth Observations for Environmental Sustainability for the Next Decade

Editors

**Yuei-An Liou**
**Steven C. Reising**
**Yuriy Kuleshov**
**Chung-Ru Ho**
**Kim-Anh Nguyen**

MDPI • Basel • Beijing • Wuhan • Barcelona • Belgrade • Manchester • Tokyo • Cluj • Tianjin

*Editors*

Yuei-An Liou
National Central University (NCU)
Taiwan

Steven C. Reising
Colorado State University
USA

Yuriy Kuleshov
Australian Bureau of Meteorology
Australia

Chung-Ru Ho
National Taiwan Ocean University
Taiwan

Kim-Anh Nguyen
National Central University
Taiwan

*Editorial Office*
MDPI
St. Alban-Anlage 66
4052 Basel, Switzerland

This is a reprint of articles from the Special Issue published online in the open access journal *Remote Sensing* (ISSN 2072-4292) (available at: https://www.mdpi.com/journal/remotesensing/special_issues/EOES_rs).

For citation purposes, cite each article independently as indicated on the article page online and as indicated below:

LastName, A.A.; LastName, B.B.; LastName, C.C. Article Title. *Journal Name* **Year**, *Volume Number*, Page Range.

**ISBN 978-3-0365-1982-1 (Hbk)**
**ISBN 978-3-0365-1983-8 (PDF)**

© 2021 by the authors. Articles in this book are Open Access and distributed under the Creative Commons Attribution (CC BY) license, which allows users to download, copy and build upon published articles, as long as the author and publisher are properly credited, which ensures maximum dissemination and a wider impact of our publications.

The book as a whole is distributed by MDPI under the terms and conditions of the Creative Commons license CC BY-NC-ND.

# Contents

**About the Editors** . . . . . . . . . . . . . . . . . . . . . . . . . . . . . . . . . . . . . . . . . . . . . . . . . . . . . . . . . . . . . . . . . . vii

**Yuei-An Liou, Yuriy Kuleshov, Chung-Ru Ho, Kim-Anh Nguyen and Steven C. Reising**
Preface: Earth Observations for Environmental Sustainability for the Next Decade
Reprinted from: *Remote Sensing* 2021, 13, 2871, doi:10.3390/rs13152871 . . . . . . . . . . . . . . . 1

**Swapan Talukdar, Pankaj Singha, Susanta Mahato, Shahfahad, Swades Pal, Yuei-An Liou and Atiqur Rahman**
Land-Use Land-Cover Classification by Machine Learning Classifiers for Satellite Observations —A Review
Reprinted from: *Remote Sensing* 2020, 12, 1135, doi:10.3390/rs12071135 . . . . . . . . . . . . . . . 5

**Yonghong Zhang, Taotao Ge, Wei Tian and Yuei-An Liou**
Debris Flow Susceptibility Mapping Using Machine-Learning Techniques in Shigatse Area, China
Reprinted from: *Remote Sensing* 2019, 11, 2801, doi:10.3390/rs11232801 . . . . . . . . . . . . . . . 29

**Ke Xiong, Basanta Raj Adhikari, Constantine A. Stamatopoulos, Yu Zhan, Shaolin Wu, Zhongtao Dong and Baofeng Di**
Comparison of Different Machine Learning Methods for Debris Flow Susceptibility Mapping: A Case Study in the Sichuan Province, China
Reprinted from: *Remote Sensing* 2020, 12, 295, doi:10.3390/rs12020295 . . . . . . . . . . . . . . . . 55

**Zhi-Weng Chua, Yuriy Kuleshov and Andrew Watkins**
Evaluation of Satellite Precipitation Estimates over Australia
Reprinted from: *Remote Sensing* 2020, 12, 678, doi:10.3390/rs12040678 . . . . . . . . . . . . . . . . 75

**Yuei-An Liou and Getachew Mehabie Mulualem**
Spatio–temporal Assessment of Drought in Ethiopia and the Impact of Recent Intense Droughts
Reprinted from: *Remote Sensing* 2019, 11, 1828, doi:10.3390/rs11151828 . . . . . . . . . . . . . . . 93

**Ying-Nong Chen**
Multiple Kernel Feature Line Embedding for Hyperspectral Image Classification
Reprinted from: *Remote Sensing* 2019, 11, 2892, doi:10.3390/rs11242892 . . . . . . . . . . . . . . . 113

**Po-Chun Hsu, Chia-Ying Ho, Hung-Jen Lee, Ching-Yuan Lu and Chung-Ru Ho**
Temporal Variation and Spatial Structure of the Kuroshio-Induced Submesoscale Island Vortices Observed from GCOM-C and Himawari-8 Data
Reprinted from: *Remote Sensing* 2020, 12, 883, doi:10.3390/rs12050883 . . . . . . . . . . . . . . . . 133

**Kuo-Wei Lan, Li-Jhih Lian, Chun-Huei Li, Po-Yuan Hsiao and Sha-Yan Cheng**
Validation of a Primary Production Algorithm of Vertically Generalized Production Model Derived from Multi-Satellite Data around the Waters of Taiwan
Reprinted from: *Remote Sensing* 2020, 12, 1627, doi:10.3390/rs12101627 . . . . . . . . . . . . . . . 151

**Yuei-An Liou, Ji-Chyun Liu, Chung-Chih Liu, Chun-Hsu Chen, Kim-Anh Nguyen and James P. Terry**
Consecutive Dual-Vortex Interactions between Quadruple Typhoons Noru, Kulap, Nesat and Haitang during the 2017 North Pacific Typhoon Season
Reprinted from: *Remote Sensing* 2019, 11, 1843, doi:10.3390/rs11161843 . . . . . . . . . . . . . . . 167

**Jehyeok Rew, Yongjang Cho, Jihoon Moon and Eenjun Hwang**
Habitat Suitability Estimation Using a Two-Stage Ensemble Approach
Reprinted from: *Remote Sensing* **2020**, *12*, 1475, doi:10.3390/rs12091475 . . . . . . . . . . . . . . . . **185**

# About the Editors

**Yuei-An Liou** is a Distinguished Professor and Head of Hydrology Remote Sensing Laboratory, Center for Space and Remote Sensing Research, National Central University, Taiwan; Founder & Honorary President, Taiwan Group on Earth Observations (2016/8~); Honorary President, Vietnamese Experts Association in Taiwan (2017/1~). Dr. Liou received the awards as Foreign Member, Prokhorov Russian Academy of Engineering Sciences in 2008; Outstanding Alumni Awards, University of Michigan Alumni Association in Taiwan & National Sun Yat-sen University in 2008; Member, International Academy of Astronautics in 2014; Fellow, The Institution of Engineering and Technology in 2015; Crystal Achievement Award in 2019/11, Vietnam Academy of Science and Technology, Vietnam; Outstanding Research Award, Ministry of Science and Technology, Taiwan, 2019.

**Yuriy Kuleshov** is a Science Lead of the Climate Risk and Early Warning Systems (CREWS) at the Australian Bureau of Meteorology. Working for the Bureau from 1995, he led the Climate Change and Tropical Cyclones International Initiative and a number of climate programs of the International Climate Change Adaptation Initiative, among others. For the past two decades, he also worked for numerous expert and task teams of the World Meteorological Organization (WMO). Currently, he is Chairman of the Steering Group for the Space-based Weather and Climate Extreme Monitoring (SWCEM), implementing this WMO flagship initiative in countries of East Asia and the Pacific. Working at the Department of Satellite Remote Sensing of the Earth Environment, Academy of Sciences, USSR, in 1981–1994, he developed novel methods and microwave instruments for satellite remote sensing, including the world's first operational space-based radar of the Cosmos-1500 satellite.

**Chung-Ru Ho** received his Ph.D. in Applied Ocean Science from the University of Delaware, USA in 1994. He is currently Professor of the Department of Marine Environmental Informatics, National Taiwan Ocean University, Taiwan. He has served Deputy Director General of National Museum of Marine Science and Technology, Taiwan. Ho is currently serving as a Committee Member of Committee on Space Research (COSPAR) and International Union of Geodesy and Geophysics (IUGG). His research interests include eddy–current interactions, typhoon–ocean interactions, global change and climate variability, and ocean dynamics.

**Kim-Anh Nguyen** received her Ph.D. degree in the Graduate Institute of Hydrological and Oceanic Sciences, National Central University, Taiwan in 2017. In the last 5 years, she has authored three books, 12 articles, 8 IEEE conference papers and delivered more than 50 international conference presentations and invited speeches. She received paper awards and young scientist awards/grants from 13 international societies and conferences, including American Academy of Sciences, AGU, IEEE GRSS, APEC, EGU, and AOGS, among others. Dr. Nguyen is currently Guest Editor of a Remote Sensing Special Issue and Sustainability Special Issue and serves as a reviewer for high quality journals such as Science of the Total Environment, Ecological Indicators, Remote Sensing, TGRS, and JSTARS. Dr. Nguyen is a member of AGU, JPGU, and IEEE GRSS and has been a social media ambassador of IEEE GRSS since January 2019.

**Steven C. Reising** reising served as Assistant Professor in Electrical and Computer Engineering at the University of Massachusetts Amherst (1998–2004), where he received tenure. He served as Associate Professor at Colorado State University (2004–2011) and is currently Full Professor in Electrical and Computer Engineering (2011–present). Dr. Reising has been the Principal Investigator of 18 grants from NASA, NOAA, NSF, DoD, ONR, NPOESS, ESA, Ball Aerospace and FIRST RF Corporation. Dr. Reising has served as Associate Editor for Atmospheric Remote Sensing of MDPI Remote Sensing since 2018 and is the founding Associate Editor of IEEE GRSL (2004–2013). He currently serves as Secretary of the IEEE Geoscience and Remote Sensing Society (GRSS) and has previously served in many leadership positions, including as an elected AdCom Member of both IEEE GRSS (2003–2020) and IEEE MTT-S (2014–2019).

## Editorial

# Preface: Earth Observations for Environmental Sustainability for the Next Decade

Yuei-An Liou [1,2,*], Yuriy Kuleshov [3,4], Chung-Ru Ho [5], Kim-Anh Nguyen [1,6] and Steven C. Reising [7]

1. Center for Space and Remote Sensing Research, National Central University, No. 300, Jhongda Rd., Jhongli District, Taoyuan City 32001, Taiwan; nguyenrose@csrsr.ncu.edu.tw
2. Taiwan Group on Earth Observations, Hsinchu 32001, Taiwan
3. Australian Bureau of Meteorology, 700 Collins Street, Docklands, Melbourne, VIC 3008, Australia; yuriy.kuleshov@bom.gov.au
4. SPACE Research Centre, School of Science, Royal Melbourne Institute of Technology (RMIT) University, Melbourne, VIC 3000, Australia
5. Department of Marine Environmental Informatics, National Taiwan Ocean University, Keelung 32001, Taiwan; b0211@mail.ntou.edu.tw
6. Institute of Geography, Vietnam Academy of Science and Technology, 18 Hoang Quoc Viet Rd., Cau Giay, Hanoi 100000, Vietnam
7. Electrical and Computer Engineering Department, Colorado State University, 1373 Campus Delivery, Fort Collins, CO 80523-1373, USA; steven.reising@ColoState.edu
* Correspondence: yueian@csrsr.ncu.edu.tw; Tel.: +886-3-4227151 (ext. 57631); Fax: +886-3-4254908

Evidence of the rapid degradation of the Earth's natural environment has grown in recent years. Sustaining our planet has become the greatest concern faced by humanity. Of the 17 Sustainable Development Goals (SDGs) in the 2030 Agenda for Sustainable Development, Earth observations have been identified as major contributors to nine of them: 2 (Zero Hunger), 3 (Good Health and Well-Being), 6 (Clean Water and Sanitation), 7 (Affordable and Clean Energy), 11 (Sustainable Cities and Communities), 12 (Sustainable Consumption and Production), 13 (Climate Action), 14 (Life Below Water), and 15 (Life on Land). Achieving the SDGs by turning knowledge into action is the critical challenge for scientists and other subject matter experts throughout the world. This monograph, *Earth Observations for Environmental Sustainability for the Next Decade*, gathers original viewpoints and knowledge advances in the use of Earth observations to address a number of urgent issues of great concern for humanity, including land use/land cover (LULC) classification [1], debris-flow assessment [2,3], precipitation estimates [4], drought assessment [5], hyperspectral image classification [6], Kuroshio-induced wakes [7], sea water primary production [8], weather system (tropical cyclone) interaction [9], and habitat suitability and biodiversity conservation [10]. In this editorial, a brief overview of the collected papers is presented.

Land cover and how people use land are important determining factors that affect a wide range of key surface parameters (evaporation, transpiration, runoff, land surface temperature, etc.). Accordingly, LULC change has been recognized as one of the most important issues with wide-ranging effects, from Earth system functioning to global environmental change. In view of the significance of land cover and its change, there is a strong demand for high-quality geospatial information on LULC classification and its dynamics at different temporal and spatial scales. Remote sensing with the aid of machine learning has been instrumental for the study of LULC change. Talukdar et al. [1] examined the accuracy of various algorithms for LULC mapping to identify the best classifier for further applications. In the article, six machine learning algorithms, namely random forest (RF), support vector machine (SVM), artificial neural network (ANN), fuzzy adaptive resonance theory-supervised predictive mapping (fuzzy ARTMAP), spectral angle mapper (SAM), and Mahalanobis distance (MD) were examined. Results showed that all these classifiers had a similar accuracy level with minor variations, but the RF algorithm had the highest accuracy of 0.89, and the MD algorithm (parametric classifier) had the lowest accuracy 0.82. Further evaluations of the RF algorithm in different morphoclimatic conditions will certainly be worthwhile in the future.

Humans and their possessions are vulnerable to debris flows wherever they occur, particularly in populated areas with a harsh natural environment and deforestation. Therefore, assessment of debris-flow susceptibility (DFS) is useful for mitigating debris flow risk. Zhang et al. [2] assessed the main triggering factors of debris flows and investigated DFS in the Shigatse area of Tibet using machine learning methods. Remote-sensing data sets and geographic information system (GIS) techniques are used to obtain influential variables of topography, vegetation, human activities, and soil for local debris flows. Five machine learning methods, i.e., back-propagation neural network (BPNN), one-dimensional convolutional neural network (1D-CNN), decision tree (DT), random forest (RF), and extreme gradient boosting (XGBoost) were utilized to examine the relationship between debris-flow triggering factors and occurrence. The results revealed that the XGBoost model exhibits the best mean accuracy (0.924) on 10-fold cross-validation, and that its performance was significantly better than the other machine learning methods, although the performance of the XGBoost did not significantly differ from the 1D-CNN (0.914). These methods could potentially be used to assist in the prevention of the casualties and economic losses caused by debris flows. Furthermore, the relevant authorities can use the XGBoost model in combination with satellite remote sensing and GIS spatial data processing to create feature maps and high-precision, area-sensitive maps to provide guidance and preparation for debris-flow prevention and mitigation.

In addition, machine learning algorithms have been widely used in disaster prevention in recent years. Due to human development and global change, debris flows have become an important issue in environmental disasters. Sichuan Province in China is an area where debris flows occur frequently and cause many dangerous incidents. Xiong et al. [3] utilized four machine learning algorithms, namely logistic regression, support vector machine, random forest, and boosted regression trees, to conduct a debris-flow sensitivity analysis in Sichuan Province to understand which algorithm was the most suitable for debris flow analysis and assistance in evaluation of debris-flow hazards. Combined with the application of remote sensing and geographic information systems, the authors found that the average altitude, altitude difference, aridity index, and groove gradient played the most important roles in the assessment. The research results also showed that all four algorithms could generate accurate and effective debris-flow sensitivity maps, which could be used to provide useful data for assessing and mitigating debris-flow hazards.

An evaluation of two widely used satellite precipitation estimates—the U.S. National Oceanographic and Atmospheric Administration's (NOAA) Climate Prediction Center morphing technique (CMORPH) and the Japan Aerospace Exploration Agency's (JAXA) Global Satellite Mapping of Precipitation (GSMaP)—has been conducted over Australia using an 18-year data set (2001–2018) [4]. Overall, statistics demonstrated that satellite precipitation estimates were of high accuracy for Australia, and that gauge-blending yielded a notable increase in performance. The dependence of performance on geography, season, and rainfall intensity was also investigated. It was found that the skill of satellite precipitation detection was reduced in areas of elevated topography and where cold front rainfall was the main precipitation source. Areas where rain-gauge coverage was sparse also exhibited reduced skill. The skill of the satellite precipitation estimates was highly dependent on rainfall intensity. The highest skill was obtained for moderate rainfall rates (2–4 mm/day). Low rainfall rates were overestimated, and large rainfall rates were underestimated, both in frequency and amount. Overall, CMORPH and GSMaP datasets were evaluated as useful sources of satellite precipitation estimates over Australia.

To address drought events in Ethiopia, several techniques and data sets were analyzed to study the spatiotemporal variability of vegetation in response to a changing climate [5]. In this study, 18 years (2001–2018) of Moderate Resolution Imaging Spectroradiometer (MODIS) Terra/Aqua, normalized difference vegetation index (NDVI), la surface temperature (LST), Climate Hazards Group Infrared Precipitation with Stations (CHIRPS) daily precipitation, and the Famine Early Warning Systems Network (FEWS NET) Land Data Assimilation System (FLDAS) soil-moisture data sets were processed. Pixel-based Mann–Kendall trend analysis and the Vegetation Condition Index (VCI) were used to assess drought patterns during the crop growth season. Results indicated that the central highlands and northwestern part of Ethiopia, which have land cover dominated by cropland, had experienced a decreasing trend in both precipitation and NDVI. This study provides valuable information for identifying locations of potential concern for drought and planning for immediate action of relief measures. Furthermore, this paper presents the results of the first attempt to apply a recently developed index, the Normalized Difference Latent Heat Index (NDLI), to monitor drought conditions. NDLI successfully captures historical droughts and shows a notable correlation with climatic variables.

It has been shown previously that the discriminative capability of the general nearest feature line embedding (FLE) transformation was successful for numerous applications; however, there are certain limitations to this methodology. For example, the conventional linear-based principle component analysis (PCA) preprocessing method in FLE cannot be used to effectively extract nonlinear information. To overcome this deficiency of FLE, a novel multiple kernel FLE (MKFLE) method was proposed and applied to classify hyperspectral images [6]. The proposed MKFLE dimension-reduction framework was performed in two stages. In the first multiple-kernel PCA stage, the multiple-kernel learning method based on between-class distance and support vector machine was used to find the kernel weights. Based on these weights, a new weighted kernel function was constructed as a linear combination of valid kernels. In the second FLE stage, the FLE method, which can preserve the nonlinear manifold structure, was applied for supervised dimension reduction using the kernel function obtained in the first stage. The effectiveness of the proposed MKFLE method was evaluated using three benchmark data sets: Indian Pines, Pavia University, and Pavia City; it was demonstrated that the performance of the MKFLE was superior compared to other methods.

Island wakes may induce ocean upwelling in the lee of the island and bring nutrition to the upper ocean, increasing the chlorophyll-a concentration. The increase in chlorophyll-a concentration in the upper ocean can affect carbon cycles and hence global changes. An improved understanding of ocean current-induced island waves is needed in the study of oceanic environments. Using high-temporal-resolution imagery from the Himawari-8 satellite, the study in [7] presented the temporal variation and spatial structure of the Kuroshio-induced Green Island wakes. Green Island is a small island located near southeast Taiwan that is on the main path of the Kuroshio. Using the Himawari-8 imagery, the authors found that the structure of the wake changed quickly, and the water mixed into different wake states. The results suggested that satellite imagery can help build up an island wake database to assist with ocean sustainability.

Evaluating the accuracy of satellite remote-sensing products is very important for their further application. A match-up data set of satellite remote-sensing observations with in situ measurements is quite useful for algorithm validation. The study in [8] evaluated the primary production derived from MODIS onboard the Aqua and Terra satellites using a vertically generalized production model with in situ data on the waters around Taiwan. The authors suggested the combined primary production product from MODIS of the Aqua and Terra satellites was more accurate than that from only one satellite. Using the product, the author concluded that the China coastal water and the Kuroshio water had the highest and the lowest primary production, respectively, in the waters adjacent to Taiwan.

Exploring dual-vortex interactions between typhoons is crucial to understanding the behaviours of typhoons during their journeys [9]. The differential averaging technique, based on the Normalized Difference Convection Index (NDCI) operator and filter, depicted differences and generated a new set of clarified images. During the first set of dual-vortex interactions, Typhoon Noru (2017) experienced an increase in intensity and a U-turn in its direction after being influenced by adjacent cooler air masses and air flows.

Triple interactions between Noru–Kulap–Nesat and Noru–Nesat–Haitung were analyzed using geosynchronous satellite infrared (IR1) and IR3 water vapor (WV) images. The results demonstrated that the generalized Liou–Liu formulas for computing threshold distances between typhoons successfully validated and quantified the triple-interaction events. Through the unusual and combined effects of the consecutive dual-vortex interactions, Typhoon Noru lasted for 22 days from 19 July to 9 August 2017, and migrated approximately 6900 km. Typhoon Noru consequently became the third-longest-lasting typhoon on record for the Northwest Pacific Ocean. A comparison was made with long-lived Typhoon Rita in 1972, which experienced similar multiple Fujiwhara interactions with the other concurrent typhoons. During the first set of dual-vortex interactions, Typhoon Noru experienced an increase in intensity and a U-turn in its direction after being influenced by adjacent cooler air masses and air flows. Thereafter, in spite of a distance of 2000–2500 km separating Typhoon Noru and the newly formed Typhoon Nesat, the influence of mid-air flows and jet flows caused an "indirect interaction" between these typhoons. Evidence of this second interaction included the intensification of both typhoons and changes in their track directions. The third interaction occurred subsequently between Tropical Storm Haitang and Typhoon Nesat.

Many species' habitats have significantly declined or become extinct in recent decades for various reasons. It is vital to detect potential habitats based on habitat-suitability analyses to enhance biodiversity conservation. The study in [10] proposed a novel scheme for assessing habitat suitability based on a two-stage ensemble approach. First, a deep neural network (DNN) model was constructed to predict habitat suitability based on environmental data. Second, an ensemble model employing various methods for habitat suitability estimation was developed based on observational and environmental data. Crowdsourced databases were utilized, and observational and environmental data were used for four amphibian species and seven bird species in South Korea. The authors demonstrated that the proposed scheme provided a more accurate estimation of habitat suitability compared to previous approaches. For example, the proposed scheme achieved a true skill statistic (TSS) score of 0.886, which was higher than previous approaches (TSS 0.725 ± 0.010).

**Funding:** This research received no external funding.

**Conflicts of Interest:** The authors declare no conflict of interest.

## References

1. Talukdar, S.; Singha, P.; Mahato, S.; Shahfahad, P.S.; Liou, Y.; Rahman, A. Land-Use Land-Cover Classification by Machine Learning Classifiers for Satellite Observations—A Review. *Remote Sens.* **2020**, *12*, 1135. [CrossRef]
2. Zhang, Y.; Ge, T.; Tian, W.; Liou, Y. Debris Flow Susceptibility Mapping Using Machine-Learning Techniques in Shigatse Area, China. *Remote Sens.* **2019**, *11*, 2801. [CrossRef]
3. Xiong, K.; Adhikari, B.; Stamatopoulos, C.; Zhan, Y.; Wu, S.; Dong, Z.; Di, B. Comparison of Different Machine Learning Methods for Debris Flow Susceptibility Mapping: A Case Study in the Sichuan Province, China. *Remote Sens.* **2020**, *12*, 295. [CrossRef]
4. Chua, Z.; Kuleshov, Y.; Watkins, A. Evaluation of Satellite Precipitation Estimates over Australia. *Remote Sens.* **2020**, *12*, [CrossRef]
5. Liou, Y.; Mulualem, G. Spatio–temporal Assessment of Drought in Ethiopia and the Impact of Recent Intense Droughts. *Remote Sens.* **2019**, *11*, 1828. [CrossRef]
6. Chen, Y.-N. Multiple Kernel Feature Line Embedding for Hyperspectral Image Classification. *Remote Sens.* **2019**, *11*, 2. [CrossRef]
7. Hsu, P.; Ho, C.; Lee, H.; Lu, C.; Ho, C. Temporal Variation and Spatial Structure of the Kuroshio-Induced Submesoscale Island Vortices Observed from GCOM-C and Himawari-8 Data. *Remote Sens.* **2020**, *12*, 883. [CrossRef]
8. Lan, K.; Lian, L.; Li, C.; Hsiao, P.; Cheng, S. Validation of a Primary Production Algorithm of Vertically Generalized Production Model Derived from Multi-Satellite Data around the Waters of Taiwan. *Remote Sens.* **2020**, *12*, 1627. [CrossRef]
9. Liou, Y.; Liu, J.; Liu, C.; Chen, C.; Nguyen, K.; Terry, J. Consecutive Dual-Vortex Interactions between Quadruple Typhoons Noru, Kulap, Nesat and Haitang during the 2017 North Pacific Typhoon Season. *Remote Sens.* **2019**, *11*, 1843. [CrossRef]
10. Rew, J.; Cho, Y.; Moon, J.; Hwang, E. Habitat Suitability Estimation Using a Two-Stage Ensemble Approach. *Remote Sens.* **2020**, *12*, 1475. [CrossRef]

*Review*

# Land-Use Land-Cover Classification by Machine Learning Classifiers for Satellite Observations—A Review

Swapan Talukdar [1], Pankaj Singha [1], Susanta Mahato [1], Shahfahad [2], Swades Pal [1], Yuei-An Liou [3,*] and Atiqur Rahman [2]

1. Department of Geography, University of Gour Banga, NH12, Mokdumpur, Malda-732103, India; swapantalukdar65@gmail.com (S.T.); pankajsingha2014@gmail.com (P.S.); mahatosusanta2011@gmail.com (S.M.); swadespalgeo@ugb.ac.in (S.P.)
2. Department of Geography, Faculty of Natural Sciences, Jamia Millia Islamia, MMAJ Marg, Jamia Nagar, New Delhi-110025, India; shahfahad179766@st.jmi.ac.in (S.); arahman2@jmi.ac.in (A.R.)
3. Center for Space and Remote Sensing Research, National Central University, 300 Jhongda Road, Jhongli District, Taoyuan City 32001, Taiwan
* Correspondence: yueian@csrsr.ncu.edu.tw; Tel.: +886-3-4227151 (ext. 57631)

Received: 25 February 2020; Accepted: 30 March 2020; Published: 2 April 2020

**Abstract:** Rapid and uncontrolled population growth along with economic and industrial development, especially in developing countries during the late twentieth and early twenty-first centuries, have increased the rate of land-use/land-cover (LULC) change many times. Since quantitative assessment of changes in LULC is one of the most efficient means to understand and manage the land transformation, there is a need to examine the accuracy of different algorithms for LULC mapping in order to identify the best classifier for further applications of earth observations. In this article, six machine-learning algorithms, namely random forest (RF), support vector machine (SVM), artificial neural network (ANN), fuzzy adaptive resonance theory-supervised predictive mapping (Fuzzy ARTMAP), spectral angle mapper (SAM) and Mahalanobis distance (MD) were examined. Accuracy assessment was performed by using Kappa coefficient, receiver operational curve (RoC), index-based validation and root mean square error (RMSE). Results of Kappa coefficient show that all the classifiers have a similar accuracy level with minor variation, but the RF algorithm has the highest accuracy of 0.89 and the MD algorithm (parametric classifier) has the least accuracy of 0.82. In addition, the index-based LULC and visual cross-validation show that the RF algorithm (correlations between RF and normalised differentiation water index, normalised differentiation vegetation index and normalised differentiation built-up index are 0.96, 0.99 and 1, respectively, at 0.05 level of significance) has the highest accuracy level in comparison to the other classifiers adopted. Findings from the literature also proved that ANN and RF algorithms are the best LULC classifiers, although a non-parametric classifier like SAM (Kappa coefficient 0.84; area under curve (AUC) 0.85) has a better and consistent accuracy level than the other machine-learning algorithms. Finally, this review concludes that the RF algorithm is the best machine-learning LULC classifier, among the six examined algorithms although it is necessary to further test the RF algorithm in different morphoclimatic conditions in the future.

**Keywords:** land use/land cover (LULC); Earth observations; machine learning algorithm; random forest; artificial neural network

## 1. Introduction

Knowledge of land-use/land-cover (LULC) change is essential in a number of fields based on the use of Earth observations, such as urban and regional planning [1,2], environmental vulnerability and impact assessment [3–7], natural disasters and hazards monitoring [8–13] and estimation of soil erosion and salinity, etc. [14–17]. Quantitative assessment and prediction of LULC dynamics are the most efficient means to manage and understand the landscape transformation [18]. Mapping LULC change has been identified as an essential aspect of a wide range of activities and applications, such as in planning for land use or global warming mitigation [19,20]. Consequently, assessment in LULC change is inevitably required for a variety of purposes for the welfare of human beings in the context of rapid and uncontrolled population growth along with economic and industrial development, especially in developing countries with intensified LULC changes [20–23]. These changes have a series of impacts on both human society and environment in many ways like increasing flood and drought vulnerability, environmental degradation, loss of ecosystem services, groundwater depletion, landslide hazards, soil erosion and others [14,15,24–27].

Several techniques have been developed to map LULC patterns and dynamics from the satellite observations, including traditional terrestrial mapping, as well as satellite-based mapping. Terrestrial mapping, known as a field survey, is a direct way of mapping in which the map can be produced at various scales incorporating information with different levels of precision, although it is a manpower-based, time- and money-consuming way to map large areas [28]. Moreover, there is a chance of subjectivity in mapping. On the other hand, the satellite- and aerial photograph-based mapping of LULC are cost-effective, spatially extensive, multi-temporal, and time-saving [29]. Earlier, the spatial resolution of satellite data was comparatively less than that of the maps prepared through terrestrial surveys. With the advancement of remote-sensing (RS) techniques and microwave sensors, satellites provide data at various spatial and temporal scales [30–32]. RS provides the opportunity for rapid acquisition of information on LULC at a much reduced price compared to the other methods like ground surveys [33,34]. The satellite images have the advantages of multi-temporal availability as well as large spatial coverage for the LULC mapping [35,36]. In the past few decades, studies on mapping, monitoring and forecasting of LULC dynamics have been carried out using medium- and low-resolution observations from satellites, such as Landsat, Satellite Pour l'Observation de la Terre, or Satellite for observation of Earth (SPOT), Indian Remote Sensing (IRS) Satellite Resourcesat, Advanced Spaceborne Thermal Emission and Reflection Radiometer (ASTER), Moderate Resolution Imaging Spectroradiomete (MODIS) and others [18,31,37–40]. With the advancement of hyperspectral satellite sensors, the importance of RS has increased many times in the research field and for planning purposes.

Recently, the application of machine-learning algorithms on remotely-sensed imageries for LULC mapping has been attracting considerable attention [41,42]. The machine-learning techniques have been categorized into two sub-types; supervised and unsupervised techniques [43,44]. The supervised classification techniques include support vector machine (SVM), random forest (RF), spectral angle mapper (SAM), fuzzy adaptive resonance theory-supervised predictive mapping (Fuzzy ARTMAP), Mahalanobis distance (MD), radial basis function (RBF), decision tree (DT), multilayer perception (MLP), naive Bayes (NB), maximum likelihood classifier (MLC), and fuzzy logic [45,46], while the unsupervised classification techniques include Affinity Propagation (AP) cluster algorithm, fuzzy c-means algorithms, K-means algorithm, ISODATA (iterative self-organizing data) etc. [41,47].

Over the last decade, more advanced methods, such as artificial neural networks (ANN), SVM, RF, decision tree, and other models, have gained exceptional attention in remote sensing-based applications, such as LULC classification. Thus, numerous studies on the LULC modelling have been carried out using different machine-learning algorithms [14,48–51] as well as comparing the machine-learning algorithms [52–54]. Furthermore, a few studies have been carried out to identify the best suited and accurate algorithm among used machine-learning classifiers for LULC mapping [52–55]. Each machine-learning technique has different types of accuracy levels. It has been found that ANN, SVM, and RF generally provide better accuracy as compared with the other traditional classifier techniques [56], while SVM and RF are the best techniques for the LULC classification compared to all other machine-learning techniques [57,58]. However, the sensor characteristics and image data-related factors, such as spatial and temporal resolution, processing software and hardware, etc. determine the accuracy of LULC classification [59].

Several studies found that the LULC classification using medium- and low-resolution observations from satellites has several spectral and spatial limitations that affect its accuracy [24,60–62]. Therefore, researchers have been applying machine-learning algorithms to reduce the aforementioned limitations and obtain high-precision LULC images. Furthermore, all machine-learning techniques do not always produce a high-precision LULC map because good results depend on the machine-learning model set-up, training samples and input parameters. Up to the present, numerous studies have been conducted on land-use classification using machine-learning algorithms [20,63], but the performance of models is not well examined. In this article, we utilized six machine-learning techniques to understand which method can produce a high-precision LULC map based on accuracy statistics.

## 2. Materials and Methods

### 2.1. Study Area

We selected a stretch of riparian landscape of the river Ganga from Rajmahal to Farakka barrage in India emphasizing three major dynamic river islands (locally, charland) dominated by patches. LULC classification in relatively stable areas is easier than highly dynamic landscape like charland and such work is undertaken by many scholars. How far the advanced methods are useful for delineating LULC units in such a dynamic area was given emphasis with different approaches of accuracy assessment. Successful application of one method in different similar sites proves its usability. Hence, to test the precision of the applied methods three such patches from the study stretch were used. The study area covers parts of Jharkhand and the West Bengal states of India. More precisely, it covers some parts of Sahibganj District of Jharkhand and Malda and the Murshidabad districts of West Bengal (Figure 1). The topography of the study area is dominated by alluvial plain, which is formed by the sediments deposited by Ganga, Mahananda and Kalindi rivers. The elevation of the region varies between approximately 12 meters to 90 meters. The regular flooding makes the region suitable for agriculture with seasonal water scarcity. The climate of the region is of sub-humid monsoon type (Koppen-Cwg) with average annual rainfall more than 1500 mm and temperature ranges from 10 and 38 °C. The rapidly increasing population causes large-scale landscape transformation by the expansion of agricultural land and human settlement along with well-defined riverbank erosion and flood hazards. Riverbank erosion has caused significant changes in the LULC pattern in the area for a long time. The frequent flooding and riverbank erosion have caused large-scale displacement of the human settlements in the region during the first decade of the 21st century [64].

### 2.2. Materials

In this work, the Landsat 8 Operational Land Imager (OLI) image (path/row 139/43) downloaded from the United States Geological Survey (USGS) website (https://earthexplorer.usgs.gov)) has been used to map the LULC using different machine-learning algorithms (Figure 2). Six first-order LULC classes have been identified based on a comprehensive literature survey and expert-based knowledge about the study area (Table 1). The acquisition date of the Landsat data downloaded was 03 October 2019. Furthermore, the Google Earth image and field-based observations have been used for the accuracy assessment of the LULC maps prepared.

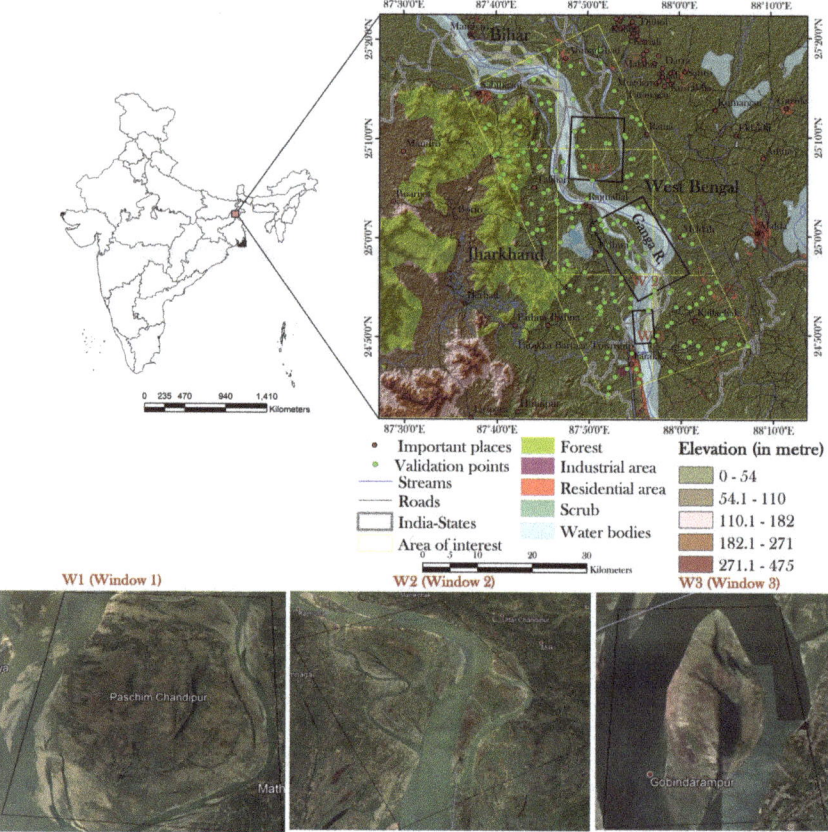

**Figure 1.** Location map and details of the study area.

**Table 1.** Description of the land-use/land-cover (LULC) classes identified.

| Class Name | Class Description | Class Description Example |
|---|---|---|
| Agricultural land | Area covered by agricultural crops | |
| Built up area | Area covered by settlement, road | |
| Sand bar | Land on the river bed | |
| Fallow land | Area without vegetation | |
| Vegetation | Area covered by forest, sparse trees, mango orchard | |
| River and wetlands | Area covered by water | |

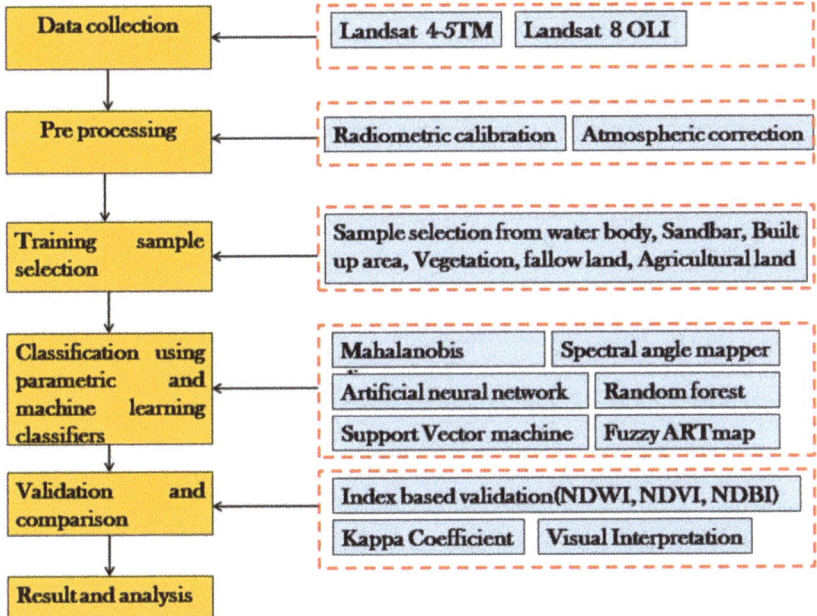

**Figure 2.** Flowchart of the methodology.

## 2.3. Methods for Land-Use/Land-Cover (LULC) Modelling

The LULC classification was performed using the six most popular machine-learning classifiers. The descriptions of the parameters for optimizing the models and software used to perform the LULC classification are given in Table 2.

**Table 2.** Optimized parameters for different classifiers used for the LULC modelling.

| S. No. | Methods | Software Used for Modelling | Optimized Parameters |
|---|---|---|---|
| 1 | Artificial neural network | TerrSet Geospatial monitoring and modelling system | Hidden layer-1, input layer-1, output layer-1, nodes-6, learning rate-0.01, momentum factor-0.5, sigmoid constant-1 |
| 2 | Support vector machine | Environment for Visualizing Images (ENVI 5.3) | Kernel type-radial basis function, gamma in kernel function-1, penalty parameter-100, pyramid level-1, pyramid reclassification threshold-0.90 |
| 3 | Fuzzy ARTMAP | TerrSet Geospatial monitoring and modelling system | F1 layer neurons-12, F2 layer neurons-385, map field layer neurons-6, choice parameter for $ART_a$-0.01, learning rate and vigilance parameter for $ART_a$- 1 and 0.98, learning rate and vigilance parameter for $ART_b$- 1 and 1, iteration 3338 |
| 4 | Spectral Angle mapper | Environment for Visualizing Images (ENVI 5.3) | Wavelength units-micrometers, Y data multiplier-1, set maximum angle (radiance)-single value, maximum angles (radians)-0.100 |
| 5 | Random Forest | R programming language (R 3.5.3) | - |
| 6 | Mahalanobis Distance | Environment for Visualizing Images (ENVI 5.3) | - |

2.3.1. Methods for Machine-Learning Classifiers

1. Artificial Neural Network (ANN)

The ANN is the most widely applied machine-learning technique, which can be efficiently used in non-linear phenomena such as parameter retrieval [65–67], LULC changes with the ability to work on big data analysis. It is currently one of the most used non-parametric classification techniques [68]. It does not depend on any assumption of generally distributed data [69].

The ANN is a forward structure black-box model, which is trained by back propagation algorithm (supervised training algorithm). The ANN is functioned like a human brain or nervous system containing nerve fibres with many interconnections through other axons [70]. It can learn and produce meaningful results from examples, even when the input data having error or complexity and incomplete. Therefore, it can simulate exactly like the human nervous system. However, the ANN has one input layer, at least one hidden layer and one output layer. Each layer is formed by neurons (like brain nerves) (Figure 3). These neurons are non-linear processing units. However, all the neurons in a layer are interconnected to all other neurons in the adjacent layers and formed networks. In addition, the connection between neurons in successive layers are weighted. This process (transferring information from one neuron to another or one layer to other layer) is called forward connection. This automatic learning is accomplished through a dynamic adjustment of network inter-connection associated with each neuron [66].

One of most important algorithms that ANN usually uses is the back propagation algorithm, which is a gradient-decent algorithm. The main function of it is to minimize the error between the actual network outputs and the outputs of training input/output pairs [71]. The network repeatedly receives the numbers of input/output pairs and the error is propagated from the output back to the input layer. The learning rate and update rule renew the weights of the backward paths [72,73]. In addition, the default processing unit, training and learning rate cannot uniquely specify the ANN. Therefore, the trial and error process of changes of model parameters can only be the best way to obtain better result. In this review article, the multilayer perceptron (MLP) ANN architecture used in the LULC classification is modelled using a layered feed-forwarded model in the TerrSet Geospatial Modelling Software.

The MLP architecture can be explained mathematically. In MLP architecture, the input layer comprises the $n_0$ neurons, which collect a normalized set of input variables of $x_i$ ($i = 1, 2 \ldots \ldots n_0$). The second layer is also known as the hidden layer that contains the $n_1$ neurons and receives a set of variables of $y_j$ ($j = 1, 2 \ldots \ldots \ldots n_1$), which are the output of the first layer. Each of the layers receive a bias value of 1 in each of the neurons that rectify their outputs. The third or output layer consists of the $n_2$ neurons with number equal to output variables of $z_k$ ($K = 1, 2 \ldots ., n_2$). A continuous non-linear mapping is performed in the $n_0$ neurons of $x_i$ variables in the output layer to the $y_i$ variables in the hidden layer after summing them up using an activation function. The parameter of this function is also defined as weights of neurons in each hidden layer for each result of neurons of the input layer [74]. One of the most common methods for ANN training is the back-propagation algorithm defined by minimizing the cost function as presented in Equation (1).

$$m = \frac{1}{2} \sum_{i=1}^{n} (a_i - b_i)^2 \qquad (1)$$

where $n$ represents the number of classes, $a_i$ denotes the expected output, and $b_i$ is the response of designed ANN from the $i$ neuron of the total $n$ neurons in the output layer.

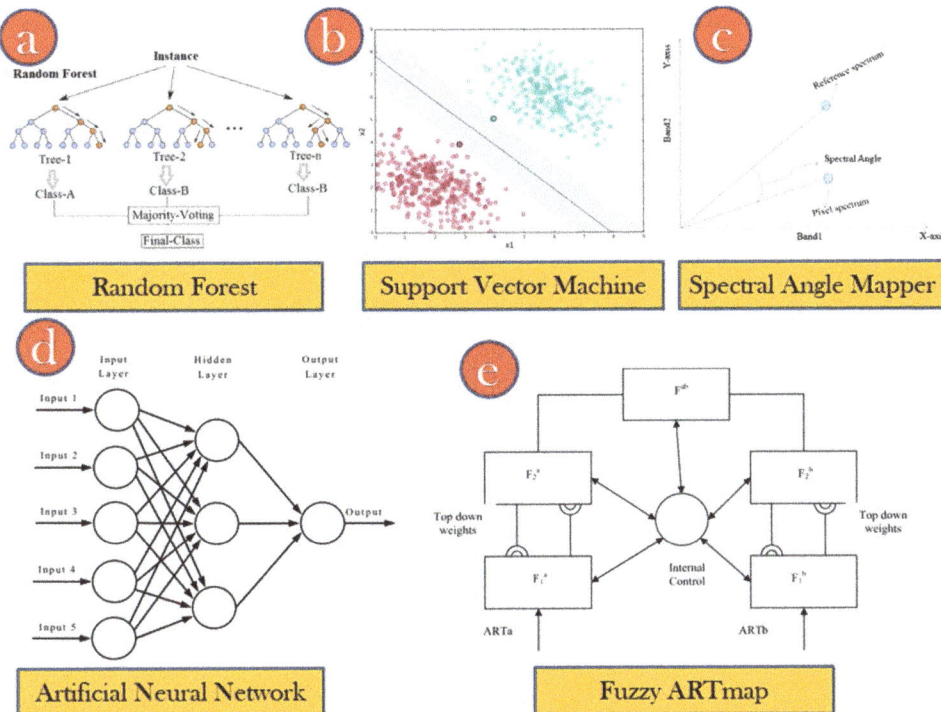

**Figure 3.** Schematic architecture of land use land cover classifiers (**a**) random forest, (**b**) support vector machine, (**c**) spectral angle mapper, (**d**) artificial neural network, and (**e**) fuzzy ARTMAP.

2. Support Vector Machine (SVM)

SVM is a non-parametric supervised machine learning technique and initially aimed to solve the binary classification problems [41,75]. It is based on the concept of structural risk minimization (SRM), which maximizes and separates the hyper-plane and data points nearest the spectral angle mapper (SAM) of the hyper-plane. It separates data points into various classes using a hyper-spectral plane. In this process, the vectors ensure that the width of the margin will be maximized [76]. SVM can support multiple continuous and categorical variables as well as linear and non-linear samples in different class membership. The training samples or bordering samples that delineate the margin or hyper-plane of SVM are known as support vectors [46]. In remote sensing, the polynomial and radial basis function (RBF) kernel has been used most commonly [41], but for LULC classification RBF is the most popular technique and gives better accuracy than the other traditional methods.

The original SVM method has been launched with a set of data, and its objective is to find the hyper-plane that can separate the datasets into a number of classes, as the aim of SVM is to find the optimal separating hyper-plane from the available hyper-planes [77]. Furthermore, the SVM algorithm needs a proper kernel function to establish the hyper-planes accurately and minimize the classification errors [78]. The essential part of the SVM technique is the kernel type used. The functionality of the SVM mainly depends on the kernel size, and the similarity of a smooth surface depends on the more significant kernel density. For simulated and real-world hyperspectral satellite data, the genetically optimized SVM using the support vectors shows the best performance [79]. The primary function of SVM is to find the optimal boundary, which will increase the separation between the entire support vectors. The RBF and polynomial function kernel were performed on ENVI software version 5.3 for LULC mapping.

3. Fuzzy ARTMAP (FA)

The fuzzy ARTMAP technique is based on the similarity of the fuzzy subset calculation as well as the adaptive and vibrant category selection through the feature space search. The structure of fuzzy ARTMAP includes two modules, i.e., $ART_a$ and $ART_b$. These two modules can be further sub-divided into two sub-modules in the function (attention and orientation subsystem). The attention subsystem has several functions. For example, it deals with the modules, establishes the exact internal illustration, and creates fine-tuning for the modules. In contrast, the orientation subsystem is used for dealing with the newly appeared module [80]. Each module of the fuzzy ARTMAP consists of three layers, namely $F_0$ as input layer, $F_1$ as comparison layer and $F_2$ as recognition layer. These characteristics of fuzzy ARTMAP are identical to the artificial neural network. Furthermore, each layer has its respective neuron units M, M, N as well as the control connections associated with the layers. $F_1$ is used for the detection of features and it has adequate nodes for the mode coding, while the nodes of $F_2$ show the categories concerning the input.

Based on the comprehensive investigation of fuzzy ARTMAP as well as the characteristics of remote-sensing data, a simplified fuzzy ARTMAP algorithm has been applied using the Terrset software. It comprises two layers in which the first is used for the feature data input and the second for the classification of remote-sensing data. In the first layer, the numbers of neurons are equal to the feature dimensions of the data, while in the second layer, the numbers of neurons are decided by the user as per the trial and process results [81]. The fuzzy ARTMAP firstly calculates the comparison between the new pattern and the existing active pattern. Then all active values are arranged in ascending order to degree of matching and compared with warning values. If the warning values are exceeded by the matching degree, the pattern of the training sample will be the same as the output layer. The fuzzy ARTMAP combines the pattern of output layer neurons and uses the weight between the output-input layers and the radius. If all the output layer neurons do not meet the matching requirements, a new output layer neuron will be built to store the new pattern and, thus, the results of classification become more accurate with more output layers.

4. Random Forest

RF is a new non-parametric ensemble machine-learning algorithm developed by Breiman [82]. The RF algorithm has been widely applied for solving the environmental problems, like water resource management and natural hazard management. It can handle a variety of data, like satellite imageries, and numerical data [83]. It is an ensemble learning method based on a decision tree, which combines with massive ensemble regression and classification trees. For setting up the RF model, two parameters are needed and called the base of the method. These parameters are (1) the number of trees, which can be explained by 'n-tree' and (2) many features in each split, which can be explained by 'm-try.' Classification trees provide an individual choosing power or vote and accurate classification in regulating the majority vote from trees in the entire forest.

In recent times, several studies have shown a satisfactory performance for LULC classification using RF in the field of remote-sensing applications [42,52,57]. A vast number of trees of this method provide better accuracy in the field of image classification [84] and land-use modelling. Breiman [82] stated that using more trees compared to required trees is an unnecessary and time-consuming process, but it does not hamper the entire model. Furthermore, Feng et al. [85] selected 200 decision trees in their study and noted that the performance of RF was accurate. The RF technique has been benefited with the two more powerful algorithms: bagging and random, which are called the powerhouse of this method. In our study, the 'randomForest' package in R has been used to produce the LULC map. As suggested by Feng et al. [85], 200 decision trees have been used with 3 input features (m-try) in our study.

2.3.2. Method for Parametric Classifier

1. Mahalanobis Distance (MD)

Supervised image classification is performed to detect the quantitative approach in the remote-sensing image. The prime goals of supervised classification are to segment the spectral domain into the areas that match the ground cover interest classes for a particular purpose. The Mahalanobis distance (MD) supervised image classification algorithm was developed by an Indian applied statistician Mahalanobis in the 1930s [86]. In MD classification, training data are given to specify the spectral classes of the pixel based on the user-defined classes. MD classification is same as maximum likelihood classification where statistics have been used for each class and it considers only equal class coefficients. The MD method measures the distance between two or more than two correlated variables. In mathematical term, MD distance is equal to Euclidean distance (ED) when the covariance matrix is the unit matrix. The small value of MD increases the chance of an observation being closer to the group's centre. For each feature vector, the MD ($D_k^2$) towards class means is calculated as in the following equation:

$$D_k^2 = (x_i - \bar{x}_k)^T S_k^{-1} (x_i - \bar{x}_k) \qquad (2)$$

where $x_i$ is the vector showing the pixel of image data, $\bar{x}_k$ is the sample mean vector of the $k$th class, $S_k^{-1}$v is the variance/covariance matrix for class $i$; and $T$ represents transpose of the matrix.

### 2.3.3. Method for Non-Parametric Classifier

1. Spectral Angle Mapper (SAM)

SAM is an auto-generated supervised spectral classifier machine learning technique that is used to determine the spectral similarity between the given image spectra and reference spectra in an $n$ (here $n$ denotes the spectral band number) dimensional space using the calculation of the angle between the spectra [87]. In recent times, a large number of bands have been used in hyper-spectral remote sensing to identify the different objects accurately and the SAM is able to analyse all bands together. Reference spectra refer to the spectra that can be taken either by field investigation or directly from satellite images. For LULC classification, reference spectra can be taken as a signature from the satellite image [88].

In SAM, only angular information can be used to identify the pixel spectra. Thus, SAM uses only angular information to identify the pixel spectra, which assumes that an observed reflectance spectrum in a vector format is a multidimensional space with the number of dimensions equal to the number of bands. The difference between image spectra and reference spectra is shown as the level of angle where a small angle indicates high similarity and a high angle indicates low similarity. The maximum threshold limits of tolerance of angle are not classified. Hence, it is better to define a threshold angle limit (in radians) under which a pixel cannot be classified. In our study, SAM has been applied using the ENVI 5.3 image processing environment for LULC classification. This technique is comparatively intensive for illumination and albedo conditions while calibrating reflectance information. The SAM is auto-generated supervised classification.

### 2.3.4. Similarity Test among the Classifiers

For representing the difference of performance of the algorithms for delineating LULC, similarity ratio (SR) is computed. It is simply the ratio between proportions of area of a given LULC computed by two algorithms. SR = 1 signifies the absolute similarity of the areal proportion of LULC computed by two algorithms. A value >1 or <1 means growing dissimilarity.

### 2.3.5. Accuracy Assessment and Correlation among the Classifiers

The post-classification accuracy assessment has been considered as the most vital part of validating the LULC maps produced by the models [61,89]. The high-precision LULC map can generate fundamental grounds for successful planning and management. The statistics only can tell about the accuracy assessment and the Kappa coefficient is a statistical technique that has been applied in the present study for assessing the accuracy. Monserud and Leemans [90] suggested five levels of agreement, poor or very poor, fair, good, very good, and excellent corresponding to the values lower than 0.4, from 0.4 to 0.55, from 0.55 to 0.70, from 0.70 to 0.85, and higher than 0.85, respectively, between images and ground reality. Thus, the Kappa coefficient has been calculated using 200 randomly selected sample ground control points in order to evaluate the accuracy of LULC maps produced by using different algorithms (the random points are shown in Figure 1). The sample points have been selected from the field observation and using Google Earth Pro for the remote and inaccessible areas.

The receiver operating characteristics (RoC) curve graph was plotted to validate the performance of LULC classifiers for detecting the different features of LULC. The graph was plotted between sensitivity and specificity being on y and x axes, respectively. The sensitivity of a model represents the proportion of correctly predicted positive pixels (i.e., the pixels belonging to a particular LULC class were correctly predicted or identified), while the specificity refers to the proportion of correctly predicted negative pixels (i.e., the pixels not belonging to a particular LULC class was correctly predicted or identified). The sensitivity and specificity were calculated following Equations (3) and (4):

$$Sensitivity = \frac{a}{a+c} \quad (3)$$

$$Specificity = \frac{d}{b+d} \quad (4)$$

where $a$ represents true positive, $d$ refers to true negative, $b$ means false positive, and $c$ represents false negative.

The area under curve (AUC) of the RoC curve depict the performance of classifiers for predicting the LULC. The value of AUC ranges from 0–1, while the AUC close to 1 represents the high degree of model performances.

The root mean square error (RMSE) was computed to evaluating the performance of machine learning classifiers using the observed and prediction sample points. The RMSE was calculated by using Equation (5). The lower the RMSE, the higher the accuracy of LULC prediction:

$$RMSE = \frac{\sum_{i=1}^{n}(Observed_i - Predicted_1)^2}{n} \quad (5)$$

where $n$ represents the number of sample points.

The "index-based technique" has been introduced to evaluate and select the best machine-learning technique for LULC mapping. Thus, three satellite data-derived indices; normalized differential vegetation index (NDVI), normalized differential water index (NDWI) and normalized differential built-up index (NDBI), have been calculated for this purpose. Each index has been classified based on a manual threshold. For better visualization, LULC classes (water, vegetation-agricultural land, built-up area) and threshold-based NDWI, NDVI, NDBI have been masked out from the study area using the selected three windows. The closeness of the area between the index-derived area and classifier-derived LULC area could be considered as a good result and vice versa. Then, we used correlation matrix among the area of land use classes of six LULC models and satellite data-derived indices to statistically validate the index-based methods:

$$NDVI = \frac{(IR\ band - R\ band)}{(IR\ band + R\ band)} \quad (6)$$

$$NDWI = \frac{(Green\ band - NIR\ band)}{(Green\ band + NIR\ band)} \quad (7)$$

$$NDBI = \frac{(MIR - NIR)}{(MIR + NIR)} \quad (8)$$

We also used a visual interpretation procedure to evaluate the accuracy assessment of LULC models. Furthermore, Karl Pearson's coefficient of correlation technique was applied to understand the association among the results of area coverage of land use classes obtained from the six LULC models. Higher correlation coefficient values indicate conformity of the models.

## 3. Results

### 3.1. LULC Classification

The spatial analysis of the LULC map shows that the built-up area, and rivers and wetland are more prominent and clearer in the outputs of SVM and random forest classifiers, while they are least prominent in the output of SAM. On the other hand, the fallow land and agricultural land are more prominent in the output of ANN, followed by fuzzy ARTMAP and Mahalanobis distance classifiers. The vegetation cover and sand bar are fairly classified in all classifiers. In RF and SVM, they are excellently classified (Figure 4). Overall, maximum coverage of built-up land was classified by the SVM and fuzzy ARTMAP methods, whereas least coverage of built-up land was classified by SAM and random forest. On the other hand, the highest coverage of vegetation is found by the SAM classifier, followed by RF and SVM classifiers, while the least coverage is found by ANN. The coverage of fallow land is completely reciprocal to the vegetation cover and the ANN classifier has the highest coverage, followed by fuzzy ARTMAP and MD classifiers, while SAM has the least coverage (Table 3). The coverage of rivers and wetland and sand bar are almost equally classified in all classifiers.

Table 3. Percentage of areal coverage of different LULC classes.

| Land Use Class | Agriculture Land (%) | Fallow Land (%) | Sand Bar (%) | Settlement (%) | Vegetation (%) | River and Wetlands (%) | Total (%) |
|---|---|---|---|---|---|---|---|
| Artificial neural network (ANN) | 10.88 | 18.29 | 1.3 | 14.03 | 44.07 | 11.43 | 100 |
| Fuzzy adaptive resonance theory-supervised predictive mapping (ARTMAP) | 6.72 | 14.34 | 1.74 | 19.59 | 46.31 | 11.3 | 100 |
| Mahalanobis distance (MD) | 8.25 | 13.65 | 1.2 | 17.66 | 45.48 | 13.75 | 100 |
| Support vector machine (SVM) | 12.27 | 10.4 | 1.5 | 17.95 | 48.31 | 9.57 | 100 |
| Random forest (RF) | 18.2 | 5.37 | 0.94 | 12.68 | 53.92 | 8.89 | 100 |
| Spectral angle mapper (SAM) | 12.54 | 6.16 | 1.84 | 9.99 | 58.62 | 10.86 | 100 |
| Average (%) | 11.47 | 11.36 | 1.42 | 15.32 | 49.45 | 10.96 | |
| Standard deviation (SD) | 4.01 | 5.02 | 0.34 | 3.67 | 5.65 | 1.69 | |
| Coefficient of variation (CV) (%) | 34.93 | 44.15 | 23.95 | 24.02 | 11.43 | 15.44 | |

Table 3 shows the percentage share of each LULC class with respect to the total land coverage in the study area for each classifier. Vegetation cover is the most dominant land-use class in the region classified by all classifiers. It covers about half of the total land surface, while the sand bar has the least share in the total land surface area. The percentage share of vegetation cover in total area varies from 44.07% by ANN to 58.62% by SAM. The built-up area (from 9.99% by SAM to 19.59% by fuzzy ARTMAP) and fallow land (from 5.37% by RF to 18.29% by ANN) and agricultural land (from 6.72% by fuzzy ARTMAP to 18.20% by RF) are at the second, third and fourth positions in terms of areal share, while sand bar has the least percentage share with respect to the total surface area by all the classifiers used (from 0.94% by RF to 1.84% by SAM).

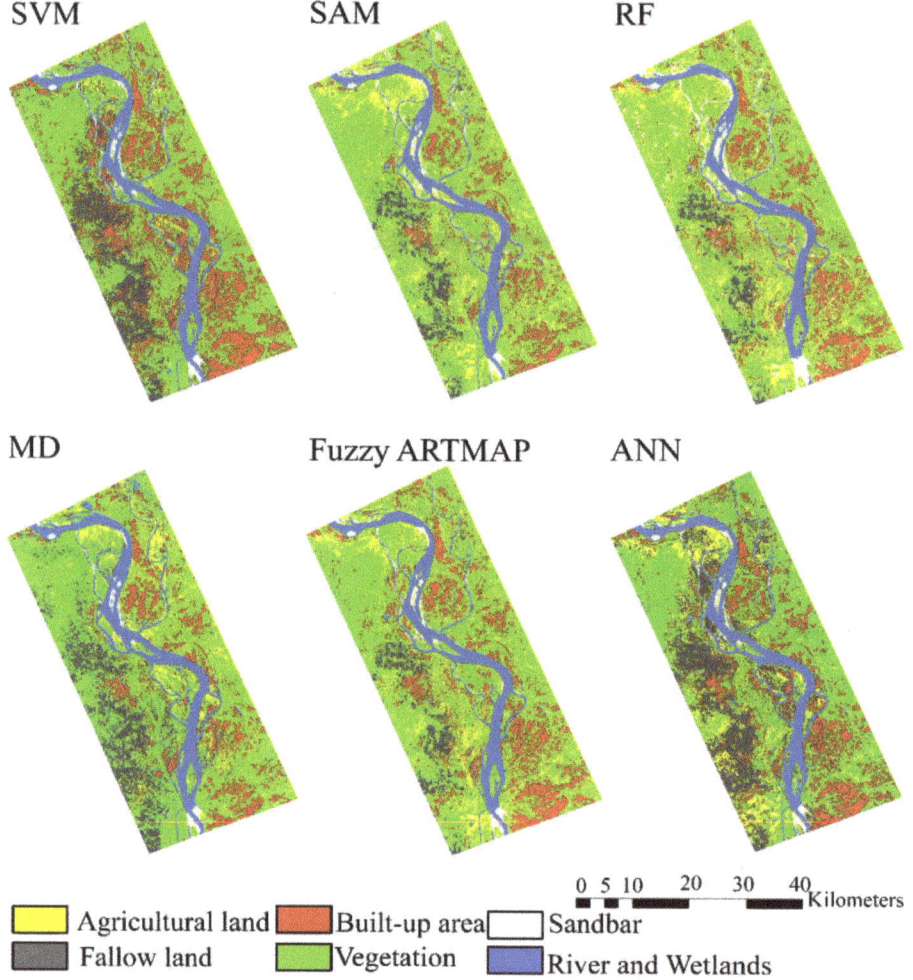

**Figure 4.** LULC map with different machine-learning techniques.

Computed standard deviation (SD) and coefficient of variation (CV) among the percentage share of area in a single LULC class by different classifiers are also displayed in Table 3. This vividly exhibits that vegetation, river and water bodies are classified more accurately as all the classifiers accounted for quite uniform areas with very low coefficient of variation. In contrast, fallow land and agricultural land are less well classified as all the classifiers accounted areal extent with considerable differences. Based on the result of similarity test, it can be stated that fuzzy ARTmap and MD, fuzzy ARTmap and SVM methods are quite similar in their performance. The difference is found to be maximum between ANN and SAM algorithms (Table 4).

Table 4. Similarity ratio matrix of the proportion of LULC between two algorithms.

| Land Use Class | ANN | Fuzzy ARTMAP | MD | SVM | RF | SAM |
|---|---|---|---|---|---|---|
| Artificial neural network (ANN) | 1 | 1.61 [i], 1.28 [ii], 0.72 [iii], 0.95 [iv], 1.01 [v], 1.11 [vi] | 1.31 [i], 1.34 [ii], 0.79 [iii], 0.97 [iv], 0.83 [v], 1.05 [vi] | 0.88 [i], 1.75 [ii], 0.78 [iii], 0.91 [iv], 1.19 [v], 1.10 [vi] | 0.59 [i], 3.04 [ii], 1.11 [iii], 0.82 [iv], 1.29 [v], 1.37 [vi] | 0.86 [i], 2.97 [ii], 1.40 [iii], 0.75 [iv], 1.05 [v], 1.41 [vi] |
| Fuzzy adaptive resonance theory-supervised predictive mapping (ARTMAP) | | 1 | 0.81 [i], 1.05 [ii], 1.11 [iii], 1.02 [iv], 0.83 [v], 0.96 [vi] | 0.54 [i], 1.38 [ii], 1.09 [iii], 0.96 [iv], 1.18 [v], 1.03 [vi] | 0.34 [i], 2.67 [ii], 1.54 [iii], 0.86 [iv], 1.27 [v], 1.34 [vi] | 0.5 [i], 2.33 [ii], 1.96 [iii], 0.79 [iv], 1.04 [v], 1.32 [vi] |
| Mahalanobis distance (MD) | | | 1 | 0.67 [i], 1.31 [ii], 0.98 [iii], 0.94 [iv], 1.44 [v], 1.07 [vi] | 0.45 [i], 2.54 [ii], 1.39 [iii], 0.84 [iv], 1.55 [v], 1.35 [vi] | 0.66 [i], 2.21 [ii], 1.77 [iii], 0.78 [iv], 1.27 [v], 1.34 [vi] |
| Support vector machine (SVM) | | | | 1 | 0.67 [i], 1.94 [ii], 1.42 [iii], 0.9 [iv], 1.08 [v], 1.2 [vi] | 0.98 [i], 1.69 [ii], 1.8 [iii], 0.82 [iv], 0.88 [v], 1.23 [vi] |
| Random forest (RF) | | | | | 1 | 1.45 [i], 0.87 [ii], 1.27 [iii], 0.92 [iv], 0.82 [v], 1.07 [vi] |
| Spectral angle mapper (SAM) | | | | | | 1 |

i = agriculture land, ii = Fallow land, iii = Settlement, iv = vegetation, v = River and wetlands, vi = Average of all land-use classes.

### 3.2. Validation of the LULC Classification

The overall accuracy (in percentage) using Kappa coefficient (K) for all the classifiers is shown in Table 5. The RF classifier has been detected as the highly accurate LULC model with Kappa coefficient of 0.89 among all the classifiers followed by ANN (K = 0.87), SVM (K = 0.86), fuzzy ARTMAP (K = 0.85), SAM (K = 0.84) and MD (K = 0.82). RF, ANN and SVM models exhibit excellent agreement and the other models show very good agreement between classified LULC map and ground reality. All the models can be treated as useful but the RF algorithm can be recommended as the best suited classifier of LULC. However, the agricultural land and river and wetland were classified better by using the RF and ANN algorithms (user's accuracy of RF and ANN: 94%, 92%) in comparison to the other algorithms. Similarly, most of the LULC classes were well classified by using RF, fuzzy ARTmap and ANN (See details in supplementary Table S1). The computed area under curve (AUC) of ROC and RMSE stated in Table 5 also yield the same result as identified when using the Kappa coefficient.

Table 5. Overall accuracy level of the classifiers used in this study.

| Methods | Kappa Coefficient (K) | Area Under Curve (AUC) | Root Mean Square Error (RMSE) |
|---|---|---|---|
| ANN | 0.87 | 0.89 | 0.09 |
| MD | 0.82 | 0.83 | 0.28 |
| Fuzzy ARTMAP | 0.85 | 0.86 | 0.17 |
| SVM | 0.86 | 0.87 | 0.11 |
| RF | 0.89 | 0.91 | 0.006 |
| SAM | 0.84 | 0.85 | 0.23 |

The areas of the three spectral indices; normalized differential water index (NDWI), normalized differential vegetation index (NDVI), and normalized differential built-up index (NDBI) have been computed and compared with the areas of water body, vegetation-agricultural land and built-up of LULC maps by using the six classifiers. Results show that the RF classifier performs better than the other classifiers because the total area of three spectral indices is strongly correlated with the area of three LULC classes (Table 6). It is thus considered as the best-fit classifier for preparing LULC in the present study area. Table 6 shows the departure of the area between spectral indices (NDWI, NDVI, and NDBI) and maximum likelihood (ML) algorithms of three LULC units (Figure 5). The total area of NDVI-based vegetation and agricultural land in the three windows is 155.87 km$^2$ and the departure of the area is −0.68 km$^2$. It is very close to the area detected by the RF classifier (156.55 km$^2$). The NDWI-based water body area is 79.57 km$^2$ and departure of area is 0.78km$^2$ in the selected window. It is the most similar with the water body area computed by the RF classifier (78.79 km$^2$) (Table 6). A similar result is found in the case of NDBI-based built-up area.

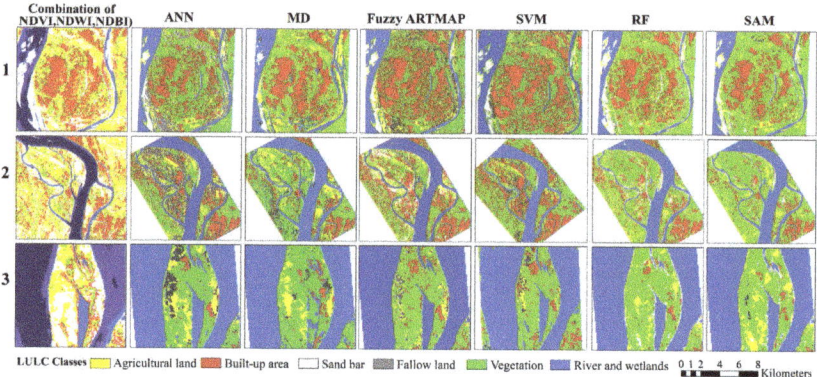

**Figure 5.** Validation of LULC of different classifiers with satellite data-derived indices (normalized differential vegetation index (NDVI), normalized differential water index (NDWI), and normalized differential built-up index (NDBI)).

The earlier analysis of closeness between the LULC models and satellite data-derived indices is based on the comparison of the absolute values, but it does not assure robustness. Similarity in area coverage does not always refer to the identical geographical location of any feature. Therefore, we conducted correlation matrix analysis between all LULC models and satellite data-derived indices. Figure 6a presents the correlation between the vegetation and agricultural land by the six classifiers and NDVI in window 1, where the maximum correlation (0.99) can be found by the RF classifier and NDVI at 0.001% significance level, followed by ANN (0.98), Fuzzy ARTMAP (0.97) and SVM (0.97), while the least correlation can be found by the Mahalanobis distance (0.87) with NDVI. The highest correlation (0.96) between water body by RF and NDWI can be found in window 2 at a significance level of 0.001% (Figure 6b), followed by Fuzzy ARTMAP (0.95), SVM (0.94) and ANN (0.93), while the least correlation (0.88) can be detected between SAM and NDWI (Figure 6b). The highest correlation (1) is found between built-up area by RF classifier and NDBI at a significant level of 0.001% followed by Fuzzy ARTMAP (1) at 0.01% of significance level, SVM (0.99) and ANN (0.93) in window 3 (Figure 6c). The lowest correlation (0.91) between SAM and NDBI has been detected (Figure 6c).

Table 6. Area of different LULC computed by the spectral indices and departure of area from the computed areas of the LULC by the Machine Learning (ML) algorithms.

| LULC | | ANN | MD | Fuzzy ART | SVM | RF | SAM | Spectral Indices |
|---|---|---|---|---|---|---|---|---|
| Vegetation and Agricultural Land | Window 1 | 57.52 (3.72) | 62.75 (−1.51) | 55.83 (5.41) | 50.47 (10.77) | 54.07 (7.17) | 46.87 (14.37) | 61.24 |
| | Window 2 | 71.57 (14.98) | 51.66 (34.89) | 73 (13.55) | 66.69 (19.79) | 94.5 (−7.95) | 68.44 (18.11) | 86.55 |
| | Window 3 | 7.26 (0.81) | 7.96 (0.11) | 7.15 (0.92) | 8.25 (−0.18) | 7.98 (0.09) | 7.95 (0.12) | 8.07 |
| | Total | 136.35 (19.52) | 122.37 (33.5) | 135.98 (19.89) | 125.41 (30.46) | 156.55 (−0.68) | 123.26 (32.61) | 155.87 |
| Water Body | Window 1 | 16.65 (−0.46) | 21.26 (−5.07) | 11.36 (4.83) | 19.7 (−3.51) | 15.9 (0.29) | 12.96 (3.23) | 16.19 |
| | Window 2 | 50.94 (−1.06) | 56.34 (−6.46) | 50.9 (−1.02) | 57.67 (−7.79) | 49.41 (0.47) | 45.6 (4.28) | 49.88 |
| | Window 3 | 13.78 (−0.28) | 13.68 (−0.18) | 13.72 (0.22) | 16.35 (−2.85) | 13.48 (0.02) | 13.04 (0.46) | 13.5 |
| | Total | 81.37 (−1.8) | 91.28 (−11.71) | 75.98 (3.59) | 93.72 (−14.15) | 78.79 (0.78) | 71.6 (7.97) | 79.57 |
| Built-up Area | Window 1 | 24.66 (−0.63) | 25.35 (−1.32) | 33.12 (−9.09) | 32.56 (−8.53) | 21.98 (2.05) | 20.4 (3.63) | 24.03 |
| | Window 2 | 24.46 (6.54) | 20.35 (10.65) | 38.9 (−7.9) | 52.1 (−21.1) | 28.36 (2.64) | 52.1 (−21.1) | 31 |
| | Window 3 | 0.55 (0.06) | 0.55 (0.06) | 1.19 (−0.58) | 1.29 (−0.68) | 0.51 (0.1) | 0.37 (0.24) | 0.61 |
| | Total | 49.67 (5.97) | 46.25 (9.39) | 73.21 (−17.57) | 85.95 (−30.31) | 50.85 (4.79) | 72.87 (−17.23) | 55.64 |

Values within parenthesis indicate the departure of area computed in spectral indices and respective MLs.

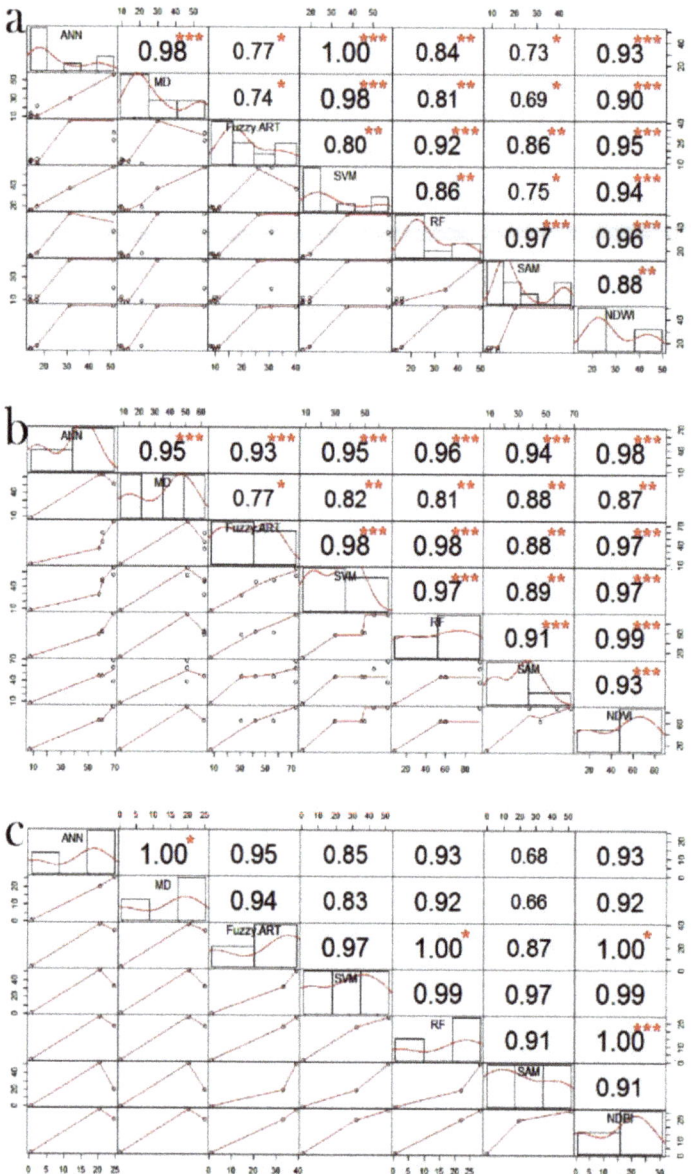

**Figure 6.** The correlation among the six LULC classifiers and satellite data-derived indices (NDWI, NDVI and NDBI) for (**a**) window 1, (**b**) window 2, and (**c**) window 3. (*, ** and *** are the significant levels at 0.001, 0.01, and 0.05, respectively).

We validated the LULC modelling using six classifiers by comparing the high-resolution images provided by Google Earth. The aforementioned three sites were selected for comparison (Figure 5). The water bodies and wetlands have been classified very well and they can be matched with the Google Earth images in three sites, whereas both vegetation and agricultural land have been prominently visualized by all classifiers. In the cases of RF, fuzzy ARTMAP and SVM, the land-use features were classified very well as proved by comparing with the Google Earth images. Although it is very difficult to model a built-up area with 30-meter resolution images, machine-learning algorithms with good training sites can perform predictions very well. Therefore, in the present study, some classifiers like MD and fuzzy ARTMAP generated more built-up area than the other classifiers and even the reality. However, the ANN, RF and SAM have classified built-up area as found in Google Earth images and even similar to what we have found in the field survey.

## 4. Discussion

### 4.1. Variation in LULC in the Output of Classifiers Used

Several studies reported that the areas of LULC classes are not equal in all the classification techniques, whether machine-learning algorithms or traditional classification techniques are adopted [54,91,92]. In this study, the variation is also found in the results of six classifiers (Figure 5). The area under any land-use class of a classifier does not exactly match with the area under the same land-use class of another different classifier. The area under each land-use class for the same region also varied in the different satellite data due to the atmospheric, illumination and geometric variations [92]. However, the differences in area under LULC classes of different classifiers occur due to the differences in the parameter optimization of the models, techniques and the accuracy differences in the algorithms used [42,92]. Furthermore, a few studies reported that the machine-learning techniques do not have significant difference in the results [42,52]. In our study, we also found similar kinds of result with small variation in a few land-use classes but not in every case. The coefficient of variation showed significant difference of area computed under various LULC classes. The chi square value clearly exhibited that as the difference in result produced in applied classifiers is significantly high, it is not due to random chance. Hence, we need to justify the suitability of any one or two model(s) adopted here.

### 4.2. Comparison of Accuracy Assessment of Different Classifiers with the Literature

The accuracy of a classification varied with methods, techniques, time and space [41,52,93,94]. Several studies reported minor to moderate fluctuation in the accuracy of the LULC classification using different classifiers [95–98]. The accuracy assessment in this study shows a small variation among the outputs of the classifiers used in the present case. The accuracy of a LULC classification does not only vary with classifier but also with space and time. This is possibly due to the influence of atmospheric, surface and illumination variations [53,61]. The Kappa coefficient is the most popular technique used to analyse the accuracy. The result shows that the maximum accuracy has been observed in the case of the RF classifier (0.89). Previous studies like Adam et al. [42] and Ma et al. [57] noted that the accuracy levels were 0.93 and 0.90, respectively, for the RF classifier. On the other hand, the minimum accuracy has been found in the case of MD (0.82). The previous studies on LULC classification using MD classifier reported the accuracy level of 0.89 [99], which is higher than the result of the present study. A small difference is found between the previous study and the present study on the accuracy levels of ANN, SAM, fuzzy ARTMAP and SVM [75,100,101]. In this study, the validation of LULC models using satellite image-derived indices is novel and the findings show that RF has modelled LULC in a very good manner, followed by Fuzzy ARTMA, SVM and ANN. However, on the basis of Kappa coefficient, index-based accuracy assessment and empirical observations, it can be concluded that RF is the best classifier for LULC classification. A number of studies from literature also noted that SVM and RF have the highest accuracy in LULC classification (Table 7), while SAM and MD have the lowest accuracy levels. Furthermore, Li et al. [53] noted that the accuracy of SVM and RF has very little difference, but the difference increases between either SVM and ANN or RF and ANN. The result shows that the difference between accuracies of RF and SVM is more than that between RF and ANN.

Table 7. Suitability of classifier in LULC modelling as per the previous works.

| S. No. | Methods Used for LULC Classification | Best Method | Study area | Authors |
|---|---|---|---|---|
| 1 | Random forest (RF), K-nearest neighbor (KNN), Support vector machine (SVM) | SVM | Red river delta, Vietnam | [93] |
| 2 | RF, SVM | RF | Eastern suburbs of Deyang city, Chaina | [57] |
| 3 | naïve Bayes (NB), Decision trees J-48, RF, Multilayer perception, SVM | SVM | Brazialian Tropical Savana | [52] |
| 4 | SVM, NB, Decision trees (DT), KNN | SVM | Haidian District of Beijing, Chaina | [100] |
| 5 | Maximum likelihood classifier (MLC), SVM, Artificial neural network (ANN) | ANN | Walnutcreek, Lowa, USA | [77] |
| 6 | SVM, ANN, Classification and regression tree (CART) | SVM | Albemarle-Pamlico Estuary System, | [63] |
| 7 | Bagging, Boosting, RF, Classification tree | Boosting and RF | Cape cod, Massachusetts, USA | [102] |
| 8 | MLC, ANN, SVM | SVM | Koh Tao, western Gulf of Thailand | [101] |
| 9 | RF, SVM | RF | Eastern Coast of KwaZulu-Natal, South Africa | [42] |
| 10 | DT, RF, ANN, SVM | RF | Granada, Spain | [93] |
| 11 | MLC, SVM, DT | DT | Kibale Sub-county, Eastern Uganda | [103] |
| 12 | NB, AdaBoost, ANN, RF, SVM | RF | Riau, Jambi and West Sumatra, Indonesia | [104] |
| 13 | SVM, RF, ANN | ANN | North Western part of Karkonosze National Park, Poland | [105] |
| 14 | MLC, SVM | SVM | Johor, Malaysia | [106] |
| 15 | SVM, MLC, ANN | SVM | Klang District, Malaysia | [107] |
| 17 | RF, MLC | MLC | Sihu Township of Yun-Ling, Taiwan | [108] |
| 18 | ANN, SVM | ANN | Abbottabad, Pakistan | [109] |
| 19 | ANN, SVM, Rotation Forest, RF, Meta Classifier | RF and Meta Classifier | Gomukh, Uttarakhand, India | [110] |

## 5. Conclusions

This study was conducted to examine the accuracy of different machine-learning classifiers for LULC mapping for satellite observations. The aim was to suggest the best classifier. Six machine-learning algorithms were applied on the Landsat 8 (OLI) data for the LULC classification. Accuracy assessments were undertaken by using the Kappa coefficient, an index-based technique and empirical observations. Results suggest that the area under each LULC class varies under different classifiers. The maximum variation is observed for the agricultural and fallow lands, while the minimum for the water bodies and wetlands. Such variation requires a need to prove the best suited classifier.

Furthermore, the Kappa coefficient and index-based analysis also show variation in the accuracy of each LULC classifier. The variation in the accuracy of the classifiers used is found to be minor, but this minor variation has very important significance in the area of LULC mapping and planning. Both the Kappa coefficient and index-based analysis show that the RF has the highest accuracy of all classifiers applied in this study. To justify the result, previous literature on this was taken into consideration and most of the studies concluded that either RF or ANN is the best classifier. Although the previous studies found a higher accuracy for RF and ANN than this study, this study concludes that RF is the best machine-learning classifier for LULC modelling in the highly dynamic charland-dominated areas.

Furthermore, numerous studies suggested that the accuracy of LULC mapping varies with time and location. Thus, for future research, it is suggested to analyse the accuracy of the classifiers for different morphoclimatic and geomorphic conditions.

**Supplementary Materials:** The following are available online at http://www.mdpi.com/2072-4292/12/7/1135/s1, Table S1: The accuracy assessment of LULC mapping of six classifiers using Kappa coefficient.

**Author Contributions:** Conceptualization, S.T.; methodology, S.T., S.P.; software, S.T.; validation, S.T., S.P. and S.M.; formal analysis, S.T., S.P. and Y.-A.L.; investigation, S.T., P.S.; resources, S.T., S.; data curation, S.T., P.S; writing—original draft preparation, S., S.M., S.P., S.T. and A.R.; writing—review and editing, Y.-A.L., S.P., S.M., S.T. and A.R.; visualization, S.T., P.S. and S.M.; supervision, Y.-A.L., S.P., A.R. and S.T.; project administration, S.T., S.M., S.P., Y.-A.L. and P.S.; funding acquisition, Y.-A.L. All authors have read and agreed to the published version of the manuscript.

**Funding:** This work was funded by the Ministry of Science and Technology under Grant MOST 108-2111-M-008 -036 -MY2, and Grant number 108-2923-M-008 -002 -MY3.

**Conflicts of Interest:** The authors declare no conflict of interest.

## References

1. Hashem, N.; Balakrishnan, P. Change analysis of land use/land cover and modelling urban growth in Greater Doha, Qatar. *Ann. Gis* **2015**, *21*, 233–247. [CrossRef]
2. Rahman, A.; Kumar, S.; Fazal, S.; Siddiqui, M.A. Assessment of land use/land cover change in the North-West District of Delhi using remote sensing and GIS techniques. *J. Indian Soc. Remote Sens.* **2012**, *40*, 689–697. [CrossRef]
3. Liou, Y.A.; Nguyen, A.K.; Li, M.H. Assessing spatiotemporal eco-environmental vulnerability by Landsat data. *Ecol. Indic.* **2017**, *80*, 52–65. [CrossRef]
4. Nguyen, K.A.; Liou, Y.A. Mapping global eco-environment vulnerability due to human and nature disturbances. *MethodsX* **2019**, *6*, 862–875. [CrossRef]
5. Nguyen, K.A.; Liou, Y.A. Global mapping of eco-environmental vulnerability from human and nature disturbances. *Sci. Total Environ.* **2019**, *664*, 995–1004. [CrossRef] [PubMed]
6. Talukdar, S.; Pal, S. Wetland habitat vulnerability of lower Punarbhaba river basin of the uplifted Barind region of Indo-Bangladesh. *Geocarto Int.* **2018**, 1–30. [CrossRef]
7. Nguyen, A.K.; Liou, Y.A.; Li, M.H.; Tran, T.A. Zoning eco-environmental vulnerability for environmental management and protection. *Ecol. Indic.* **2016**, *69*, 100–117. [CrossRef]
8. Che, T.; Xiao, L.; Liou, Y.A. Changes in glaciers and glacial lakes and the identification of dangerous glacial lakes in the Pumqu River Basin, Xizang (Tibet). *Adv. Meteorol.* **2014**, *2014*, 903709. [CrossRef]
9. Dao, P.D.; Liou, Y.A. Object-based flood mapping and affected rice field estimation with Landsat 8 OLI and MODIS data. *Remote Sens.* **2015**, *7*, 5077–5097. [CrossRef]
10. Liou, Y.A.; Kar, S.K.; Chang, L. Use of high-resolution FORMOSAT-2 satellite images for post-earthquake disaster assessment: A study following the 12 May 2008 Wenchuan Earthquake. *Int. J. Remote Sens.* **2010**, *31*, 3355–3368. [CrossRef]
11. Liou, Y.A.; Sha, H.C.; Chen, T.M.; Wang, T.S.; Li, Y.T.; Lai, Y.C.; Lu, L.T. Assessment of disaster losses in rice paddy field and yield after Tsunami induced by the 2011 great east Japan earthquake. *J. Mar. Sci. Technol.* **2012**, *20*, 618–623.
12. Zhang, Y.; Ge, T.; Tian, W.; Liou, Y.A. Debris Flow Susceptibility Mapping Using Machine-Learning Techniques in Shigatse Area, China. *Remote Sens.* **2019**, *11*, 2801. [CrossRef]
13. Talukdar, S.; Pal, S. Effects of damming on the hydrological regime of Punarbhaba river basin wetlands. *Ecol. Eng.* **2019**, *135*, 61–74. [CrossRef]
14. Talukdar, S.; Singha, P.; Mahato, S.; Praveen, B.; Rahman, A. Dynamics of ecosystem services (ESs) in response to land use land cover (LU/LC) changes in the lower Gangetic plain of India. *Ecol. Indic.* **2020**, *112*, 106121.
15. Chen, Z.; Wang, L.; Wei, A.; Gao, J.; Lu, Y.; Zhou, J. Land-use change from arable lands to orchards reduced soil erosion and increased nutrient loss in a small catchment. *Sci. Total Environ.* **2019**, *648*, 1097–1104. [CrossRef] [PubMed]
16. Braun, A.; Hochschild, V. A SAR-Based Index for Landscape Changes in African Savannas. *Remote Sens.* **2017**, *9*, 359. [CrossRef]

17. Nguyen, K.A.; Liou, Y.A.; Tran, H.P.; Hoang, P.P.; Nguyen, T.H. Soil salinity assessment by using near-infrared channel and Vegetation Soil Salinity Index derived from Landsat 8 OLI data: A case study in the Tra Vinh Province, Mekong Delta, Vietnam. *Prog. Earth Planet. Sci.* **2020**, *7*, 1–16. [CrossRef]
18. Mas, J.F.; Lemoine-Rodríguez, R.; González-López, R.; López-Sánchez, J.; Piña-Garduño, A.; Herrera-Flores, E. Land use/land cover change detection combining automatic processing and visual interpretation. *Eur. J. Remote Sens.* **2017**, *50*, 626–635. [CrossRef]
19. Reis, S. Analyzing Land Use/Land Cover Changes Using Remote Sensing and GIS in Rize, North-East Turkey. *Sensors* **2008**, *8*, 6188–6202. [CrossRef]
20. Dutta, D.; Rahman, A.; Paul, S.K.; Kundu, A. Changing pattern of urban landscape and its effect on land surface temperature in and around Delhi. *Environ. Monit. Assess.* **2019**, *191*, 551. [CrossRef]
21. Hoan, N.T.; Liou, Y.A.; Nguyen, K.A.; Sharma, R.C.; Tran, D.P.; Liou, C.L.; Cham, D.D. Assessing the Effects of Land-Use Types in Surface Urban Heat Islands for Developing Comfortable Living in Hanoi City. *Remote Sens.* **2018**, *10*, 1965. [CrossRef]
22. Rahman, A.; Aggarwal, S.P.; Netzband, M.; Fazal, S. Monitoring Urban Sprawl Using Remote Sensing and GIS Techniques of a Fast Growing Urban Centre, India. *IEEE J. Sel. Top. Appl. Earth Obs. Remote. Sens.* **2011**, *4*, 56–64. [CrossRef]
23. Kumari, B.; Tayyab, M.; Hang, H.T.; Khan, M.F.; Rahman, A. Assessment of public open spaces (POS) and landscape quality based on per capita POS index in Delhi, India. *SN Appl. Sci.* **2019**, *1*, 368.
24. Pal, S.; Talukdar, S. Assessing the role of hydrological modifications on land use/land cover dynamics in Punarbhaba river basin of Indo-Bangladesh. *Environ. Dev. Sustain.* **2018**, *22*, 363–382. [CrossRef]
25. Cerovski-Darriau, C.; Roering, J.J. Influence of anthropogenic land-use change on hillslope erosion in the Waipaoa River Basin, New Zealand. *Earth Surf. Process. Landf.* **2016**, *41*, 2167–2176. [CrossRef]
26. Pal, S.; Kundu, S.; Mahato, S. Groundwater potential zones for sustainable management plans in a river basin of India and Bangladesh. *J. Clean. Prod.* **2020**, *257*, 120311. [CrossRef]
27. Mahato, S.; Pal, S. Groundwater potential mapping in a rural river basin by union (OR) and intersection (AND) of four multi-criteria decision-making models. *Nat. Resour. Res.* **2019**, *28*, 523–545. [CrossRef]
28. Langat, P.K.; Kumar, L.; Koech, R.; Ghosh, M.K. Monitoring of land use/land-cover dynamics using remote sensing: A case of Tana River Basin, Kenya. *Geocarto Int.* **2019**. [CrossRef]
29. Hoffmann, J. The future of satellite remote sensing in hydrogeology. *Hydrogeol. J.* **2005**, *13*, 247–250. [CrossRef]
30. Liou, Y.-A.; Wu, A.-M.; Lin, H.-Y. FORMOSAT-2 Quick Imaging. In *Optical Payloads for Space Missions*; Qian, S.-E., Ed.; Wiley: Oxford, UK, 2016; 1008p, ISBN 9781118945148.
31. Wentz, E.A.; Nelson, D.; Rahman, A.; Stefanov, W.L.; Roy, S.S. Expert system classification of urban land use/cover for Delhi, India. *Int. J. Remote Sens.* **2008**, *29*, 4405–4427. [CrossRef]
32. Scaioni, M.; Longoni, L.; Melillo, V.; Papini, M. Remote Sensing for Landslide Investigations: An Overview of Recent Achievements and Perspectives. *Remote Sens.* **2014**, *6*, 9600–9652. [CrossRef]
33. Chen, Z.; Wang, J. Land use and land cover change detection using satellite remote sensing techniques in the mountainous Three Gorges Area, China. *Int. J. Remote Sens.* **2010**, *31*, 1519–1542. [CrossRef]
34. Pal, M.; Mather, P.M. Assessment of the effectiveness of support vector machines for hyperspectral data. *Future Gener. Comput. Syst.* **2004**, *20*, 1215–1225. [CrossRef]
35. Wittke, S.; Yu, X.; Karjalainen, M.; Hyyppä, J.; Puttonen, E. Comparison of two dimensional multitemporal Sentinel-2 data with three-dimensional remote sensing data sources for forest inventory parameter estimation over a boreal forest. *Int. J. Appl. Earth Obs. Geoinf.* **2019**, *76*, 167–178. [CrossRef]
36. Viana, C.M.; Girão, I.; Rocha, J. Long-Term Satellite Image Time-Series for Land Use/Land Cover Change Detection Using Refined Open Source Data in a Rural Region. *Remote Sens.* **2019**, *11*, 1104. [CrossRef]
37. Gurjar, S.K.; Tare, V. Estimating long-term LULC changes in an agriculture-dominated basin using CORONA (1970) and LISS IV (2013–14) satellite images: A case study of Ramganga River, India. *Environ. Monitor. Assess.* **2019**, *191*, 217. [CrossRef] [PubMed]
38. Toure, S.I.; Stow, D.A.; Shih, H.C.; Weeks, J.; Lopez-Carr, D. Land cover and land use change analysis using multi-spatial resolution data and object-based image analysis. *Remote Sens. Environ.* **2018**, *210*, 259–268. [CrossRef]
39. Usman, M.; Liedl, R.; Shahid, M.A.; Abbas, A. Land use/land cover classification and its change detection using multi-temporal MODIS NDVI data. *J. Geogr. Sci.* **2015**, *25*, 1479–1506. [CrossRef]

40. Stefanov, W.L.; Netzband, M. Assessment of ASTER Land Cover and MODIS NDVI Data at Multiple Scales for Ecological Characterization of an Arid Urban Center. *Remote Sens. Environ.* **2005**, *99*, 31–43. [CrossRef]
41. Maxwell, A.E.; Warner, T.A.; Fang, F. Implementation of machine-learning classification in remote sensing: An applied review. *Int. J. Remote Sens.* **2018**, *39*, 2784–2817. [CrossRef]
42. Adam, E.; Mutanga, O.; Odindi, J.; Abdel-Rahman, E.M. Land-use/cover classification in a heterogeneous coastal landscape using Rapid Eye imagery: Evaluating the performance of random forest and support vector machines classifiers. *Int. J. Remote Sens.* **2014**, *35*, 3440–3458. [CrossRef]
43. Wu, L.; Zhu, X.; Lawes, R.; Dunkerley, D.; Zhang, H. Comparison of machine learning algorithms for classification of LiDAR points for characterization of canola canopy structure. *Int. J. Remote Sens.* **2019**, *40*, 5973–5991. [CrossRef]
44. Halder, A.; Ghosh, A.; Ghosh, S. Supervised and unsupervised landuse map generation from remotely sensed images using ant based systems. *Appl. Soft Comput.* **2011**, *11*, 5770–5781. [CrossRef]
45. Ma, L.; Liu, Y.; Zhang, X.; Ye, Y.; Yin, G.; Johnson, B.A. Deep learning in remote sensing applications: A meta-analysis and review. *ISPRS J. Photogramm. Remote Sens.* **2019**, *152*, 166–177. [CrossRef]
46. Shih, H.C.; Stow, D.A.; Tsai, Y.H. Guidance on and comparison of machine learning classifiers for Landsat-based land cover and land use mapping. *Int. J. Remote Sens.* **2019**, *40*, 1248–1274. [CrossRef]
47. Camps-Valls, G.; Benediktsson, J.A.; Bruzzone, L.; Chanussot, J. Introduction to the issue on advances in remote sensing image processing. *IEEE J. Sel. Top. Signal Process.* **2011**, *5*, 365–369. [CrossRef]
48. Zhang, C.; Sargent, I.; Pan, X.; Li, H.; Gardiner, A.; Hare, J.; Atkinson, P.M. Joint Deep Learning for land cover and land use classification. *Remote Sens. Environ.* **2019**, *221*, 173–187. [CrossRef]
49. Teluguntla, P.; Thenkabail, P.S.; Oliphant, A.; Xiong, J.; Gumma, M.K.; Congalton, R.G.; Huete, A. A 30-m Landsat-derived cropland extent product of Australia and China using random forest machine learning algorithm on Google Earth Engine cloud computing platform. *ISPRS J. Photogramm. Remote Sens.* **2018**, *144*, 325–340. [CrossRef]
50. Pal, M. Random forest classifier for remote sensing classification. *Int. J. Remote Sens.* **2005**, *26*, 217–222. [CrossRef]
51. Civco, D.L. Artificial neural networks for land-cover classification and mapping. *Int. J. Geogr. Inf. Sci.* **1993**, *7*, 173–186. [CrossRef]
52. Camargo, F.F.; Sano, E.E.; Almeida, C.M.; Mura, J.C.; Almeida, T. A comparative assessment of machine-learning techniques for land use and land cover classification of the Brazilian tropical savanna using ALOS-2/PALSAR-2 polarimetric images. *Remote Sens.* **2019**, *11*, 1600. [CrossRef]
53. Li, X.; Chen, W.; Cheng, X.; Wang, L. A comparison of machine learning algorithms for mapping of complex surface-mined and agricultural landscapes using ZiYuan-3 stereo satellite imagery. *Remote Sens.* **2016**, *8*, 514. [CrossRef]
54. Rogan, J.; Franklin, J.; Stow, D.; Miller, J.; Woodcock, C.; Roberts, D. Mapping land-cover modifications over large areas: A comparison of machine learning algorithms. *Remote Sens. Environ.* **2008**, *112*, 2272–2283. [CrossRef]
55. Jamali, A. Evaluation and comparison of eight machine learning models in land use/land cover mapping using Landsat 8 OLI: A case study of the northern region of Iran. *SN Appl. Sci.* **2019**, *1*, 1448. [CrossRef]
56. Carranza-García, M.; García-Gutiérrez, J.; Riquelme, J.C. A framework for evaluating land use and land cover classification using convolutional neural networks. *Remote Sens.* **2019**, *11*, 274. [CrossRef]
57. Ma, L.; Li, M.; Ma, X.; Cheng, L.; Du, P.; Liu, Y. A review of supervised object-based land-cover image classification. *ISPRS J. Photogramm. Remote Sens.* **2017**, *130*, 277–293. [CrossRef]
58. Mountrakis, G.; Im, J.; Ogole, C. Support vector machines in remote sensing: A review. *ISPRS J. Photogramm. Remote Sens.* **2011**, *66*, 247–259. [CrossRef]
59. Deng, J.S.; Wang, K.; Deng, Y.H.; Qi, G.J. PCA-based land-use change detection and analysis using multitemporal and multisensor satellite data. *Int. J. Remote Sens.* **2008**, *29*, 4823–4838. [CrossRef]
60. Yang, C.; Wu, G.; Ding, K.; Shi, T.; Li, Q.; Wang, J. Improving land use/land cover classification by integrating pixel unmixing and decision tree methods. *Remote Sens.* **2017**, *9*, 1222. [CrossRef]
61. Manandhar, R.; Odeh, I.O.; Ancev, T. Improving the accuracy of land use and land cover classification of Landsat data using post-classification enhancement. *Remote Sens.* **2009**, *1*, 330–344. [CrossRef]
62. Latifovic, R.; Olthof, I. Accuracy assessment using sub-pixel fractional error matrices of global land cover products derived from satellite data. *Remote Sens. Environ.* **2004**, *90*, 153–165. [CrossRef]

63. Pal, S.; Ziaul, S.K. Detection of land use and land cover change and land surface temperature in English Bazar urban centre. *Egypt. J. Remote Sens. Space Sci.* **2017**, *20*, 125–145. [CrossRef]
64. Iqbal, S. Flood and Erosion Induced Population Displacements: A Socio-economic Case Study in the Gangetic Riverine Tract at Malda District, West Bengal, India. *J. Human Ecol.* **2010**, *30*, 201–211. [CrossRef]
65. Atkinson, P.M.; Tatnall, A.R. Introduction neural networks in remote sensing. *Int. J. Remote Sens.* **1997**, *18*, 699–709. [CrossRef]
66. Schuman, C.D.; Birdwell, J.D. Dynamic artificial neural networks with affective systems. *PLoS ONE* **2013**, *8*, e80455. [CrossRef] [PubMed]
67. Liou, Y.-A.; Tzeng, Y.C.; Chen, K.S. A neural network approach to radiometric sensing of land surface parameters. *IEEE Trans. Geosci. Remote Sens.* **1999**, *37*, 2718–2724. [CrossRef]
68. Liou, Y.-A.; Liu, S.-F.; Wang, W.-J. Retrieving soil moisture from simulated brightness temperatures by a neural network. *IEEE Trans. Geosci. Remote Sens.* **2001**, *39*, 1662–1673.
69. Lu, D.; Weng, Q. A survey of image classification methods and techniques for improving classification performance. *Int. J. Remote Sens.* **2007**, *28*, 823–870. [CrossRef]
70. Dixon, B.; Candade, N. Multispectral landuse classification using neural networks and support vector machines: One or the other, or both? *Int. J. Remote Sens.* **2008**, *29*, 1185–1206. [CrossRef]
71. Yilmaz, I.; Kaynar, O. Multiple regression, ANN (RBF, MLP) and ANFIS models for prediction of swell potential of clayey soils. *Expert Syst. Appl.* **2011**, *38*, 5958–5966. [CrossRef]
72. Moody, J.; Darken, C.J. Fast learning in networks of locally-tuned processing units. *Neural Comput.* **1989**, *1*, 281–294. [CrossRef]
73. Shafizadeh-Moghadam, H.; Hagenauer, J.; Farajzadeh, M.; Helbich, M. Performance analysis of radial basis function networks and multi-layer perceptron networks in modelling urban change: A case study. *Int. J. Geogr. Inf. Sci.* **2015**, *29*, 606–623. [CrossRef]
74. Ghassemieh, M.; Nasseri, M. Evaluation of stiffened end-plate moment connection through optimized artificial neural network. *J. Softw. Eng. Appl.* **2012**, *5*, 156–167. [CrossRef]
75. Huang, C.; Davis, L.S.; Townshend, J.R.G. An assessment of support vector machines for land cover classification. *Int. J. Remote Sens.* **2002**, *23*, 725–749. [CrossRef]
76. Bouaziz, M.; Eisold, S.; Guermazi, E. Semiautomatic approach for land cover classification: A remote sensing study for arid climate in Southeastern Tunisia. *Euro Mediterr. J. Environ. Integr.* **2017**, *2*, 24. [CrossRef]
77. Srivastava, P.K.; Han, D.; Rico-Ramirez, M.A.; Bray, M.; Islam, T. Selection of classification techniques for land use/land cover change investigation. *Adv. Space Res.* **2012**, *50*, 1250–1265. [CrossRef]
78. Awad, M.; Khanna, R. Support vector machines for classification. In *Efficient Learning Machines*; Apress: Berkeley, CA, USA, 2015; pp. 39–66.
79. Mathur, A.; Foody, G.M. Multiclass and binary SVM classification: Implications for training and classification users. *IEEE Geosci. Remote Sens. Lett.* **2008**, *5*, 241–245. [CrossRef]
80. Mannan, B.; Roy, J.; Ray, A.K. Fuzzy ARTMAP supervised classification of multi-spectral remotely-sensed images. *Int. J. Remote Sens.* **1998**, *19*, 767–774. [CrossRef]
81. Gopal, S. Fuzzy ARTMAP—A neural classifier for multispectral image classification. In *Spatial Analysis and GeoComputation*; Springer: Berlin/Heidelberg, Germany, 2006; pp. 209–237.
82. Breiman, L. Random forests. *Mach. Learn.* **2001**, *45*, 5–32. [CrossRef]
83. Abdullah, A.Y.M.; Masrur, A.; Adnan, M.S.G.; Baky, M.; Al, A.; Hassan, Q.K.; Dewan, A. Spatio-temporal patterns of land use/land cover change in the heterogeneous coastal region of Bangladesh between 1990 and 2017. *Remote Sens.* **2019**, *11*, 790. [CrossRef]
84. Liaw, A.; Wiener, M. Classification and Regression by randomForest. *R News* **2002**, *2*, 18–22.
85. Feng, Q.; Gong, J.; Liu, J.; Li, Y. Flood mapping based on multiple endmember spectral mixture analysis and random forest classifier—The case of Yuyao, China. *Remote Sens.* **2015**, *7*, 12539–12562. [CrossRef]
86. Mohan, B.S.S.; Sekhar, C.C. Class-Specific Mahalanobis Distance Metric Learning for Biological Image Classification. In *Image Analysis and Recognition—9th International Conference, ICIAR 2012, Aveiro, Portugal, 25–27 June 2012*; Campilho, A., Kamel, M., Eds.; Lecture Notes in Computer Science; Springer: Berlin/Heidelberg, Germany, 2012; Volume 7325, pp. 240–248.
87. Petropoulos, G.P.; Vadrevu, K.P.; Xanthopoulos, G.; Karantounias, G.; Scholze, M. A Comparison of Spectral Angle Mapper and Artificial Neural Network Classifiers Combined with Landsat TM Imagery Analysis for Obtaining Burnt Area Mapping. *Sensors* **2010**, *10*, 1967–1985. [CrossRef] [PubMed]

88. Li, D.; Ke, Y.; Gong, H.; Li, X. Object-Based Urban Tree Species Classification Using Bi-Temporal WorldView-2 and WorldView-3 Images. *Remote Sens.* **2015**, *7*, 16917–16937. [CrossRef]
89. Hurskainen, P.; Adhikari, H.; Siljander, M.; Pellikka, P.K.E.; Hemp, A. Auxiliary datasets improve accuracy of object-based land use/land cover classification in heterogeneous savanna landscapes. *Remote Sens. Environ.* **2019**, *233*, 111354. [CrossRef]
90. Monserud, R.A.; Leemans, R. Comparing global vegetation maps with the Kappa statistic. *Ecol. Model.* **1992**, *62*, 275–293. [CrossRef]
91. Abdi, A.M. Land cover and land use classification performance of machine learning algorithms in a boreal landscape using Sentinel-2 data. *GISci. Remote Sens.* **2019**, 1–20. [CrossRef]
92. Erbek, F.S.; Özkan, C.; Taberner, M. Comparison of maximum likelihood classification method with supervised artificial neural network algorithms for land use activities. *Int. J. Remote Sens.* **2004**, *25*, 1733–1748. [CrossRef]
93. Noi, P.T.; Kappas, M. Comparison of Random Forest, k-Nearest Neighbor, and Support Vector Machine Classifiers for Land Cover Classification Using Sentinel-2 Imagery. *Sensors* **2018**, *18*, 18.
94. Rodriguez-Galiano, V.F.; Chica-Rivas, M. Evaluation of different machine learning methods for land cover mapping of a Mediterranean area using multi-seasonal Landsat images and Digital Terrain Models. *Int. J. Digit. Earth* **2014**, *7*, 492–509. [CrossRef]
95. Foody, G.M. Harshness in image classification accuracy assessment. *Int. J. Remote Sens.* **2008**, *29*, 3137–3158. [CrossRef]
96. Rwanga, S.S.; Ndambuki, J.M. Accuracy assessment of land use/land cover classification using remote sensing and GIS. *Int. J. Geosci.* **2017**, *8*, 611. [CrossRef]
97. Islam, K.; Jashimuddin, M.; Nath, B.; Nath, T.K. Land use classification and change detection by using multi-temporal remotely sensed imagery: The case of Chunati wildlife sanctuary, Bangladesh. *Egypt. J. Remote Sens. Space Sci.* **2018**, *21*, 37–47. [CrossRef]
98. Leyk, S.; Uhl, J.H.; Balk, D.; Jones, B. Assessing the accuracy of multi-temporal built-up land layers across rural-urban trajectories in the United States. *Remote Sens. Environ.* **2018**, *204*, 898–917. [CrossRef] [PubMed]
99. Carvalho Júnior, O.A.; Guimarães, R.F.; Gillespie, A.R.; Silva, N.C.; Gomes, R.A. A new approach to change vector analysis using distance and similarity measures. *Remote Sens.* **2011**, *3*, 2473–2493. [CrossRef]
100. Qian, Y.; Zhou, W.; Yan, J.; Li, W.; Han, L. Comparing machine learning classifiers for object-based land cover classification using very high-resolution imagery. *Remote Sens.* **2015**, *7*, 153–168. [CrossRef]
101. Szuster, B.W.; Chen, Q.; Borger, M. A comparison of classification techniques to support land cover and land use analysis in tropical coastal zones. *Appl. Geogr.* **2011**, *31*, 525–532. [CrossRef]
102. Ghimire, B.; Rogan, J.; Galiano, V.R.; Panday, P.; Neeti, N. An evaluation of bagging, boosting, and random forests for land-cover classification in Cape Cod, Massachusetts, USA. *GIScience Remote Sens.* **2012**, *49*, 623–643. [CrossRef]
103. Otukei, J.R.; Blaschke, T. Land cover change assessment using decision trees, support vector machines and maximum likelihood classification algorithms. *Int. J. Appl. Earth Obs. Geoinf.* **2010**, *12*, S27–S31. [CrossRef]
104. Shiraishi, T.; Motohka, T.; Thapa, R.B.; Watanabe, M.; Shimada, M. Comparative Assessment of Supervised Classifiers for Land Use–Land Cover Classification in a Tropical Region Using Time-Series PALSAR Mosaic Data. *IEEE J. Sel. Top. Appl. Earth Obs. Remote. Sens.* **2014**, *7*, 1186–1199. [CrossRef]
105. Raczko, E.; Zagajewski, B. Comparison of support vector machine, random forest and neural network classifiers for tree species classification on airborne hyperspectral APEX images. *Eur. J. Remote Sens.* **2017**, *50*, 144–154. [CrossRef]
106. Deilmai, B.R.; Ahmad, B.B.; Zabihi, H. Comparison of two Classification methods (MLC and SVM) to extract land use and land cover in Johor Malaysia.7th IGRSM International Remote Sensing & GIS Conference and Exhibition, 22–23 April 2014, Kuala Lumpur, Malaysia. *IOP Conf. Ser. Earth Environ. Sci.* **2014**, *20*, 012052.
107. Ahmad, M.; Protasov, S.; Khan, A.M.; Hussain, R.; Khattak, A.M.; Khan, W.A. Fuzziness-based active learning framework to enhance hyperspectral image classification performance for discriminative and generative classifiers. *PLoS ONE* **2018**, *13*, e0188996. [CrossRef]
108. Lee, R.Y.; Ou, D.Y.; Shiu, Y.S.; Lei, T.C. Comparisons of using Random Forest and Maximum Likelihood Classifiers with Worldview-2 imagery for classifying Crop Types. In Proceedings of the 36th Asian Conference Remote Sensing Foster ACRS, Quezon City, Philippines, 24–28 October 2015.

109. Abbas, A.W.; Ahmad, A.; Shah, S.; Saeed, K. Parameter investigation of Artificial Neural Network and Support Vector Machine for image classification. In Proceedings of the 14th International Bhurban Conference on Applied Sciences and Technology (IBCAST), Islamabad, Pakistan, 10–14 January 2017; IEEE: Piscataway, NJ, USA, 2017; pp. 795–798.
110. Nijhawan, R.; Joshi, D.; Narang, N.; Mittal, A. A Futuristic Deep Learning Framework Approach for Land Use-Land Cover Classification Using Remote Sensing. In *Advanced Computing and Communication Technologies: Proceedings of the 11th ICACCT 2018*; Springer: Singapore, Singapore, 2018; Volume 702, p. 87.

 © 2020 by the authors. Licensee MDPI, Basel, Switzerland. This article is an open access article distributed under the terms and conditions of the Creative Commons Attribution (CC BY) license (http://creativecommons.org/licenses/by/4.0/).

Article

# Debris Flow Susceptibility Mapping Using Machine-Learning Techniques in Shigatse Area, China

Yonghong Zhang [1], Taotao Ge [1], Wei Tian [2],* and Yuei-An Liou [3],*

1   School of Automation, Nanjing University of Information Science & Technology, Nanjing 210044, China; zyh@nuist.edu.cn (Y.Z.); gtt347568@gmail.com (T.G.)
2   School of Computer and Software, Nanjing University of Information Science & Technology, Nanjing 210044, China
3   Center for Space and Remote Sensing Research, National Central University, Taoyuan 32001, Taiwan
*   Correspondence: tw@nuist.edu.cn (W.T.); yueian@csrsr.ncu.edu.tw (Y.-A.L.)

Received: 8 October 2019; Accepted: 21 November 2019; Published: 27 November 2019

**Abstract:** Debris flows have been always a serious problem in the mountain areas. Research on the assessment of debris flows susceptibility (DFS) is useful for preventing and mitigating debris flow risks. The main purpose of this work is to study the DFS in the Shigatse area of Tibet, by using machine learning methods, after assessing the main triggering factors of debris flows. Remote sensing and geographic information system (GIS) are used to obtain datasets of topography, vegetation, human activities and soil factors for local debris flows. The problem of debris flow susceptibility level imbalances in datasets is addressed by the Borderline-SMOTE method. Five machine learning methods, i.e., back propagation neural network (BPNN), one-dimensional convolutional neural network (1D-CNN), decision tree (DT), random forest (RF), and extreme gradient boosting (XGBoost) have been used to analyze and fit the relationship between debris flow triggering factors and occurrence, and to evaluate the weight of each triggering factor. The ANOVA and Tukey HSD tests have revealed that the XGBoost model exhibited the best mean accuracy (0.924) on ten-fold cross-validation and the performance was significantly better than that of the BPNN (0.871), DT (0.816), and RF (0.901). However, the performance of the XGBoost did not significantly differ from that of the 1D-CNN (0.914). This is also the first comparison experiment between XGBoost and 1D-CNN methods in the DFS study. The DFS maps have been verified by five evaluation methods: Precision, Recall, F1 score, Accuracy and area under the curve (AUC). Experiments show that the XGBoost has the best score, and the factors that have a greater impact on debris flows are aspect, annual average rainfall, profile curvature, and elevation.

**Keywords:** debris flow susceptibility; remote sensing; GIS; oversampling methods; back propagation neural network; one-dimensional convolutional neural network; decision tree; random forest; extreme gradient boosting

## 1. Introduction

Debris flows involve gravity-driven motion of solid-fluid mixtures with abrupt surge fronts, free upper surfaces, variably erodible basal surfaces, and compositions that may change with position and time [1]. They can cause great damage to the safety of people's lives and property, public facilities and ecological environment. Due to the harsh natural environment and deforestation caused by over-exploitation of human beings, Shigatse is a typical area with active debris flows in the Tibet Autonomous Region. Debris flows can cause very high damages because the study area is densely populated. Therefore, mitigating and reducing the disasters caused by debris flows are critical to the local authorities. Most of Shigatse mountainous area is inaccessible and characterized by very steep slope such that it is very difficult to carry out field surveys. The installation and maintenance of sufficient monitoring facilities in these areas are also very challenging. Therefore, zoning debris flow susceptibility (DFS) maps through spatial data can be used to prevent and mitigate casualties and economic losses caused by debris flow events.

Susceptibility mapping of debris flow is prominent for early warning and treatments of regional debris flows. DFS assessment is based on the spatial characteristics of debris flow events and relevant factors (topography, soil, vegetation, human activities and climate). It aims to estimate the spatial distribution of future debris flow probability in a given area [2]. Some studies have discussed and analyzed debris flows in the study area [3,4], focusing on the residential settlements and vicinity of roads. Assessing the susceptibility of debris flows in the whole study area is difficult due to the vast size of land (exceeding 180,000 square kilometers). The detailed spatial information on the debris flow triggering factors is also quite limited. In this case, satellite remote sensing has good application prospects because it can describe the characteristics of a large area, such as terrain, vegetation, and climate of the place where debris flow events occur. Therefore, compared with the traditional field geological survey, which requires a lot of work and resources, data from remote sensing represented in a GIS environment can fill the gap of on-site monitoring data. That is, it can be applied for the DFS researches in a more effective and economical way.

In recent years, GIS and remote sensing data have been used to conduct many studies of disasters in mountains. Researchers built their methodology analyzing data of known occurred debris flows and tested it through unknown debris flow events. Gregoretti et al. [5] proposed a GIS-based model tested against field measurements for a rapid hazard mapping. Kim et al. [6] used a high-resolution light detection and ranging (LiDAR) digital elevation model to calculate the volume of debris flows. Kim et al. [7] developed a GIS-based real-time debris flow susceptibility assessment framework for highway sections. Alharbi et al. [8] presented a GIS-based methodology for determining initiation area and characteristics of debris flow by using remote sensing data. At present, the DFS assessment methods can be mainly divided into two categories: qualitative and quantitative models. The qualitative model assigns a weight (0–1) to each debris flow triggering factor based on expert experience and knowledge or heuristics to assess the DFS [9]. Common qualitative analysis methods include fuzzy logic [10], analytic hierarchy [11] and network analysis [12] and so on. While these models have achieved a lot in the study of debris flows, they still suffer for some shortcomings, such as a high degree of subjectivity and limited applicability to specific areas [13].

Quantitative methods usually include two types: deterministic and statistical models based on physical mechanisms. Deterministic methods are used to study the physical laws of debris flows and establish the corresponding models to simulate the DFS [14]. The disadvantages of these models are in that they require detailed inspection data for each slope. Thus, they are only suitable for smaller areas. Statistical models are data-driven. The DFS assessment from them combines the past debris flow events with environmental characteristics. It is assumed that the environmental characteristics of the past debris flows events will lead to debris flows in the future. The models for the DFS quantitative assessment include information model [15], evidence weight method [16], frequency ratio [17] and so on.

In recent years, data mining and machine learning techniques have also received extensive attention because they can more accurately describe the nonlinear relationship between DFS and triggering factors [18], and there is no special requirement for the distribution of triggering factors. Machine learning algorithms are often superior to traditional statistical models [19] for the following reasons. First of all, machine learning can adapt to larger datasets, while traditional statistical learning methods are more suitable for small datasets. Secondly, machine learning has better controllability and extensibility than traditional statistical models. Moreover, traditional statistical models are in general limited to certain requirements and assumptions on data, whereas machine learning methods are not. In the past three decades, common machine learning methods used for studying DFS mapping include back propagation neural network (BPNN) [20], decision tree (DT) [21], Bayesian network [22], and support vector machine [23]. With the advancement of researches, more and more models have been developed with better fitting performance. Under such circumstances, continuous verification and evaluation are still necessary for constructing and selecting a DFS evaluation model. Therefore, comparisons among various models to investigate DFS have become hot topics in academia. Since the information about debris flow occurrence is very limited and different, stability and accurate predictive power are the primary requirements for selecting the appropriate method to achieve better modeling results.

Among machine learning methods, BPNN is widely used because it carries the excellent nonlinear fitting and complex learning abilities to extract the complex relationship between debris flow triggering factors and DFS [24]. Convolutional neural network, a classical deep learning method, has been rapidly developed in the past decade and is widely used in pattern recognition and medicine It is generally used for classification and recognition of two-dimensional images. In recent years, artificial intelligence scholars have made the convolutional neural network one-dimensional, so as to perform the speech recognition [25], fault diagnosis [26] and data classification [27]. As an end-to-end model, the one-dimensional convolutional neural network can extract and classify different characteristics of debris flows directly from raw data without expert guidance. DT is another powerful prediction model with three major advantages: the model is easy to build; the final model is easy to interpret; and the model provides clear information about the relative importance of input factors [28]. These advantages have motivated researchers to develop new DT models to better utilize the debris flow information. At the same time, integrated learning algorithms based on decision trees have also been widely concerned. Among them, the more representative ones are bagging and boosting. Kadavi et al. [29] used four integrated algorithms: Adaboost, Bagging, LogitBoost, and Multiclass classifier to calculate and plot the DFS map. They proved that the Multiclass classifier had the best performance by verifying the AUC value of the test set.

Due to the complex terrain, geology and other mountain conditions in the study area, the multi-source and multi-data are used as much as possible to characterize the terrain and geological conditions of debris flows. Although machine learning methods have been demonstrated to achieve results with satisfactory to some extent, this paper further discusses whether they can be applied to examine the DFS. Its most important contributions are described as follows. (a) We collected debris flow events data and a variety of original remote sensing data related to topographic factors, such as soil factors, human factors and vegetation factors, and performed pre-processing operations, including projection, registration and sampling based on remote sensing and GIS technology (ArcGIS v.10.2 software). (b) We obtained the characteristics of the study area where debris flow occurred and used the data generation algorithm to merge the collected debris flow events data. (c) Based on the Python language, using the keras framework and the scikit-learn module, five DFS models (BPNN, 1D-CNN, DT, RF, and XGBoost) were constructed for the training set. The applicability of these models was examined for the Shigatse region. It is notable that this is the first comparative experiment of XGBoost and 1D-CNN in the study of DFS. (d) Cross-validation methods were used to compare the performance of artificial neural networks and tree-based models to reduce the bias and variance. (e) Statistical analyses of the comparative algorithm were done to verify whether the performance is significantly different. (f) The test set was used to evaluate the models' prediction ability in combination with the five evaluation methods of classical Recall, Precision, F1 score, Accuracy, and AUC [30]. (g) At the end of the study, the tree-based "feature importance" ranking was used to evaluate the main characteristic factors affecting the DFS.

## 2. Material and Methods

### 2.1. Study Area

The study area covers an area of 182,000 square kilometers in the southwestern part of China. It is located in the southwest of the Tibet Autonomous Region (27°23′~31°49′N, and 82°00′~90°20′E).

As shown in Figure 1, the Shigatse area is mainly located between the central Himalayas and central part of the Gangdise-Nyqinqin Tanggula Mountains. Its elevation is high in the northern part and southern part, including the southern Tibetan Plateau and Yarlung Zangbo River basin. The overall terrain of the Shigatse region is complex and diverse, mainly consisting of mountains, wide valleys and lake basins with a maximum elevation of over 8700 m. The study area belongs to semi-arid monsoon climate zone of inland plateau. It is featured with dry climate, less precipitation, rainy season coincident with hot season, and annual average sunshine hours of 3240 h.

The transportation mainly includes three main lines: China-Nepal (Zhongni) Highway, 318 National Road and Largo Railway passing through the study area. The geological disasters in the study area are serious, mainly including debris flows, rock collapses, and landslides. Among them, the debris flow is the most common one. A large number of debris flows exist in many parts of the study region. They directly threaten the safety of the three major transportation lines and residents' lives and properties. According to the collected data and previous studies [31], the debris flows in the study area are mostly caused by heavy rain.

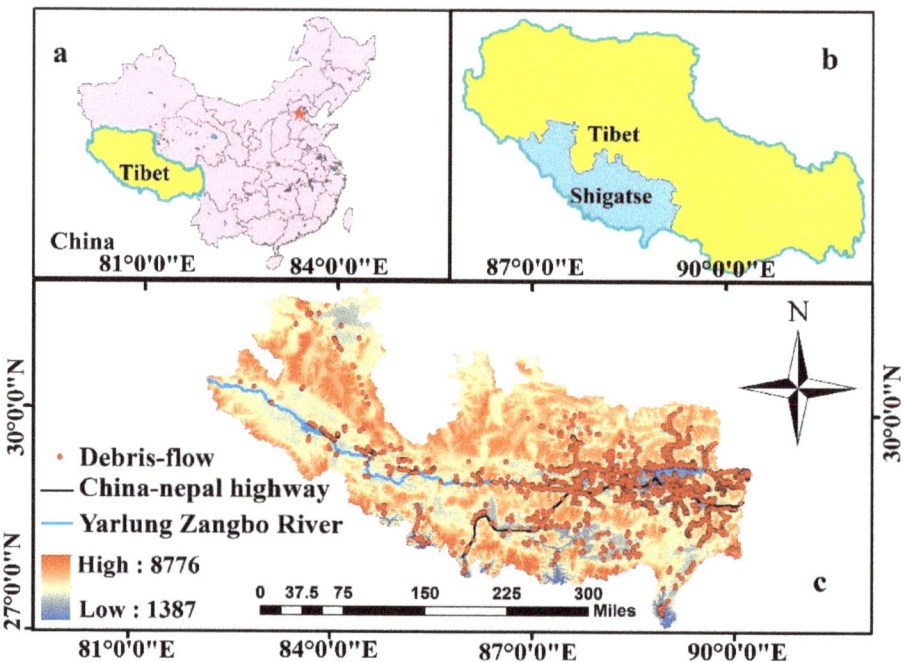

**Figure 1.** Location of the study area. Site maps of (**a**) China, (**b**) Tibet Autonomous Region, and (**c**) the study area.

2.1.1. Debris Flow Dataset

Collection and analysis of debris flow event datasets are prerequisites for the DFS assessment. There are 1944 debris flow sites in the study area from 1998 to 2008. Each case includes information obtained from field disaster investigation, such as time, debris flow susceptibility level, and geographic location. The information on debris flows is provided by the Tibet Meteorological Bureau. These events can be viewed through the geological cloud portal [32].

## 2.1.2. Debris Flow Triggering Factors

It is significant to analyze the environmental characteristics of the debris flow events for the DFS estimation. Due to the complexity of the environment and various development stages of debris flows, the causes of debris flows are controversial. Researchers have done a lot of studies on the relationship between debris flows and triggering factors, such as topography, soil, climate, and human activities. Therefore, we have classified 15 environmental factors into five categories as shown in Table 1.

Table 1. Data layer related to debris flows susceptibility (DFS) in the study area.

| Category | Factor | Scale | Data |
|---|---|---|---|
| Topography | Slope<br>Aspect<br>Elevation<br>Plan curvature<br>Profile curvature<br>Total curvature | 30 m | SRTM DEM |
| Anthropogenic | Land cover<br>Distance to road | 30 m | Land cover map<br>Road vector |
| Soil | Soil type<br>Soil texture<br>Soil Erosion | 1000 m | Spatial distribution data of soil texture<br>Spatial distribution data of soil erosion<br>Spatial distribution data of soil types |
| Vegetation | NDVI | 500 m | MODIS |
| Climate | Rainfall | 25,000 m | TRMM |

Topographic factors that include elevation, slope, aspect, and curvature are extracted from the Shuttle Radar Topography Mission Digital Elevation Model (SRTM DEM) using the ArcGIS platform [33]. The vegetation coverage is represented by the normalized difference vegetation index (NDVI), calculated from the obtained 2000–2008 MODIS images and averaged to generate the thematic layer of the annual average NDVI. Rainfall data are collected from the Tropical Rainfall Measurement Task (TRMM) [34]. We use a rainfall dataset (No: 3B42v7) with a time interval of three hours and a spatial resolution of 0.25 degree during 1998–2008 to construct a thematic layer of annual average precipitation. The 15 types of land use information layers are provided by National Earth System Science Data Sharing Infrastructure, National Science & Technology Infrastructure of China (http://www.geodata.cn) [35,36]. In addition, the road vector data provided by OpenStreetMap (OSM) (https://www.openstreetmap.org/#map=11/22.3349/113.76000) is used to calculate the distance from the road. Soil factors are provided by the Resource and Environmental Science Data Center (RESDC) of the Chinese Academy of Sciences.

Higher resolution is conducive to the topographic analysis of single-ditch debris flow, but in this work, our research focuses on the use of multiple attribute factors to analyze the disaster susceptibility of the entire study area. Golovko [2] and Ahmed [9] believe that 30M resolution Digital Elevation Model (DEM) can be used for the analysis of the susceptibility to mountain disasters. Therefore, DEM data with a pixel size of 30 m is used (Figure 2a). The slope angle is a fundamental factor calculated by the DEM data and the range of it obtained by statistics is wide (0–89°) (Figure 2b). The aspect of the slope is another key factor affecting the DFS. Because the slope surface in different directions is exposed to the wind and rain in different degrees. The aspect thematic layers are reclassified into nine categories: flat (−1), north (337.5–360°, 0–22.5°), north-east (22.5–67.5°), east (67.5–112.5°), south-east (112.5–157.5°), south (157.5–202.5°), south-west (202.5–247.5°), west (247.5–292.5°), and north-west (292.5–337.5°) (Figure 2c). The second derivative of the slope, i.e., the curvature, helps us understand the characteristics of the basin runoff and erosion processes. In this study, three curvature functions are used to show the shape of the terrain (Figure 2d–f). They are the curvature of the profile, the curvature of the plane, and the total curvature of the surface defined as the curvature of the maximum slope, the curvature of the contour, and the combination of the curvatures, respectively.

Human activities affect the geographical environment, which in turn influences the occurrence of debris flows. The land cover thematic map shows how human production can change natural land, and 14 land use types including farmland and forest can be identified (Figure 2g). The DFS assessment often takes the distance from the road into account, because the road construction and maintenance cause certain change and damage to the local morphology. This variable is calculated by using the Euclidean distance calculation technique in the spatial analysis tool of ArcGIS 10.2 (Figure 2h).

The vegetation coverage is one of the important parameters to evaluate the DFS. NDVI extracted from remote sensing images is a commonly used vegetation index for inferring the vegetation density. It is very sensitive to the presence of chlorophyll on vegetation surface (Figure 2i). We calculated the NDVI value using the following formula:

$$NDVI = (NIR - RED)/(NIR + RED) \tag{1}$$

where NIR and RED represent the near-infrared and red-band, respectively, and they are the second and first channels of the MODIS image. The NDVI value ranges between −1 and 1. The negative value indicates that the ground cover is an object highly reflective to visible light such as clouds, snow, water, etc. 0 means bare land. A positive value represents a vegetation coverage area and it increases with the vegetation coverage density.

Debris flows are usually influenced by changes in humidity caused by rainfall infiltration. Permeability can be expressed by soil type (Figure 2j), soil texture (Figure 2k–m) and soil erosion (Figure 2n). Since the particle distribution determines the shape of soil water characteristic curve and affects the soil hydraulic characteristics, the soil type and texture have a great influence on the DFS. Most of the study area is covered by alpine soil, including grass mat soil, cold soil, and frozen soil. According to statistics, most of the alpine soil is brown and has a strong acidity. The alpine soil is mainly composed of silt, sand and clay fine sand, and has fast permeable and low moisturized ability. Soil erosion is sometimes used as a synonym for soil and water loss, and areas with severe erosion are susceptible to debris flows. The external causes of soil erosion are mainly hydro, wind, and freeze-thaw. Clearly, fragile soil characteristics accompanied by concentrated rainfall usually result in debris flows.

Rainfall is the main factor leading to debris flows. The study area is affected by the monsoon climate with rare precipitation and an annual average precipitation less than 1300 mm (Figure 2o). However, statistical analyses of the geological hazard points occurring in the study area show that heavy rain and continuous rainfall are important external factors leading to geological disasters in the Shigatse area.

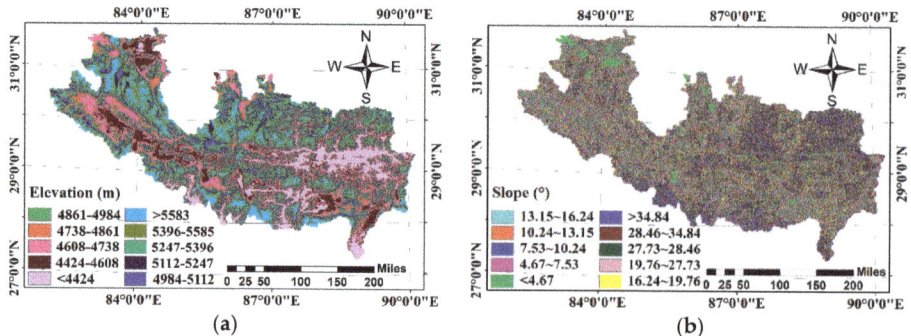

**Figure 2.** *Cont.*

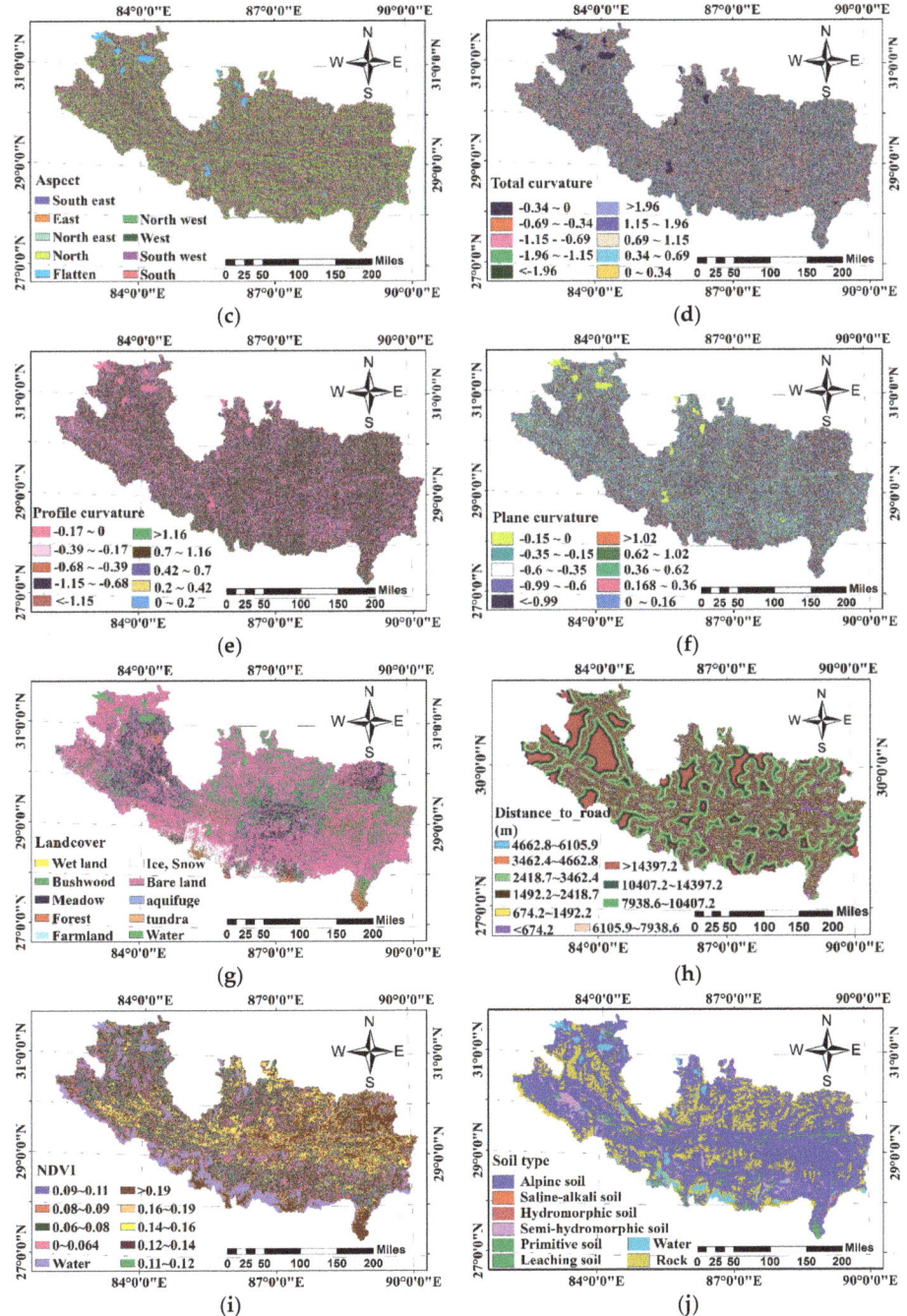

**Figure 2.** *Cont.*

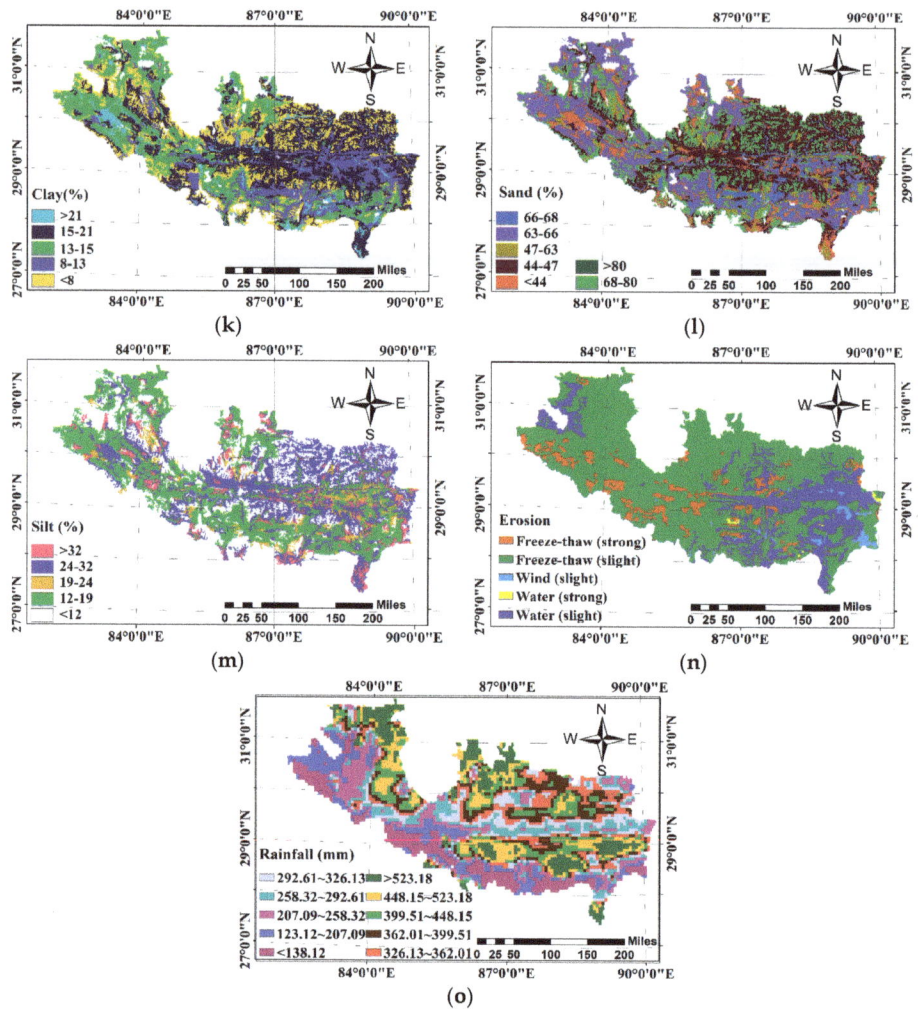

**Figure 2.** Spatial distribution of debris flow characteristics; (**a**) elevation, (**b**) slope angel, (**c**) aspect, (**d**) total curvature, (**e**)profile curvature, (**f**) plan curvature, (**g**) landcover, (**h**) distance to road, (**i**) NDVI, (**j**) soil type, (**k**) sand, (**l**) silt, (**m**) clay, (**n**) erosion, and (**o**) rainfall.

## 2.2. Methods

The main purpose of our research is to fit the relationship between the triggering factors and occurrence of debris flows. The problem can be expressed as a multi-class classification. Given a set of input quantities, the classification model attempts to label the DFS level for each pixel in the study area. The input quantities to the models are the triggering factors of the debris flow events that were collected by the local Chinese Geological Survey researchers after many years of field investigation. According to the researchers' investigation of the debris flow sites, we obtain the values of 15 triggering factors at the corresponding positions through the value extraction function of ArcGIS v10.2 software. That is, the input of the model is a one-dimensional vector form [×1, ×2, ... , ×15]. The output value of the model is the DFS level, indicating the occurrence probability of debris flows. The division criteria of regional DFS are based on the detailed survey and specification of landslide collapse debris flows by the China Geological Survey as shown in Table 2.

Table 2. DFS level classification.

| DFS Level | Frequency of Debris Flow |
|---|---|
| Very low | No debris flow occurs within 100 years |
| Low | Debris flow occurs once within 10–100 years |
| Medium | Debris flow occurs once in within1–10 years |
| High | 1-10 debris flow events occurred with a year |

According to statistics, the number of moderately susceptible units in the study area is much higher than that of the other susceptible grades (Figure 3). Therefore, before constructing a debris flow assessment model, the oversampling technique Borderline-SMOTE algorithm should be used to eliminate the classification imbalance in the dataset. The number of each debris flow susceptibility level after oversampling is shown in Figure 3. The original dataset is divided into training sets and test sets according to a percentage of 75 and 25%, respectively. The training set of the debris flow triggering factors is used to learn the ability to fit the actual DFS classification, and the validation set is used for adjusting the model parameters to prevent over-fitting or under-fitting. In this study, five DFS models have been established using DT, BPNN, 1D-CNN, RF, and XGBoost. Among them, the DT and BPNN have been the most commonly used machine learning models in the past few decades. The one-dimensional convolutional neural network (1D-CNN) has achieved remarkable results in one-dimensional signal processing, such as fault diagnosis and speech recognition. RF is based on the DT. It is a typical integrated algorithm in machine learning algorithms. The XGBoost is also a tree-based integration model, which counts on the residuals generated by the last iteration. To the best of our knowledge, the XGBoost and 1D-CNN have not been used for the DFS. Based on the above introductions, the research framework for the DFS in Shigatse is shown in Figure 4.

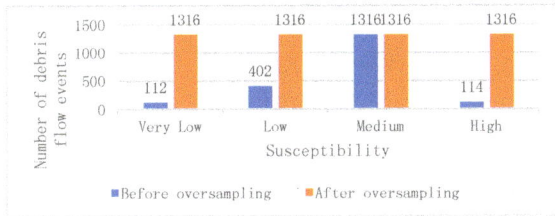

Figure 3. Statistics of debris flow events with different susceptibility.

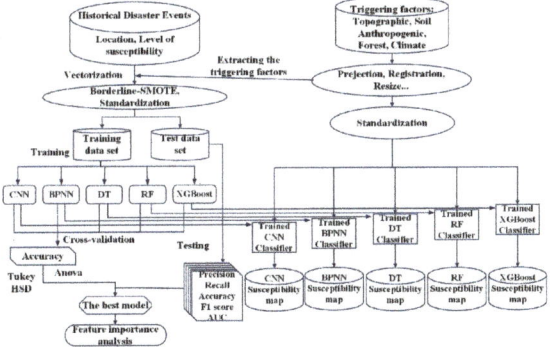

Figure 4. The research framework for DFS mapping.

In addition, we have also implemented other traditional machine learning algorithms, such as support vector machine, logistic regression, and naive Bayesian model, but the results are disappointing. Therefore, these methods are not introduced here. The following part is a brief introduction to the data sampling generation algorithm and five classifiers used in this paper.

2.2.1. Borderline-SMOTE

It is well known that in the model training process, when a certain class in the classified data set is of a high proportion, the classifier performance will be seriously affected. Synthetic Minority Oversampling Technique is often referred to as SMOTE that has been improved for its application in solving data imbalance problems [37]. It is used to artificially generate vector data to achieve the consistency among each category in the dataset. In the study, it is common that most units are with moderate susceptibility. In the classification process, the scarcity of the category data with fewer samples (the minority class) is one of the main factors for over-fitting and inaccuracy. This paper chooses the boundary-based SOMTE algorithm (Borderline-SMOTE) to handle the imbalance of the data. Specifically, the k-nearest neighbor algorithm is used to calculate the nearest neighbor sample in the minority sample set in the training set. The boundary sample set is constructed according to whether the majority class in the nearest neighbor sample set is dominant. The k-nearest neighbors of the sample $T_i$ in the boundary sample set are calculated, and the sample $T_j$ is randomly selected. The SMOTE algorithm is used to randomly insert the feature vector between the selected neighbor samples and the initial vector. The SMOTE algorithm is shown in Equation (2),

$$T_{new} = T_i + random(0,1) * |T_i - T_j| \qquad (2)$$

Finally, the generated new sample is added to the original sample set.

2.2.2. Back Propagation Neural Network

Back propagation neural network (BPNN) is a mathematical model that simulates the processing information of the biological nervous system. The BPNN, as the most classic part of artificial neural network, usually has three or more network structures, including input layer, output layer and hidden layer. The main structural functions of the BPNN are the forward propagation of the signal and the back propagation of the error. The neurons between the layers are fully connected, while the neurons of each layer are not connected to each other. The network is a gradient descent method, using gradient search technology to minimize the cost function of the network. It has strong nonlinear mapping ability and is especially suitable for dealing with the intricate relationship between debris flow triggering factors and DFS susceptibility. The general operation of the network is as follows. The input sample leaves the input layer and enters the hidden layer. After being activated by the transfer function (such as Tanh, Relu, Sigmoid and Tanh used in this article.), it passes to the next hidden layer until the output layer. The output formula for each layer is as follows:

$$z_i = f_\theta(\sum_j w_{ij} x_j - b_i) \qquad (3)$$

where $f(\theta)$ represents the transfer function; $\theta = \{w, b\}$ represents the network parameter; $w$ is the weight; and $b$ is the threshold.

2.2.3. One-Dimensional Convolutional Neural Network

As a feedforward neural network, one-dimensional neural network (1D-CNN) is inspired by the mammalian visual cortex receptive field. The network perceives the local features and integrates the local features in high-dimensional space, and finally obtains global information. The basic structure of the convolutional neural network includes an input layer, alternating convolution layers, pooling layers, and a fully connected layer. The convolutional layer captures the information of the local connections in the input features through the convolution kernel and reduces the parameters of the model using the weight sharing principle. The convolution formula is:

$$x_j^l = f(\sum_{i=1}^{M} x_i^{l-1} * k_{ij}^l + b_j^l)$$ (4)

where $f()$ represents the transfer function, $x_j^l$ represents the j-th feature map of convolutional layer $l$, $M$ represents the number of feature maps, $x_i^{l-1}$ represents the $i$th feature map of the l-1 layer, $*$ represents convolution operation, $k_{ij}^l$ represents trainable convolution kernel, and $b_j^l$ represents bias. The shape and number of one-dimensional convolution kernels can largely determine the feature-extraction ability of the overall network. The shape of the convolution kernel affects the fineness of feature extraction. The number of convolution kernels corresponds to the size of the feature layer, affecting multiple ways of feature extraction and the computational scale of the network. The pooling layer combines multiple adjacent nodes to merge similar features and performs down-sampling operation on the feature layer extracted by the convolutional layer, thereby reducing training parameters and preventing network over-fitting to enhance the generalization ability of the network. At present, the main pooling methods include maximum pooling, mean pooling, and L2-norm pooling. After the convolutional layer and pooling layer are located, the fully connected layer trains the weights and biases of the convolution kernels in the entire convolutional neural network based on the back-propagation principle. The fully connected layer structure is similar to the BP neural network mentioned in the previous section, which has a hidden layer and uses the Softmax activation function to complete the classification. The structure of the entire network is shown in Figure 5.

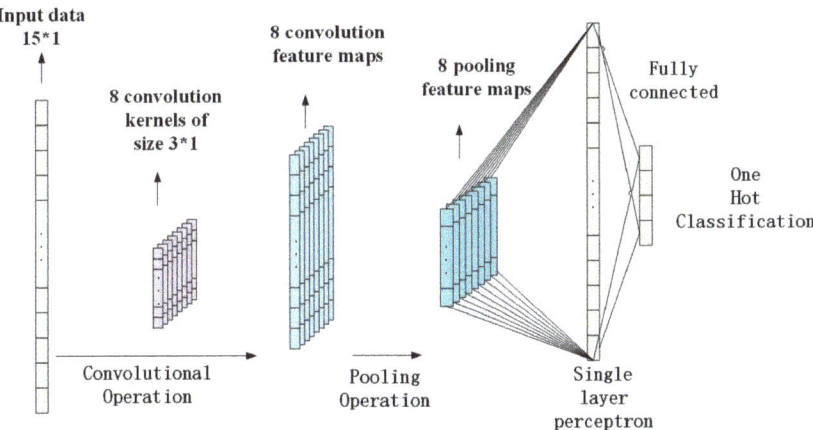

**Figure 5.** One-dimensional neural network structure used in this research.

### 2.2.4. Decision Tree

DT is a common machine learning algorithm similar to the tree structure, often used to find pattern structures in data. It aims to establish correct decision rules and determine the relationship between feature attributes and target variables without expert experience. Usually DT contains a root node, multiple internal nodes, and a set of leaf nodes from top to bottom. The leaf node corresponds to the prediction result, and the node contains all samples. The key to DT learning is to divide the best attributes. At present, the algorithms for constructing DT mainly include CART, C4.5 and ID3. This study uses a CART algorithm with better performance and efficiency [38]. CART uses the Gini coefficient to divide the node properties and establish a DT by selecting the attributes that minimize the Gini coefficient after dividing the nodes. The Gini index is shown in Equation (5) where $k$ is the category and $t$ is the node. Finally, pruning techniques are used to deal with the over-fitting problem of the model. Upon completing the entire algorithm, we can clearly understand the internal decision-making mechanism and thus get a more objective knowledge of debris flow triggering factors.

$$Gini(t) = 1 - \sum_{k}[p(c_k|t)]^2 \tag{5}$$

2.2.5. Random Forest

As an integrated classification algorithm of machine learning, RF aims to improve the flexibility and accuracy of classification trees. In RF, a large number of trees are generated by constructing a base DT on multiple bootstrap sample sets of data during the running of the algorithm. Because the feature attributes of each node are randomly selected, the node characteristics are effectively reduced without increasing the deviation. Each feature subset is much smaller than the total number of features in the input space so that each DT is decorrelated. Finally, the output of the classification task is predicted by a simple voting method. RF has been constructed with a number of DTs. It has been greatly improved compared with a single DT. However, RF is as complex as the single basic DT. Therefore, RF is also a fairly effective integrated learning algorithm.

2.2.6. XGBoost

XGBoost, also known as extreme Gradient Boosting, is a gradient-enhanced integration algorithm based on classification trees or regression trees. It works the same way as Gradient Boosting, but adds features similar to Adaboost. The algorithm combines multiple DT models in an iterative way to form a classification model with a better structure and higher precision. It has achieved impressive results in many international data mining competitions and won more than two championships in the Kaggle competition. In the DFS analysis experiment, the XGBoost can classify the DFS level according to the environmental characteristics of the Shigatse region and rank the importance of the triggering factors.

The XGBoost uses both the first and second derivatives to perform Taylor expansion on the loss function and adds a regular term to it. Therefore, while considering the model accuracy, the model complexity is also well controlled. Finally, the predictive power of the model is trained by minimizing the total loss function [39]. The objective function of the model can be expressed as Equation (6):

$$J(f_t) = \sum_{i=1}^{n} L\left(y_i, \hat{y}_i^{(t-1)} + f_t(x_i)\right) + \Omega(f_t) + C \tag{6}$$

where $i$ represents the $i$th sample, $\hat{y}_i^{(t-1)}$ represents the predicted value of the $(t-1)$th model for the sample $i$, $f_t(x_i)$ represents the newly added $t$th model, $\Omega(f_t)$ represents the regular term, C represents some constant terms, and the outermost L() represents the error.

The optimizer aims to calculate the structure and the leaf score of the CART tree. XGBoost accelerates existing lifting tree enhancement algorithms through the cache-aware read-ahead technology, distributed external memory computing technology and AllReduce fault-tolerant tools. It can also be trained by using a graphics processing unit to provide a very high speed boost.

In this work, we can import the XGBoost toolkit in Python. The training process controls the establishment of DT by adjusting five hyper-parameters: the number of iterations, the number of trees generated, the learning rate, the maximum depth of each tree, and the L2 regularization. The Gamma hyper-parameter limits the gain amount required for segmentation.

2.3. Evaluation Measures

DFS mapping should have the ability to effectively predict the probability of future debris flows in the study area. In order to evaluate the predictive power of several machine learning algorithms, five common evaluation methods are used to quantify model performances, including Precision, Recall, F1 score, Accuracy and AUC. Finally, 293 debris flow events are applied as a test set.

In the case of the binary classification problem, four elements, i.e., TP, TN, FP and FN, are defined as follows.

TP: True Positive. Samples belonging to the TRUE class are correctly marked as positive by the model.

TN: True Negative. Samples belonging to the TRUE class are incorrectly marked as negative by the model.

FP: False Positive. Samples belonging to the FALSE class are incorrectly marked as positive by the model.

FN: False Negative. Samples belonging to the FALSE class are correctly marked as negative by the model.

2.3.1. Precision

In the binary classification task, precision represents the ratio of the correct labeled True class samples to the total number of predicted values labeled true. The formula is as follows:

$$Precision = TP/(TP + FP) \qquad (7)$$

Precision is expressed as a weighted average of the precision of each class in a multivariate classification task.

2.3.2. Recall

The Recall rate is the ratio of the correct labeled True sample to the total number of True samples, expressed as Equation (8) in the binary classification task.

$$Recall = TP/(TP + FN) \qquad (8)$$

The Recall rate represents the weighted average of the Recall rates for each category in a multivariate classification task.

2.3.3. F1 Score

The F1 score is represented by Precision and Recall, with values between 0 and 1, which represent the worst and best, respectively. The relative contributions of accuracy and recall to the F1 score are equal. The formula is defined as follows:

$$F1\ score = 2 * (Precision * Recall)/(Precision + Recall) \qquad (9)$$

In a multivariate classification task, the F1 score represents a weighted average of F1 scores for each category.

2.3.4. Accuracy

In a multivariate classification task, accuracy represents the ratio of correctly classified samples to the total number of samples.

2.3.5. Area Under the Curve (AUC)

The AUC method is defined as the area under the receiver operating characteristic curve (ROC), which can give different distributions of each class. It can be used to judge classifiers' performance. The ROC curve is plotted as the relationship between the true positive rate (TPR) and false positive rate (FPR). TPR represents the ratio of the positive instance correctly classified to the total number of all the positive instances, as represented by Equation (10):

$$TPR = TP/(TP + FN) \tag{10}$$

FPR is the ratio of the positive instance misclassified to the total number of all the negative instances, as represented by Equation (11):

$$FPR = FP/(FP + TN) \tag{11}$$

The AUC method is also designed to evaluate the binary classification. First, we need to convert the multivariate classification task into multiple binary classification, and then separately calculate the AUC values of the respective categories. Finally, the multivariate classification result is obtained by obtaining the average of the total AUC values [40].

### 2.4. Cross-Validation

In this paper, the cross-validation method is used to complete the parameter optimization. Specifically, based on the error-based verification evaluation index, the training set is divided into $k$ pairs of mutually equal exclusion subsets, where $k - 1$ pairs are used as the training sets and the remaining subset are used as the verification set. The experiment is performed by rotating the subset $k$ times in turn, and the $k$ verification results are averaged. In this paper, the GridSearchCV module via the scikit-learn and the cv function via the XGBoost library are used to optimize the parameters of the decision tree, random forest and the XGBoost model. In the Keras framework, the cross-validation method based on the GridSearchCV module is also used to search in the parameter space, and the optimal parameter estimation of the model in the data set is given.

### 2.5. Statistical Analysis

In order to compare the performance differences between the models, we conducted a statistical analysis. One-way ANOVA can be used to test whether there is a statistically significant difference in two or more unrelated groups [41]. Model performance is evaluated by the accuracy of test results during the model training. Therefore, the accuracy group obtained by cross-validation of different algorithms is used for one-way ANOVA. The null hypothesis given by H0 tested by One-way ANOVA is as follows.

H0: The accuracy of all algorithms is not significantly different.
H1: There are significant differences in the accuracy of at least two or more algorithms.
The One-way ANOVA results in a P-value, and the P-value is the risk of rejecting the hypothesis H0.
The results can only determine whether there is a significant difference between at least one group of samples and other groups, but it is impossible to judge whether there is a difference between the two groups. Therefore, comparisons between specific groups are often performed after one-way ANOVA. The honestly significant difference (HSD) method was developed by Tukey and is favored by researchers as a simple and practical pairwise comparison technique. The main idea of HSD is to use the statistical distribution to calculate the true significant difference between two mean values and call it q-distribution [42]. This distribution gives an exact sampling distribution of the largest difference between a set of mean values in the same population. All pairwise differences were evaluated by using this distribution. This paper uses HSD as a post-hoc analysis to test the variance homogeneity of performance indicators from different algorithms.

All statistical analyses were completed by using the Statistical Package for Social Sciences (IBM SPSS Statistics for Windows Version 22.0).

## 2.6. Feature Importance

The tree-based machine learning approach in this study provides a "feature importance" toolkit for ordering index factors based on the function strength of a particular problem [43]. In the basic decision tree, feature attributes are selected for the node segmentation, and the number of times measures the importance of the attribute. For a single decision tree T, Equation (12) represents the score of importance for each predictor feature $x_l$, and the decision tree has $J - 1$ internal nodes.

$$w_l^2(T) = \sum_{t=1}^{J-1} \hat{\imath}_t^2 \qquad (12)$$

The selected feature is the one that provides maximal estimated improvement $\hat{\imath}_t^2$ in the squared error risk over that for a constant fit over the entire region. The following formula represents the importance calculation over the additive M trees.

$$w_l^2(T) = \frac{1}{M} \sum_{m=1}^{M} \hat{\imath}_t^2(T_m) \qquad (13)$$

In reality, a frequently used attribute often has a good distinguishing ability. In this study, the importance of the factors affecting the debris flows occurrences is ranked from high to low according to the characteristic attribute of the decision-making process of DFS.

## 3. Results

In this section, a specific implementation of five machine learning algorithms is introduced. Using Python as the development language, the BPNN and the 1D-CNN are constructed based on the Keras learning framework. The DT and the RF are implemented by API provided by the scikit-learn module, and the XGBoost algorithm is implemented by the Python-based code provided by its official website. The performance of DFS model depends largely on the choice, adjustment and optimization of its parameters. Therefore, the optimization of the model structure and parameters requires multiple experiments. The cross-validation method is used to complete the parameter optimization. After many experiments, the optimal parameters of the five DFS models are obtained as shown in Table 3.

Table 3. Calculated parameters of the algorithms.

| Algorithm | Parameter |
|---|---|
| BPNN | Number of iterations: 3000;<br>Learning rate: 0.01;<br>Activation function: tanh, softmax;<br>Number of nodes: input layer = 15, hidden layer = (30,30), output layer = 4;<br>Optimization function: Adam;<br>Loss: Logarithmic Loss Function;<br>Alpha:0.005 |
| DT | Criterion: gini;<br>Min_samples_split = 2;<br>Mat_depyh: 38;<br>Splitter: random |
| 1D-CNN | Convolutional Layer: Filter = 8, Kernel_size = 3, Stride = 1, activation = Relu;<br>Pooling Layer: max_pooling;<br>Fully connected layer: node =15, activation = Relu;<br>Output layer: node = 4, activation = Softmax |

Table 3. Cont.

| Algorithm | Parameter |
|---|---|
| RF | Number of iterations: 30; Max_feature = sprt Max_depth: 20; Criterion:gini; Min_samples_split = 0.8; Min_samples_leaf = 1 |
| XGBoost | Number of iterations: 39; Max_depth:15; colsample_bytree: 0.5; subsample: 0.9; Eval_metric: mlogloss; Objective:multi: softmax; Eta: 0.1; Lamda:0.2 Alpha = 0.005 Min_child_weight: 0.6; Num_class: 4 |

### 3.1. Performance Metrics Evaluation

Cross-validation produces a list of accuracy, which can be seen from the first row in Table 4. The poor accuracy value in the second line indicates the performance of the model under the non-SMOTE data set.

Table 4. Performance metrics.

| Model | BPNN | 1D-CNN | DT | RF | XGBoost |
|---|---|---|---|---|---|
| Accuracy | 0.871 ± 0.017 | 0.914 ± 0.01 | 0.816 ± 0.023 | 0.901 ± 0.011 | 0.924 ± 0.011 |
| Accuracy(non-SMOTE) | 0.664 ± 0.011 | 0.683 ± 0.023 | 0.671+0.005 | 0.684 ± 0.013 | 0.695 ± 0.01 |

In order to obtain robust verification results, we use the One-way ANOVA method to test whether there is a significant difference between the methods. The ANOVA method is used according to the five groups of accuracy, and the results are shown in Table 5.

Table 5. ANOVA results for five groups of accuracy.

| | Sum of Squares | df | Mean Square | F | Sig |
|---|---|---|---|---|---|
| Between Groups | 0.076 | 4 | 0.019 | 167.683 | 0 |
| Within Groups | 0.005 | 45 | 0 | | |
| Total | 0.081 | 49 | | | |

The F value in the table indicates the ratio of the mean square between the groups to the mean square within the group. The corresponding P value is found according to the F value through the lookup table. Sig represents the P value, which is 0 and less than 0.05, indicating that we can reject the null hypothesis H0. We can think that there are significant differences between at least two sets of models. Significant differences are calculated according to post-hoc Tukey's HSD for all pairwise comparisons between accuracies as shown in Table 6.

Table 6. Means for groups in homogeneous subsets.

| Model | N | \multicolumn{4}{c}{Subset for Alpha = 0.05} | | | |
|---|---|---|---|---|---|
| | | 1 | 2 | 3 | 4 |
| DT | 10 | 0.817 | | | |
| BPNN | 10 | | 0.871 | | |
| RF | 10 | | | 0.901 | |
| 1D-CNN | 10 | | | 0.914 | 0.914 |
| XGBoost | 10 | | | | 0.924 |

According to statistics, XGBoost performs best in terms of accuracy, and there is a significant difference ($p < 0.005$) from BPNN, DT and RF. There is no significant difference between XGBoost and CNN, but the average accuracy of XGBoost is higher than that of 1D-CNN.

### 3.2. DFS Map Construction

In this study, the relationships between the debris flow triggering factors and the DFS levels are fitted by training the BPNN, CNN, DT, RF and XGBoost models to predict the susceptibility index of each pixel in the study area, and to establish a pixel-based DFS classification map (Figure 6). The result is a raster map with each raster pixel assigned a unique susceptibility index value. The susceptibility index values are divided into four categories: 0, 1, 2 and 3, indicating very low, low, medium and high debris flow levels, respectively.

In the debris flow map constructed by the BPNN model, about 2% of the study area is not susceptible to debris flow, and the other 19.8%, 68.1%, and 9.6% are low, medium and high probability DFS levels, respectively, as shown in Figure 6a.

The debris flow susceptibility map predicted by 1D-CNN is shown in Figure 6c. Among them, the medium-prone area accounts for the vast majority of the study area. The other 11.12%, 21.02% and 3.35% are very low, Low and high probability DFS levels, respectively.

The DFS map generated by the DT model is shown in Figure 6b. 11.2% of the study area is not susceptible to debris flows, 21.8% is seldom affected by debris flows, while the medium probability debris flow area accounts for 56.5%, and the remaining 10.3% is high-probability area (Figure 6b).

The proportion of each RF susceptibility level in the DFS map fitted by the RF model is very low (2.1%), low (34.5%), medium (62.8%), and high (0.27%), as shown in Figure 6d.

Finally, based on the XGBoost model, a DFS map is generated as shown in Figure 6e. The results of DFS level distribution are very similar to those based on the random forest model. The medium susceptibility is the main debris flow level, which accounts for 52.5%; the second large area corresponds to the low susceptibility level, 37.2%. Very low and highly susceptible areas are small, accounting for 6.6% and 3%, respectively (Figure 7).

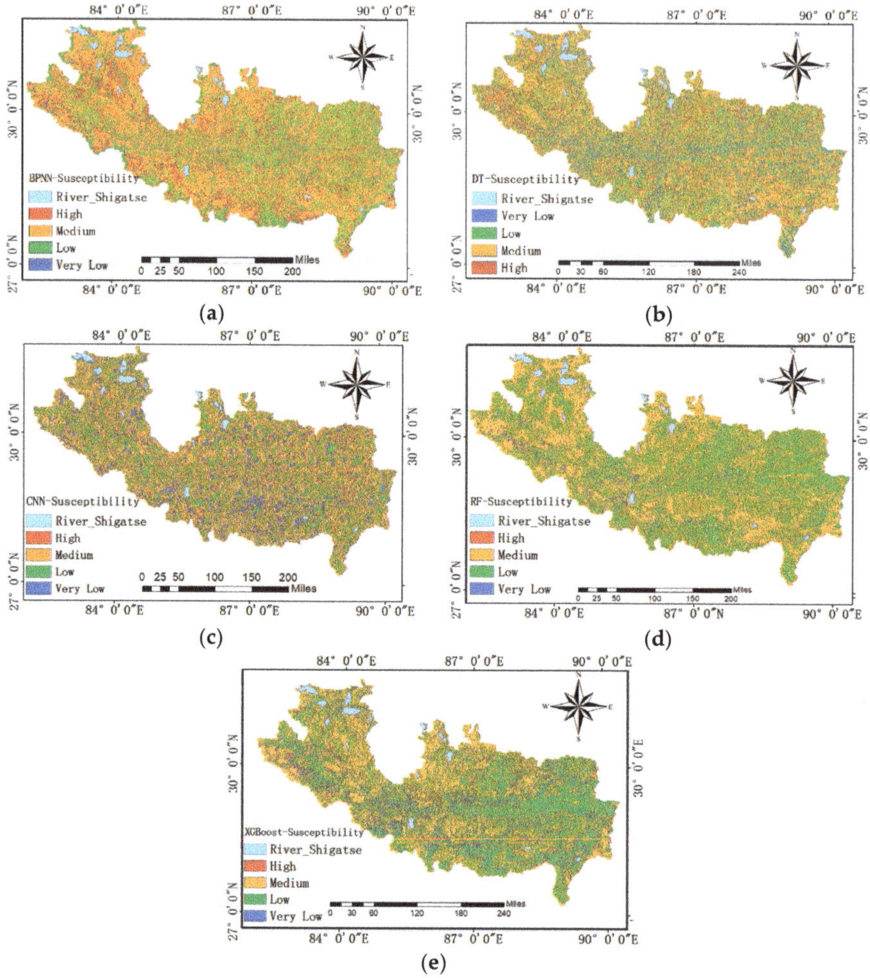

Remote Sens. **2019**, *11*, 2801

**Figure 6.** DFS maps based on the models of (**a**) BPNN, (**b**) DT, (**c**) 1D-CNN (**d**) RF, and (**e**) XGBoost in Shigatse area.

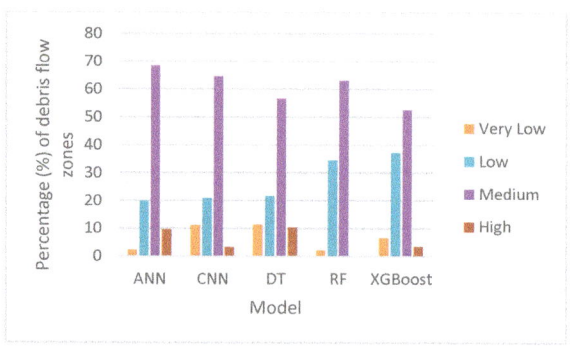

**Figure 7.** Susceptibility level distributions in DFS maps constructed by the five models.

46

## 3.3. Model Evaluation

After successfully constructing the DFS model, we evaluated its performance using five traditional evaluation methods, i.e., Recall, Precision, F1 score, Accuracy, and AUC. We also calculated the time required to forecast the entire study area. The results of each model are shown in Table 7. In particular, because the ROC curve corresponding to the AUC index of each model can directly reflect the advantages and disadvantages of the model, it is drawn in Figure 8. As seen from Table 7 and Figure 8, the following findings are listed.

(1) The values of the Recall, Precision, F1 score, Accuracy, and AUC evaluation score of the five algorithms are quite different. That is, the performances of different algorithms show great differences in the test set.
(2) Despite large differences in the evaluation index values, their differences show the same trend. That is, the algorithm is superior when each evaluation index is superior to the other algorithms.
(3) The AUC evaluation scores of the five algorithms are very high, indicating that they are excellent for evaluating the DFS in our study area. The AUC values of the BPNN, 1D-CNN, DT, RF and XGBoost are 0.946, 0.976, 0.911, 0.976 and 0.988, respectively. It can be seen that XGBoost has the best performance.
(4) From the results of the five indicators, the evaluation scores of the BPNN and DT models are similar, and the 1D-CNN, RF and XGBoost models also take approximate scores, but the former has a large gap with the latter.
(5) The models in this manuscript are all operated on Intel (R) Core (TM) i7-6800K CPU @ 3.4 GHz with 64 RAM servers. In terms of predicting the time of the entire area, the DT model takes the shortest time. XGBoost and 1D-CNN models take about the same time, and the calculation time is at a medium level. The prediction speed of the BPNN model is slow, which takes about 20 min for a single prediction. Finally, it can be seen that the RF calculation takes the longest time.

**Table 7.** Various assessment scores for five debris flow-prone models.

| Model | F1 Score | AUC | Precision | Accuracy | Recall | Prediction Time/min |
|---|---|---|---|---|---|---|
| BPNN | 0.783 | 0.946 | 0.723 | 0.795 | 0.914 | 20 |
| DT | 0.77 | 0.855 | 0.709 | 0.782 | 0.911 | 1.2 |
| RF | 0.859 | 0.976 | 0.813 | 0.852 | 0.934 | 28 |
| XGBoost | 0.9124 | 0.988 | 0.878 | 0.912 | 0.955 | 12 |
| 1D-CNN | 0.901 | 0.976 | 0.914 | 0.906 | 0.906 | 11.2 |

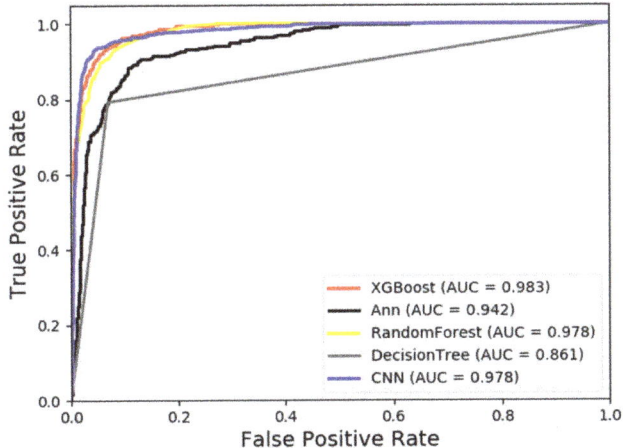

**Figure 8.** ROC curve of five models.

## 3.4. Importance of Triggering Factors

After successfully implementing the five evaluation algorithms, it is concluded that the XGBoost model performs the best in the DFS mapping, and the model can be used to fit the relationship between the debris flow triggering factors and the susceptibility. However, not all selected debris flow triggering factors have a good predictive power. Different triggering factors have different contributions to the model. Therefore, it is necessary to understand the contribution of each triggering factor. The practical significance of this research is that we can provide suggestions and references for local governments and researchers on site selection for public facilities by studying the importance of different debris flow triggering factors characteristics.

As shown in Figure 9, the ordinate represents fifteen index factors for evaluating the DFS, and the abscissa is expressed as the ratio of the number of times each feature attribute used for DT node segmentation to that all attributes used for node segmentation. It can be clearly seen that the aspect is the most important factor affecting the occurrence of debris flows, closely followed by the rainfall, and the impacts of elevation and slope curvature are ranked the third and fourth, respectively. According to these main characteristics, topographic and climatic factors are the main triggering factors for the occurrence of debris flows in Shigatse. The last few debris flow triggering factors ranked by feature importance have relatively little impact on debris flow events especially soil factors.

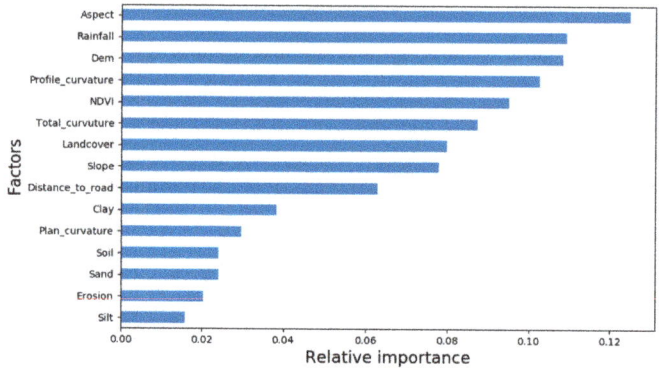

**Figure 9.** Relative importance of DFS triggering factors.

## 4. Discussion

This study aims to estimate the regional DFS by using five highly representative machine learning models, i.e., BPNN, 1D-CNN, DT, RF, and XGBoost. According to literature, such investigations are rare in Shigatse, particularly based on 1D-CNN and XGBoost.

### 4.1. Model Performance

First, as seen from Table 4, category imbalances have caused over-fitting problems on all of the above machine learning models, resulting in poor model performance. The data generated by the interpolation method is close to the original data, and we will adopt better interpolation methods to obtain more reasonable data and use other methods to alleviate the over-fitting problem of machine learning for unbalanced data [44] in the future studies.

As shown in Table 7, among the five evaluation methods (Recall, Precision, F1 score, Accuracy and AUC), the results show that there is a gap in performance among different algorithms. The performance rankings of the five models from high to low are XGBoost, 1D-CNN, RF, BPNN and DT. In addition, ANOVA and Tukey HSD test results show that XGBoost is significantly different from the RF, BPNN, and DT models. This also shows that XGBoost's generalization performance and predictive ability are significantly better than RF, BPNN and DT.

In the early days, BPNN showed excellent performance in a variety of classification tasks. However, this research only demonstrates its accuracy to outperform a single DT. XGBoost is not only better than BPNN in terms of accuracy, but also in terms of speed, because BPNN has too many parameters to be adjusted. Especially, XGBoost can generate "feature importance" that allows researchers to analyze the data and BPNN is a black box model, for which much research has been done to explain the internal structure. Although XGBoost has not been used for debris flow susceptibility analysis, some scholars in the field of mountain disaster study have similar conclusions that the boost model exceeds the accuracy of BPNN by 8% [18].

RF and XGBoost are integrated machine learning algorithms based on DT. The corresponding evaluation scores are higher than that of a single DT. Such a result shows that the integrated algorithm can make up the lack of fitting ability of a single DT. Although RF and XGBoost are both integrated machine learning algorithms, XGBoost's overall performance is better than the RF algorithm. The RF algorithm focuses on the final voting results of all DTs, which can reduce variance, while XGBoost focuses on the residuals generated by the last iteration which can reduce both variance and bias. Performance comparisons between XGBoost and RF have been commonly obtained in many research areas. Usually, XGBoost is in the leading position [39,45].

Like XGoost, 1D-CNN has not been used in debris flow susceptibility, and little literature is concerned about them. The cross-validation results show that the accuracies of XGBoost and 1D-CNN are not significantly different, but the average accuracy of XGBoost is better than that of 1D-CNN. The test performance of the two models is also led by XGBoost. The main reason for this result is that CNN can capture things like image, audio and possibly text quite well by modeling the spatial temporal locality, while tree-based models solve tabular data very well.

When considering the model classification performance comprehensively, we can find that XGBoost has the best comprehensive performance with high classification accuracy, good prediction effect and less calculation time. Therefore, the XGBoost research method should attract more attention in the future evaluation of DFS.

*4.2. Feature Importance*

Based on the selected model to construct the DFS map, the following conclusions can be drawn from Figure 4. The DFS in the study area is mainly medium and low, accounting for more than 50% of the entire study area. The feature attribute scores provided by the tree-based machine learning method are important for analyzing the cause of debris flows. The results have shown that the aspect, profile curvature, annual average rainfall and DEM are the main factors affecting the occurrence of debris flows in the study area. The other triggering factors such as vegetation cover and human activities also have a certain impact on debris flow, while the contribution of soil factors to the modeling is relatively weak. According to the evaluation results of the model feature attributes, the targeted analysis and investigation of the debris flow triggering factors in the study area can be carried out. Based on historical data statistics, analysis of the main triggering factors is conducted.

In the study area, different slope directions lead to differences in hydrothermal conditions, which in turn affect the geographical element distributions such as vegetation, hydrology, soil, and topography. Some Chinese scholars have also examined the relationship between vegetation and debris flow erosion and suggested that the slope direction largely determines the vegetation type and soil type [46]. Feature selection show that the slope aspect has the greatest impact on the distribution of debris flows. According to the actual debris flow events statistics (Figure 10a), the distribution of debris flow events in each aspect is given, and the number of debris flows on the southwest slope and the east slope are relatively large. Among them, the debris flow events on the southwest slope are the densest as well as the distribution location of highly occurrence-prone debris flow events.

The curvature of the slope describes its shape, which controls the formation of debris flow events by affecting the gravitational potential energy and water collection conditions. The feature selection results show that the profile curvature can be better used to estimate the DFS than the plane curvature and total curvature. The shape of the slope is usually linear, concave or convex, indicating the mid-term evolution of the landscape, the maturity of the landscape and the period of violence of the landscape, respectively. According to the statistic (Figure 10b), the debris flows are mainly concentrated in the area where the curvature is negative, i.e., the surface of the pixel is convex. This statistical result is consistent with the results of Guo et al. [47] on mountain debris flow events.

The overall elevation of the Shigatse region is very high and the valley is deep. Statistics on the distribution of debris flow events at different altitudes indicate that debris flow events are mainly distributed at altitudes between 3600–4600 m (Figure 10c). High-susceptibility and medium-susceptibility levels are distributed at the altitude of about 4000 m. The reason is that the region at the altitude between 3600 and 4600 m is very steep and densely populated. As a result, human activities have a huge impact on it. In addition, the area within this altitude is mainly eroded by flowing water with serious accumulation of loose materials and debris flow events particularly develop. Tang et al. [31] also got similar conclusions for the investigation of the study area.

Rainfall is the main triggering factor for debris flows. It mainly promotes the mountain debris flows by increasing soil bulk density and reducing cohesion and internal friction. The study area is mainly a rain-sparing region with annual average rainfall less than 1350 mm (Figure 10d). However, the debris flow has a very significant correlation with the rainfall season in Shigatse. Although the annual precipitation is not high, the temporal distribution is concentrated. That is, heavy rain season is also the season when debris flows frequently occur. According to statistics, debris flows in the flood season in Shigatse accounts for more than 70% of the total debris flows.

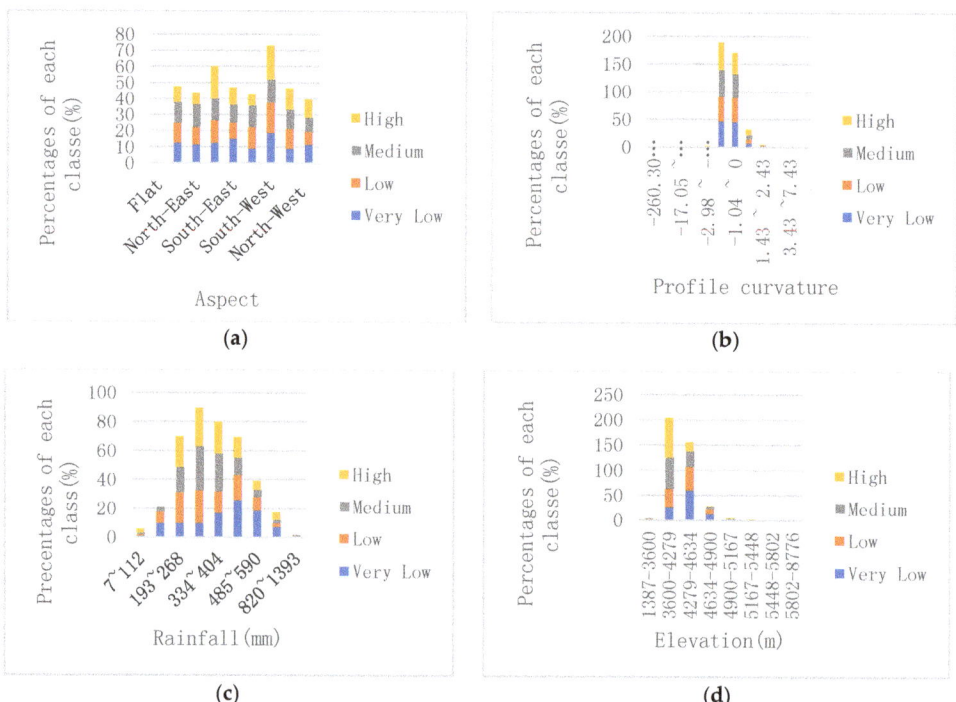

**Figure 10.** Major triggering factors data obtained from the initial region of the debris flow: (**a**) Aspect; (**b**) Profile curvature; (**c**) Rainfall; and (**d**) Elevation.

Machine learning algorithms can handle large-scale data. In addition, they are more objective than the traditional qualitative evaluation methods and can support making decisions without expert system support. However, there are some inherent problems. For example, the data preprocessing workload is large and time-consuming, and the data processing results have a great impact on the classifier.

## 5. Conclusions

In this study, multi-source satellite data and GIS are used to characterize the gestation environment of debris flows in the study area, and then input these environmental characteristics into the machine learning methods to establish the DFS model. The role and weight of the triggering factors shown by the training process are analyzed for the purpose of further studying the main causes of debris flow. In the entire research process described above, the four main findings are described as follows:

(1) Satellite remote sensing can provide data for regional DFS analysis, especially for mountainous areas such as the southwestern Tibet with steep terrain where the sites are not always accessible for investigation. Higher resolution does allow the image to better describe the terrain where the debris flow occurs [48] and potentially improve further analysis. It is important and necessary to use topographical factors, human activities, vegetation cover, climatic and soil elements provided by satellite remote sensing to estimate regional debris flow susceptibility.

(2) Five machine learning algorithms were used to construct DFS map in Shigatse. The results confirm that all five methods can be used to analyze the susceptibility of debris flows. According to the performance, XGBoost ranks the first, and 1D-CNN is the second, followed by RF, BPNN, and DT. XGBoost has the best predictive performance with the highest score among the five evaluation methods. The ANOVA method and the Tukey's HSD test showed that the accuracy of XGBoost is significantly better than those of RF, BPNN, and DT, but it is not significantly different from 1D-CNN. In terms of the time required for prediction, DT takes the least time, and the time required for 1D-CNN is moderate and close to XGBoost. RF and BPNN are slower to calculate. It is notable that this is the first comparative experiment of XGBoost and 1D-CNN in the study of DFS. The ranking of the model based on the "feature importance" indicates that the slope aspect, rainfall, profile curvature and DEM have a greater impact on the debris flows. The results of this study are significant for the local public facility construction and the residential property protection. Therefore, the XGBoost method has good prospects in estimating the DFS.

(3) By comparing the debris flow susceptibility maps of the five prediction algorithms, it is found that the prediction results of five models all show that the moderately susceptible areas account for a large proportion. This experiment has not yet explained the reasons for the different prediction results. The causes will be explored in the subsequent studies. There may be some shortcomings in the use of susceptibility in statistics as a label in experiments. In the follow-up study, we are going to use the clustering algorithm first to obtain the location where the debris flow is not easy to occur and use it together with the existing debris flow data for the classification of debris flow susceptibility. With the development of machine learning technology, we will strive to further improve the performance of the model for DFS by modifying and optimizing the algorithm.

(4) Debris flows are common in mountain areas. Five machine learning models are used to analyze the debris flow events in the study. The results show that the XGBoost model has the best predictive performance, which can be used to prevent casualties and economic losses caused by debris flows. For local land planning and land use, relevant departments can use the XGBoost model in combination with satellite remote sensing and GIS spatial data processing to create feature maps and high-precision area-sensitive maps to provide guidance and preparation for debris flow prevention and mitigation.

**Author Contributions:** Conceptualization, Y.Z. and T.G.; methodology, T.G.; software, T.G.; validation, W.T., and Y.Z.; formal analysis, T.G.; investigation, T.G.; resources, Y.Z.; data curation, W.T.; writing—original draft preparation, T.G.; writing—review and editing, Y.Z., W.T., Y.-A.L.; visualization, T.G.; supervision, W.T.; project administration, Y.Z.; funding acquisition, Y.Z.

**Funding:** This research was funded by the National Natural Science Foundation of China, grant number 41661144039, 41875027 and 41871238.

**Acknowledgments:** The authors would like to thank the Tibet Plateau Institute of Atmospheric Environment, Geospatial data cloud, Resource and Environmental Cloud Platform, earth observing system data and information system and Geoscientific Data & Discovery Publishing Systems for the data that they kindly provided. Acknowledgement for the data support from "National Earth System Science Data Sharing Infrastructure, National Science & Technology Infrastructure of China. (http://www.geodata.cn)".

**Conflicts of Interest:** The authors declare no conflict of interest.

## References

1. Iverson, R.M. Debris-flow mechanics. In *Debris-Flow Hazards and Related Phenomena*; Springer: Berlin/Heidelberg, Germany, 2005; pp. 105–134. ISBN 9783540207269.
2. Golovko, D.; Roessner, S.; Behling, R.; Wetzel, H.U.; Kleinschmit, B. Evaluation of remote-sensing-based landslide inventories for hazard assessment in Southern Kyrgyzstan. *Remote Sens.* **2017**, *9*, 943. [CrossRef]
3. LV, X.; Ding, M.; Zhang, Y.; Teng, J. Hazard assessment of mountainous disasters in Nieyou section of Sino-Nepal highway based on triangle whitening weight function. *J. Southwest Univ. Sci. Technol.* **2017**, 1. [CrossRef]
4. Sun, Y.; Chen, H.; Zhang, Z.; Zhao, Y.; Bao, P.; Bai, J. Distribution regularities of geological hazards along the g318 lhasa-shigatse section and their influence factors. *J. Nat. Disasters* **2014**, *23*, 111–119.
5. Gregoretti, C.; Stancanelli, L.M.; Bernard, M.; Boreggio, M.; Degetto, M.; Lanzoni, S. Relevance of erosion processes when modelling in-channel gravel debris flows for efficient hazard assessment. *J. Hydrol.* **2019**, *568*, 575–591. [CrossRef]
6. Kim, H.; Lee, S.W.; Yune, C.Y.; Kim, G. Volume estimation of small scale debris flows based on observations of topographic changes using airborne LiDAR DEMs. *J. Mt. Sci.* **2014**, *11*, 578–591. [CrossRef]
7. Kim, H.S.; Chung, C.K.; Kim, S.R.; Kim, K.S. A GIS-based framework for real-time debris-flow hazard assessment for expressways in Korea. *Int. J. Disaster Risk Sci.* **2016**, *7*, 293–311. [CrossRef]
8. Alharbi, T.; Sultan, M.; Sefry, S.; ElKadiri, R.; Ahmed, M.; Chase, R.; Chounaird, K. An assessment of landslide susceptibility in the Faifa area, Saudi Arabia, using remote sensing and GIS techniques. *Nat. Hazards Earth Syst. Sci.* **2014**, *14*, 1553. [CrossRef]
9. Ahmed, B.; Dewan, A. Application of bivariate and multivariate statistical techniques in landslide susceptibility modeling in Chittagong City Corporation, Bangladesh. *Remote Sens.* **2017**, *9*, 304. [CrossRef]
10. Li, Y.; Wang, H.; Chen, J.; Shang, Y. Debris flow susceptibility assessment in the Wudongde Dam area, China based on rock engineering system and fuzzy C-means algorithm. *Water* **2017**, *9*, 669. [CrossRef]
11. Liou, Y.A.; Nguyen, A.K.; Li, M.H. Assessing spatiotemporal eco-environmental vulnerability by Landsat data. *Ecol. Indic.* **2017**, *80*, 52–65. [CrossRef]
12. Sujatha, E.R.; Sridhar, V. Mapping debris flow susceptibility using analytical network process in Kodaikkanal Hills, Tamil Nadu (India). *J. Earth Syst. Sci.* **2017**, *126*, 116. [CrossRef]
13. Aditian, A.; Kubota, T.; Shinohara, Y. Comparison of GIS-based landslide susceptibility models using frequency ratio, logistic regression, and artificial neural network in a tertiary region of Ambon, Indonesia. *Geomorphology* **2018**, *318*, 101–111. [CrossRef]
14. Di Cristo, C.; Iervolino, M.; Vacca, A. Applicability of Kinematic and Diffusive models for mud-flows: A steady state analysis. *J. Hydrol.* **2018**, *559*, 585–595. [CrossRef]
15. Xu, W.B.; Yu, W.J.; Jing, S.C.; Zhang, G.P.; Huang, J.X. Debris flow susceptibility assessment by GIS and information value model in a large-scale region, Sichuan Province (China). *Nat. Hazards* **2013**, *65*, 1379–1392. [CrossRef]
16. Chen, X.; Chen, H.; You, Y.; Chen, X.; Liu, J. Weights-of-evidence method based on GIS for assessing susceptibility to debris flows in Kangding County, Sichuan Province, China. *Environ. Earth Sci.* **2016**, *75*, 70. [CrossRef]
17. Achour, Y.; Garçia, S.; Cavaleiro, V. GIS-based spatial prediction of debris flows using logistic regression and frequency ratio models for Zêzere River basin and its surrounding area, Northwest Covilhã, Portugal. *Arab. J. Geosci.* **2018**, *11*, 550. [CrossRef]
18. Oh, H.J.; Lee, S. Shallow landslide susceptibility modeling using the sata mining models artificial neural network and boosted tree. *Appl. Sci.* **2017**, *7*, 1000. [CrossRef]

19. Shirzadi, A.; Shahabi, H.; Chapi, K.; Bui, D.T.; Pham, B.T.; Shahedi, K.; Ahmad, B.B. A comparative study between popular statistical and machine learning methods for simulating volume of landslides. *Catena* **2017**, *157*, 213–226. [CrossRef]
20. Wang, L.J.; Guo, M.; Sawada, K.; Lin, J.; Zhang, J. A comparative study of landslide susceptibility maps using logistic regression, frequency ratio, decision tree, weights of evidence and artificial neural network. *Geosci. J.* **2016**, *20*, 117–136. [CrossRef]
21. Abancó, C.; Hürlimann, M. Estimate of the debris-flow entrainment using field and topographical data. *Nat. Hazards* **2014**, *71*, 363–383. [CrossRef]
22. Prenner, D.; Kaitna, R.; Mostbauer, K.; Hrachowitz, M. The value of using multiple hydrometeorological variables to predict temporal debris flow susceptibility in an alpine environment. *Water Resour. Res.* **2018**, *54*, 6822–6843. [CrossRef]
23. Jiang, W.; Rao, P.; Cao, R.; Tang, Z.; Chen, K. Comparative evaluation of geological disaster susceptibility using multi-regression methods and spatial accuracy validation. *J. Geogr. Sci.* **2017**, *27*, 439–462. [CrossRef]
24. Kang, S.; Lee, S.R.; Vasu, N.N.; Park, J.Y.; Lee, D.H. Development of an initiation criterion for debris flows based on local topographic properties and applicability assessment at a regional scale. *Eng. Geol.* **2017**, *230*, 64–76. [CrossRef]
25. Zhao, J.; Mao, X.; Chen, L. Learning deep features to recognise speech emotion using merged deep CNN. *IET Signal Process.* **2018**, *12*, 713–721. [CrossRef]
26. Pan, H.; He, X.; Tang, S.; Meng, F. An improved bearing fault diagnosis method using one-dimensional CNN and LSTM. *Strojinski Vestnik/J. Mech. Eng.* **2018**, *64*, 443–452. [CrossRef]
27. Zhao, X.; Wen, Z.; Pan, X.; Ye, W.; Bermak, A. Mixture gases classification based on multi-label one-dimensional deep convolutional neural network. *IEEE Access* **2019**, *7*, 12630–12637. [CrossRef]
28. Tsangaratos, P.; Ilia, I. Landslide susceptibility mapping using a modified decision tree classifier in the Xanthi Perfection, Greece. *Landslides* **2016**, *13*, 305–320. [CrossRef]
29. Kadavi, P.R.; Lee, C.-W.; Lee, S. Application of ensemble-based machine learning models to landslide susceptibility mapping. *Remote Sens.* **2018**, *10*, 1252. [CrossRef]
30. Nikolopoulos, E.I.; Destro, E.; Bhuian, M.; Borga, M.; Anagnostou, E. Evaluation of predictive models for post-fire debris flow occurrence in the western United States. *Nat. Hazard Earth Syst. Sci.* **2018**, *18*, 2331–2343. [CrossRef]
31. Tang, M.; Fu, T.; Zhang, W.; Yang, J. Genetic mechanism of geohazard along national highway 318 in Tibet and prevention countermeasure. *J. Highw. Transp. Res. Dev.* **2012**, *5*, 005. [CrossRef]
32. Geological Cloud Portal Home Page. Available online: http://geocloud.cgs.gov.cn/#/portal/home (accessed on 12 May 2017).
33. Marco, C.; Stefano, C.; Sebastiano, T.; Lorenzo, M. GIS tools for preliminary debris-flow assessment at regional scale. *J. Mt. Sci.* **2017**, *14*, 2498–2510. [CrossRef]
34. Djeddaoui, F.; Chadli, M.; Gloaguen, R. Desertification susceptibility mapping using logistic regression analysis in the Djelfa area, Algeria. *Remote Sens.* **2017**, *9*, 1031. [CrossRef]
35. Gong, P.; Wang, J.; Yu, L.; Zhao, Y.; Zhao, Y.; Liang, L.; Niu, Z.; Huang, X.; Fu, H.; Liu, S.; et al. Finer resolution observation and monitoring of global landcover: First mapping results with Landsat TM and ETM+ data. *Int. J. Remote Sens.* **2013**, *34*, 2607–2654. [CrossRef]
36. Li, C.; Gong, P.; Wang, J.; Zhu, Z.; Biging, G.S.; Yuan, C.; Hu, T.; Zhang, H.; Wang, Q.; Li, X.; et al. The first all-season sample set for mapping global landcover with Landsat-8data. *Sci. Bull.* **2017**, *62*, 508–515. [CrossRef]
37. Verbiest, N.; Ramentol, E.; Cornelis, C.; Herrera, F. Preprocessing noisy imbalanced datasets using SMOTE enhanced with fuzzy rough prototype selection. *Appl. Soft Comput.* **2014**, *22*, 511–517. [CrossRef]
38. Mao, Y.; Zhang, M.; Sun, P.; Wang, G. Landslide susceptibility assessment using uncertain decision tree model in loess areas. *Environ. Earth Sci.* **2017**, *76*, 752. [CrossRef]
39. Wang, S.; Dong, P.; Tian, Y. A novel method of statistical line loss estimation for distribution feeders based on feeder cluster and modified XGBoost. *Energies* **2017**, *10*, 2067. [CrossRef]
40. Wu, Y.J.; Chiang, C.T. ROC representation for the discriminability of multi-classification markers. *Pattern Recognit.* **2016**, *60*, 770–777. [CrossRef]

41. Rajaraman, S.; Antani, S.K.; Poostchi, M.; Silamut, K.; Hossain, M.A.; Maude, R.J.; Thoma, G.R. Pre-trained convolutional neural networks as feature extractors toward improved malaria parasite detection in thin blood smear images. *PeerJ* **2018**, *6*, e4568. [CrossRef]
42. Abdi, H.; Williams, L.J. Tukey's honestly significant difference (HSD) test. In *Encyclopedia of Research Design*; Salkind, N., Ed.; Sage: Thousand Oaks, CA, USA, 2010; pp. 1–5.
43. Li, C.; Zheng, X.; Yang, Z.; Kuang, L. Predicting short-term electricity demand by combining the advantages of ARMA and XGBoost in fog computing environment. *Wirel. Commun. Mob. Comput.* **2018**, *2018*, 18. [CrossRef]
44. Shimoda, A.; Ichikawa, D.; Oyama, H. Using machine-learning approaches to predict non-participation in a nationwide general health check-up scheme. *Comput. Methods Programs Biomed.* **2018**, *163*, 39–46. [CrossRef] [PubMed]
45. Zhang, L.; Ai, H.; Chen, W.; Yin, Z.; Hu, H.; Zhu, J.; Liu, H. CarcinoPred-EL: Novel models for predicting the carcinogenicity of chemicals using molecular fingerprints and ensemble learning methods. *Sci. Rep.* **2017**, *7*, 2118. [CrossRef] [PubMed]
46. Wang, Z.Y.; Gong, T.L.; Shi, W.J. Typical types of vegetation and erosion in the Yalutsangpo Basin. *Adv. Earth Sci.* **2011**, *26*, 1208–1216.
47. Guo, C.W.; Yao, L.K.; Duan, S.S.; Huang, Y.D. Distribution regularities of landslides induced by Wenchuan earthquake, Lushan earthquake and Nepal earthquake. *J. Southwest Jiaotong Univ.* **2016**, *51*, 71–77.
48. Stolz, A.; Huggel, C. Debris flows in the Swiss National Park: The influence of different flow models and varying DEM grid size on modeling results. *Landslide* **2008**, *5*, 311–319. [CrossRef]

© 2019 by the authors. Licensee MDPI, Basel, Switzerland. This article is an open access article distributed under the terms and conditions of the Creative Commons Attribution (CC BY) license (http://creativecommons.org/licenses/by/4.0/).

*Article*

# Comparison of Different Machine Learning Methods for Debris Flow Susceptibility Mapping: A Case Study in the Sichuan Province, China

Ke Xiong [1], Basanta Raj Adhikari [1,2], Constantine A. Stamatopoulos [3], Yu Zhan [4], Shaolin Wu [4], Zhongtao Dong [1] and Baofeng Di [1,4,*]

1. Institute for Disaster Management and Reconstruction & Research Center for Integrated Disaster Risk Reduction and Emergency Management, Sichuan University, Chengdu 610207, China; 2017226200010@stu.scu.edu.cn (K.X.); bradhikari@ioe.edu.np (B.R.A.); 2017226200007@stu.scu.edu.cn (Z.D.)
2. Department of Civil Engineering, Pulchowk Campus, Tribhuvan University, Lalitpur 44600, Nepal
3. Stamatopoulos and Associates Co. & Hellenic Open University, 11471 Athens, Greece; k.stam@saa-geotech.gr
4. Department of Environmental Science and Engineering, College of Architecture and Environment, Sichuan University, Chengdu 610065, China; yzhan@scu.edu.cn (Y.Z.); wushaolin@stu.scu.edu.cn (S.W.)
* Correspondence: dibaofeng@scu.edu.cn; Tel.: +86-028-8599-3558

Received: 19 December 2019; Accepted: 8 January 2020; Published: 16 January 2020

**Abstract:** Debris flow susceptibility mapping is considered to be useful for hazard prevention and mitigation. As a frequent debris flow area, many hazardous events have occurred annually and caused a lot of damage in the Sichuan Province, China. Therefore, this study attempted to evaluate and compare the performance of four state-of-the-art machine-learning methods, namely Logistic Regression (LR), Support Vector Machines (SVM), Random Forest (RF), and Boosted Regression Trees (BRT), for debris flow susceptibility mapping in this region. Four models were constructed based on the debris flow inventory and a range of causal factors. A variety of datasets was obtained through the combined application of remote sensing (RS) and geographic information system (GIS). The mean altitude, altitude difference, aridity index, and groove gradient played the most important role in the assessment. The performance of these modes was evaluated using predictive accuracy (ACC) and the area under the receiver operating characteristic curve (AUC). The results of this study showed that all four models were capable of producing accurate and robust debris flow susceptibility maps (ACC and AUC values were well above 0.75 and 0.80 separately). With an excellent spatial prediction capability and strong robustness, the BRT model (ACC = 0.781, AUC = 0.852) outperformed other models and was the ideal choice. Our results also exhibited the importance of selecting suitable mapping units and optimal predictors. Furthermore, the debris flow susceptibility maps of the Sichuan Province were produced, which can provide helpful data for assessing and mitigating debris flow hazards.

**Keywords:** debris flow; susceptibility mapping; machine learning; remote sensing; geographical information system

## 1. Introduction

Debris flow, a serious geological hazard, is defined as a mixture of water and a large number of loose materials like sediments, detritus, and muds, that cause great casualties and economic losses in mountainous areas all over the world [1–3]. Due to the complex natural conditions, South-West China is a typical area with active debris flow. About half of these debris flows took place in the high mountain zone of South-West China [4]. Geomorphological variations, heavy rainfalls, frequent seismic activities, and unreasonable land uses are responsible for triggering such a large number of debris flow in this region. Especially in the Sichuan Province, dense residential areas scattered in mountainous areas are exposed to severe risks during the flood season. For example, large-scale debris flows have caused enormous harm to the human settlements, infrastructures, and ecological security in the Danba County (2003), Dechang County (2004), and Qingping Township (2010) [5,6]. However, there is little understanding of the detection of the potentially prone areas. Thus, appropriate disaster mitigation and prevention solutions should be determined based on the debris flow susceptibility zoning.

Susceptibility maps are useful tools that show the likelihood of occurrence of an event in a specific area based on the local environmental conditions [7]. Field survey and dynamic monitoring in remote mountains are very challenging, therefore, susceptibility zoning is a prominent alternative. Plenty of susceptibility analyses have been performed and published during the last several decades [8]. Multifarious classification methods have been applied, from qualitative assessment to quantitative assessment, such as heuristic methods [9], physical methods [10–12], and data-driven methods [13,14]. Heuristic methods determine the impact of causal factors on debris flow by relying on subjective experience, and then zone susceptibility, descriptively, so that the accuracy of heuristic studies is instable due to their high subjectivity. In the physical methods, debris flow models are formulated based on mechanical principles, physical laws, and simplified physical assumptions. However, constructing physical models on a medium or large scale is quite complex. Physical models are more suitable for understanding such hazards in an individual gully rather than a whole region. In recent years, with the innovation of algorithms and the boom of data, data-driven methods, especially machine learning methods, are more popular.

Machine learning methods aim to analyze the spatial relationship between past events and causal factors by studying data characteristics and predicting the spatial probability of debris flow occurrence. Most of these methods, including Back Propagation Neural Network (BPNN), Logistic Regression (LR), Decision Tree (DT), Random Forest (RF), Boosted Regression Trees (BRT), Bayesian network (BN), and Support Vector Machines (SVM), were developed one after another. In many regions, these methods have been applied to the susceptibility mapping of landslides [15–17], gully erosion [18,19], debris flow [13,14,20], and ground subsidence [21] by integrating environmental remote sensing (RS) data under the Geographic Information System (GIS). These studies indicate that several machine-learning methods provide a good predictive performance. Unlike other algorithms, multilayer BPNN has a distinctive structure and ability to implement deep mining of data. It should be studied separately and carefully due to training difficulties. Whereas, DT and BN have been rarely used for the mapping of debris flow susceptibility, in previous investigations. Therefore, in this study, we ignored the three algorithms mentioned above.

Overall, advanced machine learning algorithms have been used for solving problems of all sorts, but only a few of these research objects were debris flows, therefore, more investigations are needed. Additionally, it is highly important to compare different machine learning methods for susceptibility mapping as each method has its own characteristics and finding an optimal method might have a large impact on real applications. Which method is most suitable for the spatial prediction of debris flows is still debated upon. Therefore, the objective of this study was to compare and analyze four machine learning methods, including LR, RF, SVM, and BRT, for debris flow susceptibility mapping. The Sichuan Province, which is well-known in China as the region with the most frequent and severe debris flow, was therefore selected as a case study. The study was implemented with the help of GIS tools and multifarious remote sensing datasets of the study area, such as environmental and sociometric factors. The results of our study can provide support and help assess and mitigate debris flow hazards.

## 2. Study Area

Southwest China's Sichuan Province is located in the upper reaches of the Yangtze River, between the latitudes of 26°03′N to 34°19′N and longitudes of 97°21′E to 108°33′E, which covers an area of 486,000 square kilometers, at altitudes ranging from 212 m to 6904 m above sea level (Figure 1). The topography of the whole province includes mountainous areas, basins, and plateaus. Terrains are complex and various. Three active faults run through the study area, namely the Longmenshan fault, the Xianshuihe fault, and the Anninghe fault. Complex geological and geomorphological characteristics play an important role to trigger heavy geological hazards in this region, because of complex interactions between the Qinghai–Tibet Plateau and the Sichuan Basin [22]. The study area belongs to the subtropical monsoon zone with an annual average rainfall of 1000 mm; over 80% of precipitation usually takes place in the monsoon season (between April and October) [5]. The average temperature in January ranges from 3 °C to 8 °C and the average temperature in July is 25 °C~29 °C [5]. Landslides, debris flow, and mountain torrents are widespread here. In particular, debris flows triggered by heavy rains pose the greatest threat in the study area.

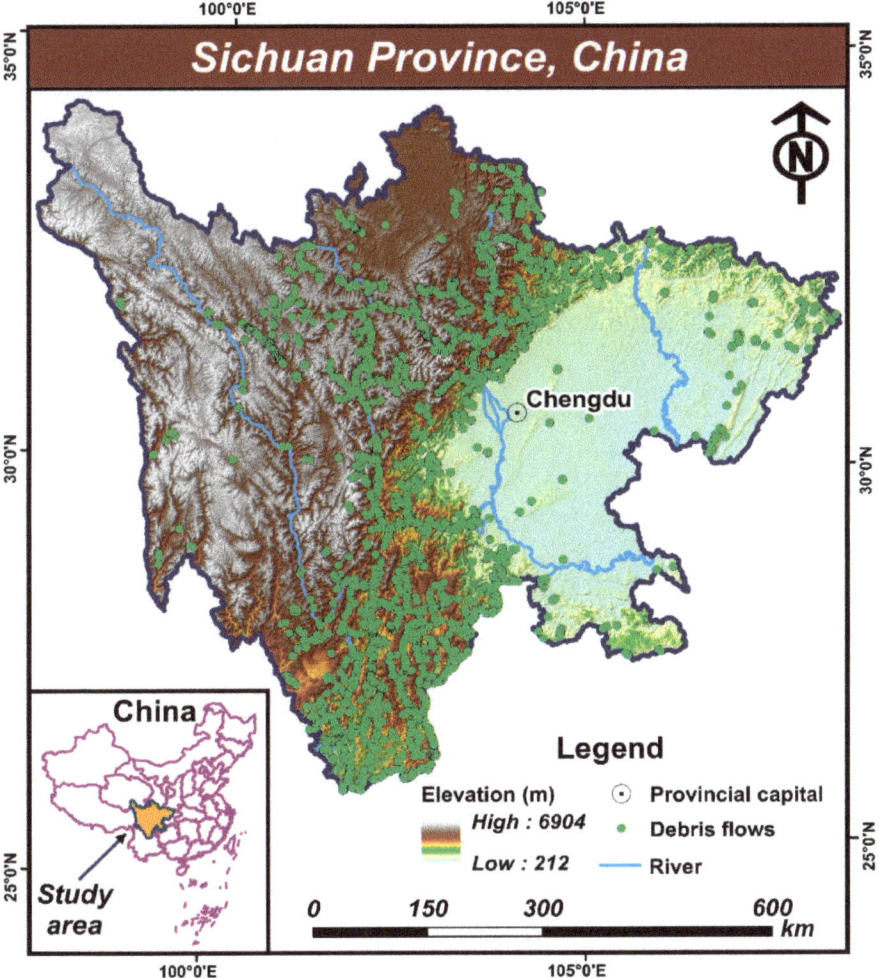

**Figure 1.** Study area and debris flow location map. Each dot represents the geographic coordinate of a debris flow.

The Sichuan Province is an important node of the "One Belt and One Road" where many important projects such as the Sichuan–Tibet Railway and the Baihetan Hydropower Station exist in the region.

To promote sustainable development of society and economy, it is always important to conduct debris flow susceptibility analyses to understand the surface dynamics and climatic variability.

## 3. Materials and Methods

To achieve debris flow susceptibility mapping, four main stages were adopted; illustrated in Figure 2. First, the debris flow inventory was prepared and the causal factors were selected. Then, they were separated into two independent groups, namely the training set and the validation set, using the validation set approach. Second, four debris flow susceptibility models were set up based on the LR, RF, SVM, and BRT algorithms. Third, we applied the constructed models to develop debris flow susceptibility maps of the study area. Finally, these models were evaluated and compared using two widely used criteria, including predictive accuracy (ACC) and the area under the receiver operating characteristic curve (AUC).

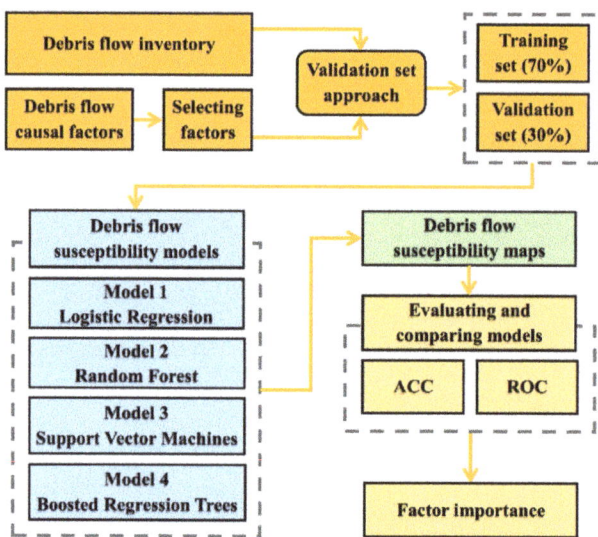

**Figure 2.** Methodological flowchart of this study.

*3.1. Preparation of Data Sets*

3.1.1. Compilation of Debris Flow Inventory

Debris flow inventory is an important prerequisite for the analyses of debris flow susceptibility because there is an assumption that past events have a great influence on the future [2]. In this study, detailed information of 3839 rainfall-triggered debris flow events in the Sichuan Province, from 1949 to 2017 was collected, based on historical records collection, aerial photographs, satellite remote sensing images interpretation, and field verification. Some of the information was obtained from government departments in Sichuan. Therefore, the inventory is reliable in both quality and completeness. The locations of debris flows are shown as points (Figure 1). As can be seen in the diagram, these events were concentrated in the mountainous and hilly areas of the mid-Sichuan region.

3.1.2. Selection of Debris Flow Causal Factors

The selection of debris flow causal factors is also an important task for susceptibility modeling and mapping. Investigators have been using diverse geo-environmental factors in previous studies and have been trying to explore their relationship with debris flows. Based on the general cause of debris flows, six clusters of factors were initially determined for modeling in this study, including topographical, geological, edaphic, meteorological, land-cover, and sociometric factors (Table 1). All factors were prepared with the help of GIS and RS. The topographical factors, including mean slope angle, slope aspect, mean altitude, altitude difference, and groove gradient, were derived from the Digital Elevation Model (DEM). Notably, the groove gradient referred to the ratio of the height difference of gully to its length and was an elemental parameter for the initiation and motion of debris flows. Geological factors, namely seismic intensity and lithology, were prepared in a GIS environment using a seismic information map and lithological composition map, respectively. Similarly, edaphic factors (soil texture and soil erosion) and meteorological factors (moisture index, aridity index, mean annual temperature, accumulated temperature of 10 °C, and annual precipitation) were acquired from the Data Center for Resources and Environmental Sciences, Chinese Academy of Sciences (RESDC) and were pre-processed by GIS technology [23]. Additionally, sociometric factors (population density and road density) were obtained from the remote sensing datasets provided by RESDC and the OpenStreetMap [24]. Land cover factors (Normalized Difference Vegetation Index and land use) constructed from remote sensing images were also commonly used [25].

Watersheds were selected as mapping units to avoid problems of raster grid-cells, such as lack of physical relations with debris flows [26]. Watershed units have significant conceptual and operational advantages [8]. We used the Hydrological Analysis Tool of ArcGIS v.10.2 software to divide the study area into watersheds. Figure 3 exhibits 2471 mapping units ranging from 8.26 km$^2$ to 1829.68 km$^2$. We excluded a few regions (e.g., plateau and plain areas) of the Sichuan Province in susceptibility mapping because of the inadequate conditions to trigger a debris flow.

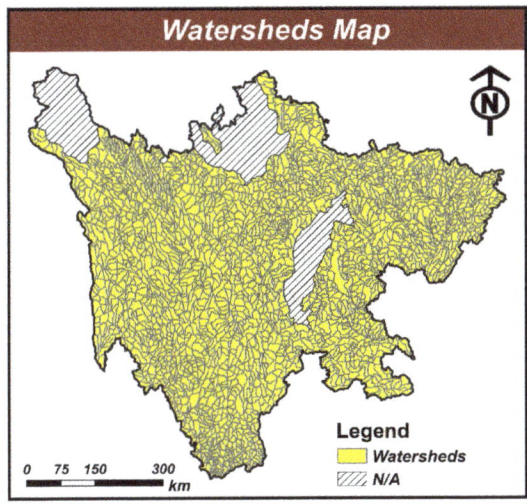

**Figure 3.** Watersheds map of the Sichuan Province, China.

The raw data of the causal factors were resampled based on the mapping units, using the Zonal Statistics Tool in the ArcGIS v.10.2 software. The following factors took the average value in the watershed—mean slope angle, mean altitude, moisture index, aridity index, mean annual temperature, accumulated temperature of 10 °C, annual precipitation, and NDVI. Moreover, other factors took the mode value in the watershed, including slope aspect, seismic intensity, lithology, soil erosion, and land use. In particular, five factors had sub-factors. (1) The seismic intensity was reclassified into five groups (<VI, VI, VII, VIII, and ≥IX), and the area of each group was also taken as a factor. (2) The lithology was constructed with five groups based on hardness (extremely soft, soft, moderate hard, hard, and extremely hard), and the area of each group was also taken as a factor. (3) The soil texture

included three factors—clay content, sand content, and silt content. (4) Soil erosion was reclassified into six groups (micro, mild, moderate, serious, drastic, and very drastic), and the area of each group was also taken as a factor. (5) Land use was interpreted as six groups (cropland, woodland, grassland, waterbody, construction land, and unused land), and the area of each type was also taken as a factor. Therefore, 42 initial causal factors were prepared.

Table 1. Initial debris flow causal factors and their sources [1].

| No. | Causal Factors | Clusters | Sources |
|---|---|---|---|
| 1 | Mean slope angle | Topographic | ASTER GDEM (Spatial resolution of 30 m × 30 m) (http://earthexplorer.usgs.gov) |
| 2 | Slope aspect | | |
| 3 | Mean altitude | | |
| 4 | Altitude difference | | |
| 5 | Groove gradient | | |
| 6 | Seismic intensity * | Geological | China seismic information (Scale of 1:4,000,000) (http://www.csi.ac.cn) |
| 7 | Lithology * | | Lithological composition map of Sichuan Province (Scale of 1:200,000) (http://www.csi.ac.cn) |
| 8 | Soil texture * | Edaphic | Spatial distribution datasets of soil texture in China (Spatial resolution of 1 km × 1 km) (http://www.resdc.cn) |
| 9 | Soil erosion * | | Spatial distribution datasets of soil erosion in China (Spatial resolution of 1 km × 1 km) (http://www.resdc.cn) |
| 10 | Moisture index (Calculated by Thornthwaite method) | Meteorological | Meteorological datasets in China (Spatial resolution of 500 m × 500 m) (http://www.resdc.cn) |
| 11 | Aridity index | | |
| 12 | Mean annual temperature | | |
| 13 | Accumulated temperature of 10 °C | | |
| 14 | Annual precipitation | | |
| 15 | Population density | Sociometric | Spatial distribution datasets of population in China (Spatial resolution of 1 km × 1 km) (http://www.resdc.cn) |
| 16 | Road density | | OpenStreetMap Data (http://planet.openstreetmap.org) |
| 17 | Normalized Difference Vegetation Index (NDVI) | Land cover | MODIS images (Spatial resolution of 500 m × 500 m) (https://modis.gsfc.nasa.gov) |
| 18 | Land use * | | The land use and land cover change database in China (Spatial resolution of 1 km × 1 km) (http://www.resdc.cn) |

[1] The factors with "*" in the table have the following sub-factors: (1) The seismic intensity was reclassified into five groups (<VI, VI, VII, VIII, and ≥IX), and the area of each group was also taken as a factor. (2) The lithology was constructed with five groups based on hardness (extremely soft, soft, moderate hard, hard, and extremely hard), and the area of each group was also taken as a factor. (3) The soil texture included three factors—clay content, sand content, and silt content. (4) Soil erosion was reclassified into six groups (micro, mild, moderate, serious, drastic, and very drastic), and the area of each group was also taken as a factor. (5) Land use was interpreted as six groups (cropland, woodland, grassland, waterbody, construction land, and unused land), and the area of each type was also taken as a factor.

An excellent debris flow susceptibility model relies on a set of suitable factors, so the above factors should be further evaluated and selected [8]. For this study, we adopted the backward variable selection method based on the RF algorithm to select the optimal factors and improve the predictive capability [27]. First, we constructed and evaluated an RF model with all factors, where the model performance and variable importance were recorded. Then, the factor with the lowest importance was eliminated and a new model was implemented. This procedure was repeated until there was only one factor left. Finally, a set of factors with the highest performance was chosen for the final prediction, and the rest were removed. In Figure 4 it can be seen that only 15 factors were selected as the optimal predictors for assessing the susceptibility of debris flow in this study.

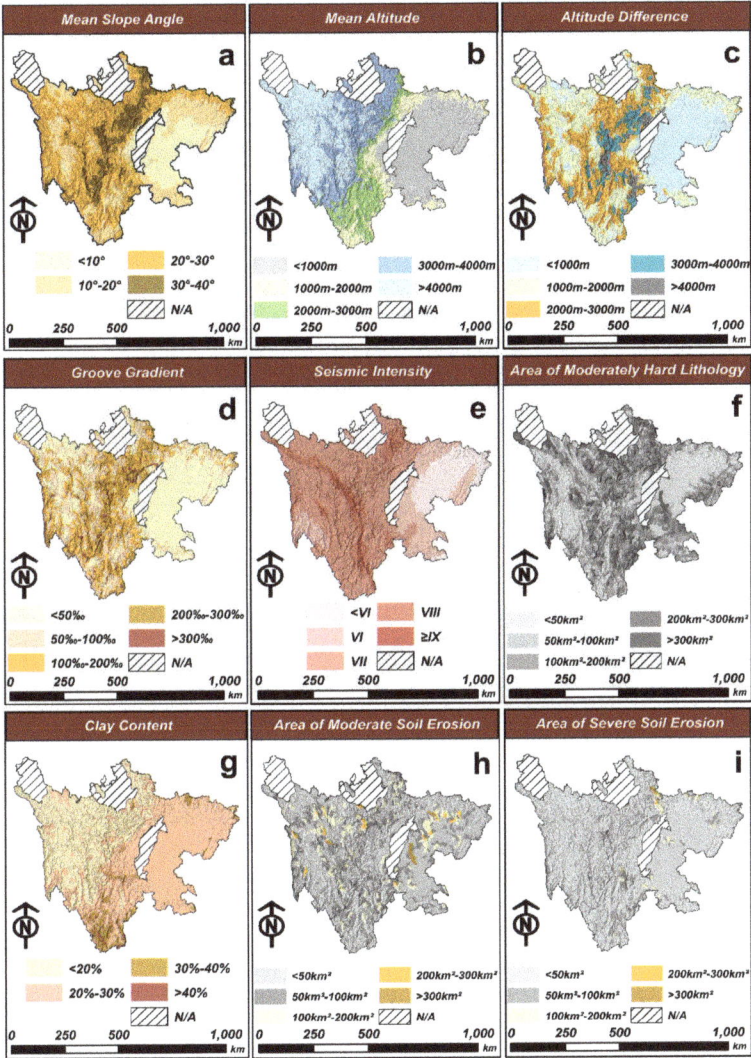

**Figure 4.** *Cont.*

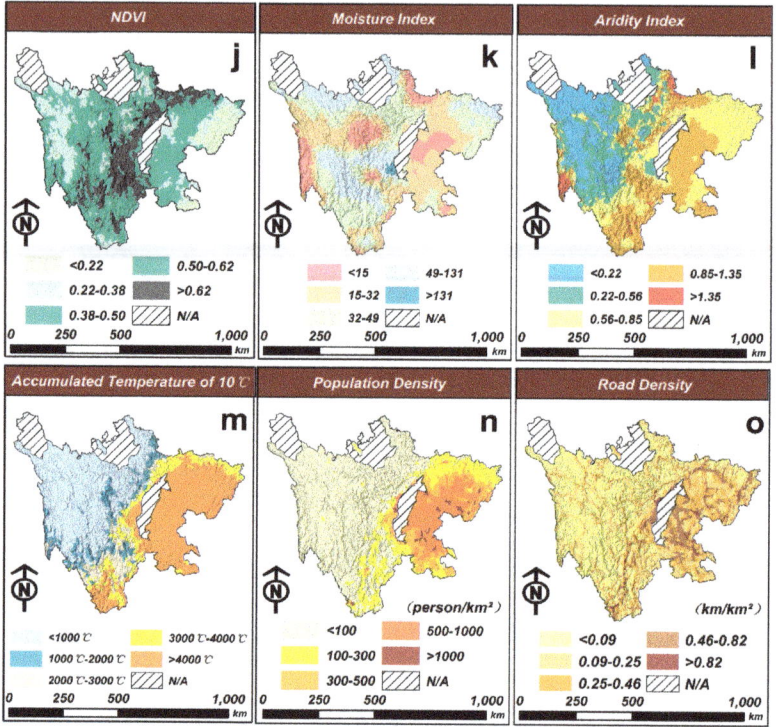

**Figure 4.** Optimal causal factors used for final debris flow susceptibility mapping. (**a**) mean slope angle, (**b**) mean altitude, (**c**) altitude difference, (**d**) groove gradient, (**e**) seismic intensity, (**f**) area of moderately hard lithology in each watershed, (**g**) clay content of each watershed, (**h**) area of moderate soil erosion in each watershed, (**i**) area of severe soil erosion in each watershed, (**j**) NDVI, (**k**) moisture index, (**l**) aridity index, (**m**) accumulated temperature of 10 °C, (**n**) population density, and (**o**) road density.

3.1.3. Partition of Data Sets

Of the 2471 watersheds in the study area, 772 watersheds were positive cases (debris flows had occurred) and the remaining 1699 watersheds were negative cases (debris flows had not occurred). According to previous studies, the size of the types of cases selected for a model should be similar [28,29]. Therefore, 772 negative cases were selected randomly along with the same number of positive cases, to train and validate the models. In general, approximately 70% of the data was randomly selected for model training, meanwhile the remaining 30% was used for model validation. This data partitioning method, namely the validation set approach, was easy to implement. However, the ratio of training/validation set needed to be chosen carefully. The inappropriate ratio might cause potential problems in the procedure of data mining, such as overfitting or deficient model training, which significantly affects the predictive performance of the model. A split of 70%–30% is a common choice adopted by many investigators for coping with this challenge [17,22,30,31]. Therefore, 1082 watersheds consisted of 541 positive cases and 541 negative cases were used to train the models, while 462 watersheds contained 231 positive cases and 231 negative cases served the output validation. The positive/negative cases were labeled as 1/0 for modeling. To obtain more robust conclusions, the sampling procedure was repeated three times. All three sample datasets participated in the model operation. Lastly, the values of causal factors were resampled for each watershed.

## 3.2. Model Construction Using Machine Learning Algorithms

After preparing datasets, we constructed and trained four debris flow susceptibility models using machine learning algorithms. The statistical tool R version 3.4.4 was used for the model training [32]. We paid more attention to adjust and optimize the parameters based on the cross-validation approach, for improving the effectiveness of the models. The model output was the occurrence probability of debris flow and was used to simulate susceptibility.

### 3.2.1. Logistic Regression (LR)

Logistic Regression (LR) is a multivariate regression algorithm that has been extensively used for the susceptibility assessment [22,33,34]. LR is suitable to understand the relationship between a binary variable (whether the debris flow will occur or not) and several causal factors, and estimate the probability of an event [35]. The logit–natural logarithm of LR can be expressed as below:

$$\log\left(\frac{p(X)}{1-p(X)}\right) = \beta_0 + \beta_1 X_1 + \cdots + \beta_p X_p \quad (1)$$

Therefore, in this study, the probability $p$ of a debris flow occurrence in each watershed could be estimated by using the following equation:

$$p(X) = \frac{e^{\beta_0 + \beta_1 X_1 + \cdots + \beta_p X_p}}{1 + e^{\beta_0 + \beta_1 X_1 + \cdots + \beta_p X_p}} \quad (2)$$

where $X = (X_1, \ldots, X_p)$ are the debris flow causal factors, $\beta_0$ represents the intercept, $(\beta_1, \ldots, \beta_p)$ are the regression coefficients. LR uses the maximum likelihood method to estimate $(\beta_1, \ldots, \beta_p)$. Finally, the probability of a debris flow occurring varies from 0 to 1.

### 3.2.2. Random Forest (RF)

Random Forest (RF) is a multivariate model that belongs to one of the ensemble-learning techniques [36]. The algorithm is also suitable for debris flow susceptibility assessment. According to the decision rules, a series of decision trees were established, and final decision (whether the debris flow will occur or not) was determined based on the majority vote [37]. When constructing these decision trees, each time a split in a tree was considered, a random sample containing $m$ causal factors was selected as the split candidates, among all factors. Forcing each split to consider only a subset of all factors helped to overcome the weakness of overfitting and improved the stability. This process was thought of as de-correlating trees, thereby, making the results more reliable. There were two important parameters, namely the number of trees and the tree depth, which needed to be tuned when modeling. Additionally, to assess factor importance, the mean decrease accuracy and mean decrease Gini were calculated [38–40].

### 3.2.3. Support Vector Machines (SVM)

Support Vector Machines (SVM) was developed in the 1990s [41] and has grown into a popular approach for classification because of its superior empirical performance in a variety of settings [30,42,43]. In this study, debris flow causal factors were mapped into a high-dimensional feature space. Then, the model struggled to detect a hyperplane to separate positive cases and negative cases, as much as possible [44]. The optimal hyperplane can be obtained by solving the following optimization problem:

$$\begin{array}{c} \underset{\beta_0,\beta_{11},\beta_{12},\cdots,\beta_{p1},\beta_{p2},\varepsilon_1,\varepsilon_2,\cdots,\varepsilon_n}{\text{maximize}} M \\ \text{subject to } y_i\left(\beta_0 + \sum_{j=1}^{p}\beta_{j1}x_{ij} + \sum_{j=1}^{p}\beta_{j2}x_{ji}^2\right) \geq M(1-\varepsilon_i) \\ \sum_{i=1}^{n}\varepsilon_i \leq C, \varepsilon_i \geq 0, \sum_{j=1}^{p}\sum_{k=1}^{2}\beta_{jk}^2 = 1 \end{array} \quad (3)$$

where $C$ is a non-negative tuning parameter, $M$ is the width of the margin, $\varepsilon_1, \varepsilon_2, \ldots, \varepsilon_n$ are slack variables. Later, to classify new data, the decision function can be written as below:

$$f(x) = \text{sgn}\left(\sum_{i=1}^{n} y_i \alpha_i K(x_i, x_j) + b\right) \quad (4)$$

where $K$ is the function that we will refer to as a kernel, $b$ represents the offset from the origin of the hyperplane, $n$ means the number of causal factors, and $\alpha_i$ are positive real constants. The radial basis kernel function was adopted in this study due to its robustness, as reported by Rahmati et al. [30] and Kavzoglu et al. [45]. The core parameters of SVM modeling included gamma and cost.

3.2.4. Boosted Regression Trees (BRT)

Boosted Regression Trees (BRT) is an approach of combining gradient boosting algorithm with classification and regression trees [46]. BRT adopts a method similar to RF to implement the debris flow susceptibility assessment. The difference is that smaller trees are typically sufficient in BRT, because of their slow learning process. Additionally, each tree in BRT is created based on the modification of previous trees, unlike the RF algorithm. The core of training the BRT model is to select the optimal value of three pivotal parameters—the shrinkage coefficient, the number of trees, and splits in each tree. They control the rate at which boosting learns, the model's scale, and the complexity of the boosted ensemble, respectively. The optimal parameters were automatically set through cross-validation.

3.3. Evaluation and Comparison Methods

In this study, two commonly used criteria, including the predictive accuracy (ACC) and receiver operating characteristic (ROC) curve were applied to quantify and compare the performance of models. ACC is a statistical metric that relies on the components of the confusion matrix [30,47]. As Table 2 shows, the confusion matrix reveals the discrepancy between the model results and the actual observed outcomes. ACC can be estimated by the following equation:

$$ACC = \frac{TP + TN}{TP + TN + FP + FN} \quad (5)$$

where $TP$ and $TN$ refer to the number of watersheds that are correctly classified, while $FP$ and $FN$ refer to the number of watersheds classified incorrectly.

Table 2. Confusion Matrix.

| Observed | Predicted | |
|---|---|---|
| | Debris-Flow | Non-Debris-Flow |
| Debris-flow | True positive (TP) | False negative (FN) |
| Non-debris-flow | False positive (FP) | True negative (TN) |

The ROC curve, which elucidates the alterations of true positive rate (TPR) and false positive rate (FPR) when the discrimination threshold changes [48,49], is also a widely used technique to measure the goodness-of-fit and the predictive power of probabilistic models. TPR is the ratio of positive cases that are correctly identified under a specific threshold value. FPR means the ratio of all negative cases that are incorrectly predicted to be positive, under the same threshold value. They also rely on the confusion matrix and can be obtained from following equations:

$$TPR = TP/(TP + FN) \qquad (6)$$

$$FPR = FP/(FP + TN) \qquad (7)$$

This popular graph visualizes the confusion matrix under various thresholds and tracks two kinds of classification errors [50]. The overall performance of debris flow susceptibility models is quantified by the area under the curve (AUC). An ideal ROC curve should be close to the upper-left corner, usually the higher the AUC value the better the model. According to the previous studies, the performance of a model based on the AUC value can be classified as several levels: 0.5~0.6 = poor, 0.6~0.7 = moderate, 0.7~0.8 = acceptable, 0.8~0.9 = excellent, and 0.9~1 = almost perfect [19,30].

## 4. Results

*4.1. Development of Debris Flow Susceptibility Maps*

The core of LR modeling was the estimation of the regression coefficients using the maximum likelihood method. During the RF modeling, the number of trees and the tree depth were determined as 1000 and 5. For the SVM model, the parameters, gamma and cost, were tuned to 1 and 10, respectively. The important parameters in the BRT model, i.e., the shrinkage coefficient, the number of trees, and splits in each tree, were identified to be 0.2, 1000, and 4, respectively. After model building and operation, we averaged the model outputs of three sample datasets to generate the results. Repeated sampling was helpful to reduce sampling error and gain more robust analysis results. Four models were applied to calculate the debris flow susceptibility index for each watershed in the Sichuan Province. According to the computed index, ranging from 0 to 1, susceptibility levels were reclassified into five categories (very low, low, moderate, high, and very high) using the natural break classification method in the GIS environment [17]. Then, susceptibility maps were produced in the GIS platform for visualization (Figure 5). The results of the assessment showed that watersheds with high and very high debris flow susceptibility were chiefly distributed in the central mountainous region of the study area. Whereas there was lower susceptibility in the western plateau districts as well as the eastern plain districts, with a gentle topography fluctuation.

Figure 6 depicts the relative distribution of the susceptibility classes calculated for each model. In the LR model, the low class had the largest proportion (22.42%). 21.57%, 17.44%, 16.35%, and 22.22% of watersheds which fell into the 'very low', 'moderate', 'high', and 'very high' susceptibility classes, respectively. For the debris flow susceptibility maps of the RF and SVM model, the percentages of each class were very similar to those acquired by the LR model. Furthermore, the percentage of very high class in the BRT model (12.59%) was small, which was lower than that based on other models. The moderate and lower debris flow susceptibilities were the main levels in the study area.

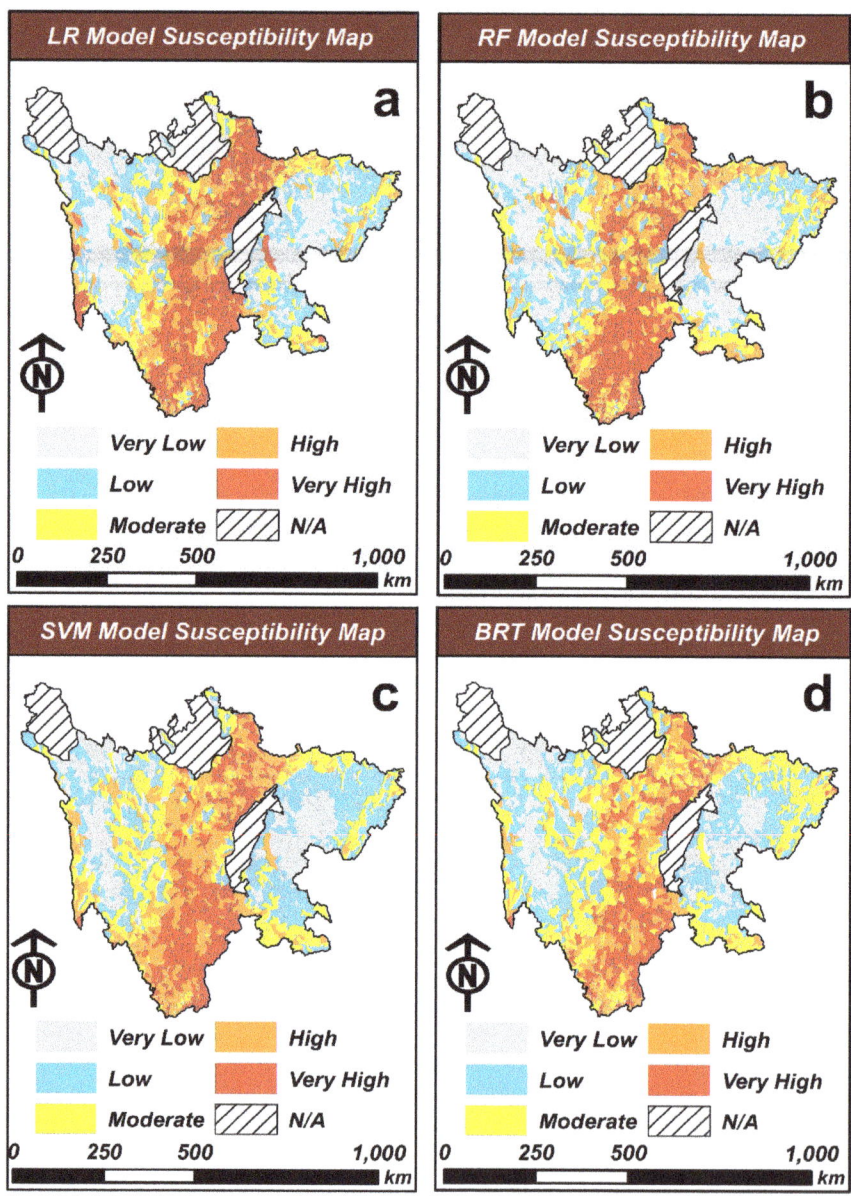

**Figure 5.** The debris flow susceptibility maps of the Sichuan Province based on the (**a**) Logistic Regression (LR), (**b**) Random Forest (RF), (**c**) Support Vector Machines (SVM), and (**d**) Boosted Regression Trees (BRT) models.

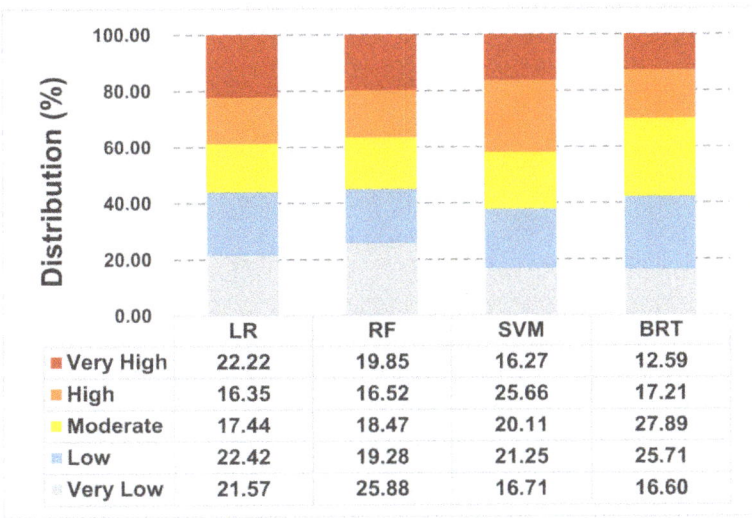

**Figure 6.** Proportions of the different debris flow susceptibility classes from the four models.

Overall, the susceptibility map intuitively describes the prone distribution of future debris flows. The establishment of a large-scale prediction system based on machine learning methods has extremely high application values and broad application prospects. More comprehensive analyses of debris flow prediction system should be conducted to guide the practice of disaster prevention and reduction in the future.

*4.2. Evaluation and Comparison of Machine Learning Models*

The performance of the four models was evaluated and compared using the criteria chosen in Section 3.3. Analyses of the ACC and AUC using the training set are shown in Figure 7a and Table 3. The highest AUC value belonged to the BRT model (AUC = 0.907), followed by the RF model (AUC = 0.870), the SVM model (AUC = 0.865), and the LR model (AUC = 0.843), respectively. Similarly, it could be found that the BRT model had the highest ACC value (0.823), other models followed it. The criteria showed a high goodness-of-fit for all models in the training step. However, performance in the training step was not enough to assess the prediction capacity of the model [51]. Therefore, we paid more attention to the performance of models in the validation set. Table 4 and Figure 7b show the ACC and AUC values on the validation set. The highest ACC and AUC values belong to the BRT model (ACC = 0.781, AUC = 0.852), followed by the RF model (ACC = 0.779, AUC = 0.849), the SVM model (ACC = 0.781, AUC = 0.849), and the LR model (ACC = 0.762, AUC = 0.829), respectively. The ACC values of these models are far above 75% and the AUC values range from excellent to almost perfect. According to these, all four machine-learning models performed well, considering the above factors for debris flow susceptibility mapping. The BRT model was superior to the rest of the models. This reiterates the fact that a data-driven classification model that learns slowly shows impressive performance [48].

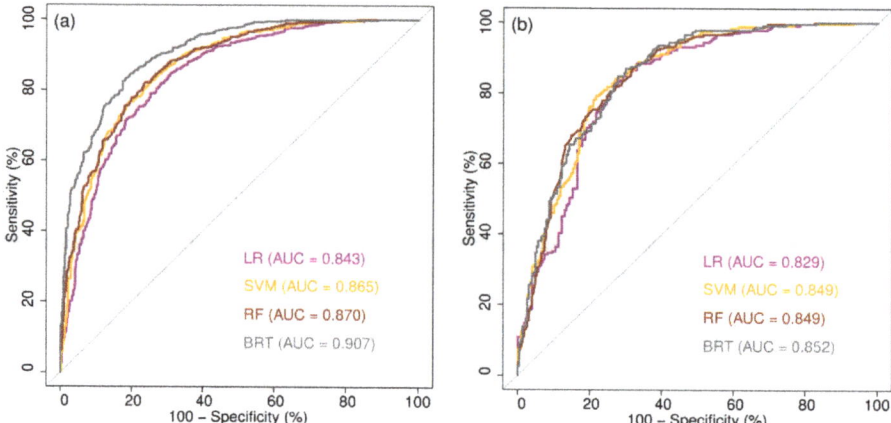

**Figure 7.** The ROC curves of four debris flow susceptibility models using (**a**) training set and (**b**) validation set.

**Table 3.** The ACC and AUC values of the four models on the training set.

| Evaluation Criteria | Models | | | |
|---|---|---|---|---|
| | LR | RF | SVM | BRT |
| ACC | 0.762 | 0.791 | 0.785 | 0.823 |
| AUC | 0.843 | 0.870 | 0.865 | 0.907 |

**Table 4.** The ACC and AUC values of the four models on the validation set.

| Evaluation Criteria | Models | | | |
|---|---|---|---|---|
| | LR | RF | SVM | BRT |
| ACC | 0.762 | 0.779 | 0.781 | 0.781 |
| AUC | 0.829 | 0.849 | 0.849 | 0.852 |

### 4.3. Assessment of Factor Importance

To evaluate the effect of factor selection, BRT was utilized. The AUC value of BRT experienced improvement after removal of unimportant factors. Therefore, a careful analysis of causal factors before modeling is indispensable. Considering the relevance and their corresponding weights, and discarding unimportant factors, result in better forecasting performance.

A variety of factors can trigger occurrences of debris flows. Under the premise that the main controlling factors of debris flow are still controversial, the assessment of factor importance is valuable for interpreting and diagnosing the contribution of different predictor variables. The relative importance of the fifteen factors used to build the models and produce the debris flow susceptibility maps are presented in Figure 8. The results are shown based on the mean decrease of the Gini index in the RF model and are expressed relative to the maximum value. The Gini index is regarded as a commonly-used measurement of total variance across all classifications, and is suitable for assessing the factor importance [48]. A large mean decrease value of the Gini index by splits over a given factor shows a significant predictor. The classification tree models (RF and BRT) have the same mechanism for assessing the relative importance of factors. While LR and SVM rank the factor importance by relying on the regression coefficients and weight vectors, respectively. One of the advantages of applying the Gini index in the RF or BRT model is that it is easier to interpret these results than the SVM or LR.

As this figure shows, we can deem that all fifteen factors have positive contributions to debris flow susceptibility modeling. The mean altitude, altitude difference, aridity index, and groove gradient have the largest mean decrease in the Gini index, followed by others. There were four topographical factors, three meteorological factors, three edaphic factors, two geological factors, two sociometric factors, and one land-cover factors. That is, the topography, meteorology, and edaphology were the most important factor clusters. Previous studies also illustrated that by explaining the general cause of debris flows—surface rock and soil gradually lose their strength because of earthquakes or weather conditions, which are potentially unstable in the steep slopes, and finally, seepage forces formed by rainfall cause them to slide, the slide distance depends on the topography and strength loss amount [52–54].

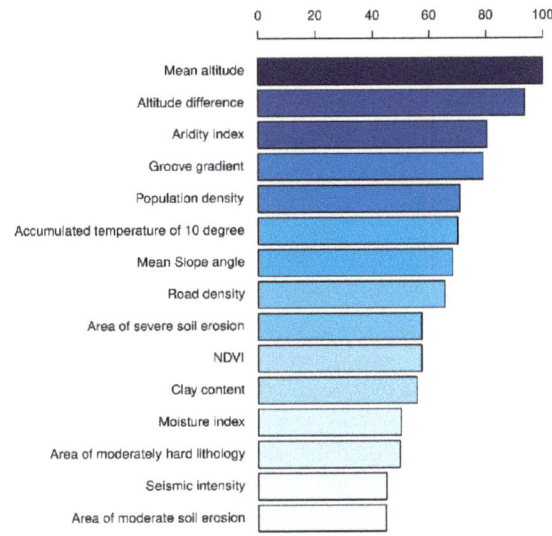

**Figure 8.** Importance of causal factors in the RF model. Values are arranged based on the mean decrease in Gini index and the expressed relative to the maximum value.

The susceptibility map derived from the BRT model has been combined with the factor maps to analyze the relationship between the causal factors and debris flow occurrences. The distribution of watersheds with high and very high susceptibility classes (737 watersheds) on four most important factor maps (mean altitude, altitude difference, aridity index, and groove gradient) is shown in Figure 9. It can be seen that there are obvious regularities. Most watersheds with high and very high susceptibility classes are highly associated with the following conditions—mean altitude varying from 2000 to 3000 m, altitude difference varying from 2000 to 3000 m, aridity index varying from 0.85 to 1.35, and groove gradient varying from 100% to 200%. According to this case, prevention and mitigation of debris flow risk should be paid more attention to in these types of highly coupled areas.

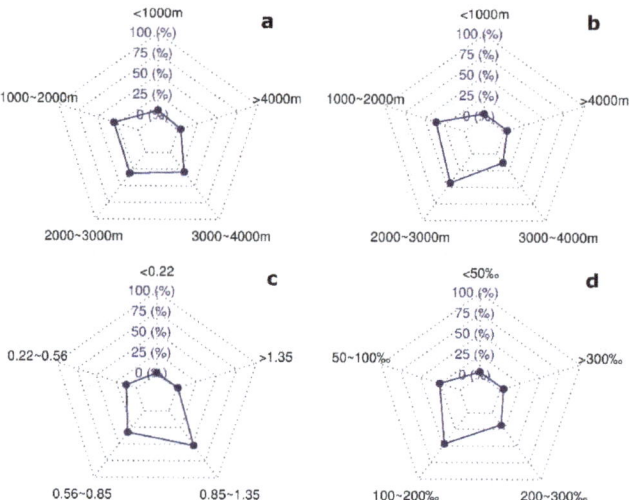

**Figure 9.** Distribution of watersheds with high and very high debris flow susceptibility derived from the BRT model on (**a**) mean altitude map, (**b**) altitude difference map, (**c**) aridity index map, and (**d**) groove gradient map.

## 5. Discussion

On the basis of the remote sensing data, GIS tools, and machine learning algorithms, debris flow susceptibility assessment of the Sichuan Province was implemented. The final four susceptibility maps did not vary considerably between the models and had a consistent spatial distribution pattern. There were obvious regional characteristics exhibited in the susceptibility maps. The transition belt of the Qinghai–Tibet Plateau to the Sichuan Basin concentrates most watersheds of high and very high debris flow susceptibility, where the topography varies enormously. Additionally, this region is coupled with dry valleys and fault zones. Severe soil erosion and frequent earthquakes provide abundant loose materials for debris flows. Similarly, we should also blame the hazard prone on the engineering dregs generated by high-intensity road and hydropower development. Through a combined analysis of factor importance, we identified the high-risk areas and major causal factors that were conducive to preferable hazard prevention. Some factors, such as NDVI and seismic intensity, were always regarded as necessary factors. However, the analyses of factor importance revealed that they were not highly important in this particular application. Hence, we inferred that some factors were site-specific. This inference was in agreement with the investigation conducted by Chen et al. [55].

In this study, all models exhibited good performance and was suitable for constructing debris flow susceptibility maps. Among them, the BRT model was the most reliable and accurate in the study area. As shown in Figure 4, proportions of the different debris flow susceptibility classes from the four models were not exactly the same. From this empirical observation, we concluded that the predictions of the BRT model tended to be optimistic, even though there was no structural evidence. There was no universal agreement on which algorithm performed best on various environments. Each machine learning method has its pros and cons, and the performance of one method is not always better than the other. Rahmati et al. [30] applied seven machine learning methods to analyze the susceptibility of gully erosion and found that the BRT model exhibited a better performance than SVM. However, Garosi et al. [19] illustrated that the BRT and ANN models obtained similar outstanding performance in their research. This might result from the lack of uniform criteria for the selection of factors. Although advanced machine learning methods have slightly different performance in various studies, they always have good predictive abilities and are suitable for the study of susceptibility. Based on the above analyses, we recommend that governors and investigators obtain an optimal susceptibility map by comparing and combining multiple models in practical applications. Therefore, further comparison and ensemble studies are necessary to guide the method selection for predicting debris flows.

## 6. Conclusions

Comparative studies of multiple machine learning models for debris flow susceptibility mapping are very useful to predict future events. The important contributions of our comparative research are summarized below. In this study, all four machine-learning models showed great performance. The BRT model obtained the optimal goodness-of-fit and predictive capability as compared to the other models, in terms of both AUC and ACC, while it was also stable and did not show any overfitting. Therefore, these models, especially BRT, show promising techniques for producing debris flow susceptibility map. This map of the study area shows that the distribution of the watersheds with high and very high susceptibility is coupled with an extreme topography transition zone. Additionally, environmental data based on RS and GIS provide important data sources for regional analyses of debris flow susceptibility. Proper selection of the optimal factors and appropriate mapping units not only improved the prediction performance of the models but also helps avoid the arbitrariness of the factors used. The topographical factors, meteorological factors, and edaphic factors played the most important role in this case. These study results provide a comprehensive perspective on debris flow susceptibility in the Sichuan Province, which are essential for policymakers to implement sustainable disaster mitigation in high debris-flow-prone areas.

**Author Contributions:** Conceptualization, K.X., and B.D.; methodology, K.X., B.R.A., and Y.Z.; software, K.X., B.R.A., Y.Z., and Z.D.; validation, K.X., C.A.S., and B.D.; formal analysis, K.X., B.R.A., C.A.S., S.W., Z.D., and B.D.; investigation, K.X.; resources, C.A.S., Y.Z., and B.D.; data curation, B.D.; writing—original draft preparation, K.X., and Z.D.; writing—review and editing, B.R.A., C.A.S., Y.Z., and B.D.; visualization, K.X., B.R.A., and S.W.; supervision, B.D.; project administration, B.D.; funding acquisition, B.D. All authors have read and agreed to the published version of the manuscript.

**Funding:** This research was funded by the National Natural Science Foundation of China, grant number 41977245, the Strategic Priority Research Program of the Chinese Academy of Sciences, grant number XDA23090502, the National Key R&D Program of China, grant number 2017YFC1502903, and the Fundamental Research Funds for the Central Universities, grant number YJ201765.

**Conflicts of Interest:** The authors declare no conflict of interest.

## References

1. Simoni, S.; Zanotti, F.; Bertoldi, G.; Rigon, R. Modelling the probability of occurrence of shallow landslides and channelized debris flows using GEOtop-FS. *Hydrol. Process.* **2008**, *22*, 532–545. [CrossRef]
2. Kang, S.; Lee, S.-R. Debris flow susceptibility assessment based on an empirical approach in the central region of South Korea. *Geomorphology* **2018**, *308*, 1–12. [CrossRef]
3. Liu, X.; Miao, C.; Guo, L. Acceptability of debris-flow disasters: Comparison of two case studies in China. *Int. J. Disaster Risk Reduct.* **2019**, *34*, 45–54. [CrossRef]
4. Zhong, D.; Xie, H.; Wei, F. *Comprehensive Regionalization of Debris Flow Risk Degree in the Upper Yangtze River*, 1st ed.; Scientific and Technical Publishers: Shanghai, China, 2010; ISBN 978-7-5478-0103-1.
5. Di, B.; Chen, N.; Cui, P.; Li, Z.L.; He, Y.P.; Gao, Y.C. GIS-based risk analysis of debris flow: An application in Sichuan, southwest China. *Int. J. Sediment Res.* **2008**, *23*, 138–148. [CrossRef]
6. Tang, C.; Van Asch, T.; Chang, M.; Chen, G.; Zhao, X.; Huang, X. Catastrophic debris flows on 13 August 2010 in the Qingping area, southwestern China: The combined effects of a strong earthquake and subsequent rainstorms. *Geomorphology* **2012**, *139*, 559–576. [CrossRef]
7. Brabb, E.E. Innovative Approaches to Landslide Hazard Mapping. In Proceedings of the Fourth International Symposium on Landslides, Canadian Geotechnical Society, Toronto, ON, Canada, 16–21 September 1984; pp. 307–324.
8. Reichenbach, P.; Rossi, M.; Malamud, B.D.; Mihir, M.; Guzzetti, F. A review of statistically-based landslide susceptibility models. *Earth Sci. Rev.* **2018**, *180*, 60–91. [CrossRef]

9. Blais-Stevens, A.; Behnia, P. Debris flow susceptibility mapping using a qualitative heuristic method and Flow-R along the Yukon Alaska Highway Corridor, Canada. *Nat. Hazards Earth Syst. Sci. Discuss.* **2015**, *3*, 3509–3541. [CrossRef]
10. Hutter, K.; Svendsen, B.; Rickenmann, D. Debris flow modeling: A review. *Contin. Mech. Thermodyn.* **1994**, *8*, 1–35. [CrossRef]
11. Aronica, G.T.; Cascone, E.; Randazzo, G.; Biondi, G.; Lanza, S.; Fraccarollo, L.; Brigandi, G. Assessment and mapping of debris flow hazard through integrated physically based models and GIS assisted methods. In Proceedings of the EGU General Assembly Conference Abstracts, Vienna, Austria, 2–7 May 2010; p. 13107.
12. Schippa, L.; Pavan, S. Numerical modelling of catastrophic events produced by mud or debris flows. *Int. J. Saf. Secur. Eng.* **2011**, *1*, 403–422. [CrossRef]
13. Liu, Y.; Di, B.; Zhan, Y.; Stamatopoulos, C.A. Debris Flows Susceptibility Assessment in Wenchuan Earthquake Areas Based on Random Forest Algorithm Model. *Mt. Res.* **2018**, *36*, 765–773. (In Chinese)
14. Zhang, Y.; Ge, T.; Tian, W.; Liou, Y.-A. Debris Flow Susceptibility Mapping Using Machine-Learning Techniques in Shigatse Area, China. *Remote Sens.* **2019**, *11*, 2801. [CrossRef]
15. Chang, T.-C.; Chao, R.-J. Application of back-propagation networks in debris flow prediction. *Eng. Geol.* **2006**, *85*, 270–280. [CrossRef]
16. Pham, B.T.; Pradhan, B.; Bui, D.T.; Prakash, I.; Dholakia, M.B. A comparative study of different machine learning methods for landslide susceptibility assessment: A case study of Uttarakhand area (India). *Environ. Model. Softw.* **2016**, *84*, 240–250. [CrossRef]
17. Roy, J.; Saha, S.; Arabameri, A.; Blaschke, T.; Bui, T.D. A Novel Ensemble Approach for Landslide Susceptibility Mapping (LSM) in Darjeeling and Kalimpong Districts, West Bengal, India. *Remote Sens.* **2019**, *11*, 2866. [CrossRef]
18. Conoscenti, C.; Angileri, S.; Cappadonia, C.; Rotigliano, E.; Agnesi, V.; Märker, M. Gully erosion susceptibility assessment by means of GIS-based logistic regression: A case of Sicily (Italy). *Geomorphology* **2014**, *204*, 399–411. [CrossRef]
19. Garosi, Y.; Sheklabadi, M.; Pourghasemi, H.R.; Besalatpour, A.A.; Conoscenti, C.; Oost, K.V. Comparison of differences in resolution and sources of controlling factors for gully erosion susceptibility mapping. *Geoderma* **2018**, *330*, 65–78. [CrossRef]
20. Carrara, A.; Crosta, G.; Frattini, P. Comparing models of debris-flow susceptibility in the alpine environment. *Geomorphology* **2008**, *94*, 353–378. [CrossRef]
21. Lee, S.; Park, I.; Choi, J.-K. Spatial Prediction of Ground Subsidence Susceptibility Using an Artificial Neural Network. *Environ. Manag.* **2012**, *49*, 347–358. [CrossRef]
22. Cao, J.; Zhang, Z.; Wang, C.; Liu, J.; Zhang, L. Susceptibility assessment of landslides triggered by earthquakes in the Western Sichuan Plateau. *CATENA* **2019**, *175*, 63–76. [CrossRef]
23. Xu, X.; Zhang, Y. Meteorological Datasets in China. Data Registration and Publishing System of Resource and Environment Science Data Center of Chinese Academy of Sciences. Available online: http://www.resdc.cn/DOI (accessed on 13 December 2017). (In Chinese)
24. Xu, X. Spatial Distribution Datasets of Population in China. Data Registration and Publishing System of Resource and Environment Science Data Center of Chinese Academy of Sciences. Available online: http://www.resdc.cn/DOI (accessed on 11 December 2017). (In Chinese).
25. Xu, X.; Liu, J.; Zhang, S.; Li, R.; Yan, C.; Wu, S. The Land Use and Land Cover Change Database in China. Data Registration and Publishing System of Resource and Environment Science Data Center of Chinese Academy of Sciences. Available online: http://www.resdc.cn/DOI (accessed on 2 July 2018). (In Chinese).
26. Guzzetti, F.; Carrara, A.; Cardinali, M.; Reichenbach, P. Landslide hazard evaluation: A review of current techniques and their application in a multi-scale study, Central Italy. *Geomorphology* **1999**, *31*, 181–216. [CrossRef]
27. Zhan, Y.; Luo, Y.; Deng, X.; Grieneisen, M.L.; Zhang, M.; Di, B. Spatiotemporal prediction of daily ambient ozone levels across China using random forest for human exposure assessment. *Environ. Pollut.* **2018**, *233*, 464–473. [CrossRef] [PubMed]
28. Nefeslioglu, H.A.; Duman, T.Y.; Durmaz, S. Landslide susceptibility mapping for a part of tectonic Kelkit Valley (Eastern Black Sea region of Turkey). *Geomorphology* **2008**, *94*, 401–418. [CrossRef]
29. Schicker, R.; Moon, V. Comparison of bivariate and multivariate statistical approaches in landslide susceptibility mapping at a regional scale. *Geomorphology* **2012**, *161–162*, 40–57. [CrossRef]

30. Rahmati, O.; Tahmasebipour, N.; Haghizadeh, A.; Pourghasemi, H.R.; Feizizadeh, B. Evaluation of different machine learning models for predicting and mapping the susceptibility of gully erosion. *Geomorphology* **2017**, *298*, 118–137. [CrossRef]
31. Arabameri, A.; Pradhan, B.; Rezaei, K.; Lee, C.-W. Assessment of Landslide Susceptibility Using Statistical- and Artificial Intelligence-Based FR–RF Integrated Model and Multiresolution DEMs. *Remote Sens.* **2019**, *11*, 999. [CrossRef]
32. Team, R.C. *R: A Language and Environment for Statistical Computing*; R Foundation for Statistical Computing: Vienna, Austria, 2018; ISBN 3-900051-07-0.
33. Greco, R.; Sorriso-Valvo, M.; Catalano, E. Logistic Regression analysis in the evaluation of mass movements susceptibility: The Aspromonte case study, Calabria, Italy. *Eng. Geol.* **2007**, *89*, 47–66. [CrossRef]
34. Bai, S.-B.; Wang, J.; Lü, G.-N.; Zhou, P.-G.; Hou, S.-S.; Xu, S.-N. GIS-based logistic regression for landslide susceptibility mapping of the Zhongxian segment in the Three Gorges area, China. *Geomorphology* **2010**, *115*, 23–31. [CrossRef]
35. Atkinson, P.; Massari, R. Generalised Linear Modelling of Susceptibility to Landsliding in the Central Apennines, Italy. *Comput. Geosci. Comput. Geosci.* **1998**, *24*, 373–385. [CrossRef]
36. Breiman, L. Random Forests. *Mach. Learn.* **2001**, *45*, 5–32. [CrossRef]
37. Micheletti, N.; Foresti, L.; Robert, S.; Leuenberger, M.; Pedrazzini, A.; Jaboyedoff, M.; Kanevski, M. Machine Learning Feature Selection Methods for Landslide Susceptibility Mapping. *Math. Geosci.* **2014**, *46*, 33–57. [CrossRef]
38. Svetnik, V.; Liaw, A.; Tong, C.; Culberson, J.C.; Sheridan, R.P.; Feuston, B.P. Random Forest: A Classification and Regression Tool for Compound Classification and QSAR Modeling. *J. Chem. Inf. Comput. Sci.* **2003**, *43*, 1947–1958. [CrossRef] [PubMed]
39. Archer, K.J.; Kimes, R.V. Empirical characterization of random forest variable importance measures. *Comput. Stat. Data Anal.* **2008**, *52*, 2249–2260. [CrossRef]
40. Pourghasemi, H.R.; Kerle, N. Random forests and evidential belief function-based landslide susceptibility assessment in Western Mazandaran Province, Iran. *Environ. Earth Sci.* **2016**, *75*, 185. [CrossRef]
41. Vapnik, V.N. *The Nature of Statistical Learning Theory*; Springer: Berlin/Heidelberg, Germnay, 1995; ISBN 0-387-94559-8.
42. Xing, Z.; Xu, Q.; Tang, M.; Nie, W.; Ma, S.; Xu, Z. Comparison of two optimized machine learning models for predicting displacement of rainfall-induced landslide: A case study in Sichuan Province, China. *Eng. Geol.* **2017**, *218*, 213–222.
43. Zhou, C.; Yin, K.; Cao, Y.; Ahmed, B.; Li, Y.; Catani, F.; Pourghasemi, H.R. Landslide susceptibility modeling applying machine learning methods: A case study from Longju in the Three Gorges Reservoir area, China. *Comput. Geosci.* **2018**, *112*, 23–37. [CrossRef]
44. Kavzoglu, T.; Sahin, E.K.; Colkesen, I. Landslide susceptibility mapping using GIS-based multi-criteria decision analysis, support vector machines, and logistic regression. *Landslides* **2014**, *11*, 425–439. [CrossRef]
45. Kavzoglu, T.; Colkesen, I. A kernel functions analysis for support vector machines for land cover classification. *Int. J. Appl. Earth Obs. Geoinf.* **2009**, *11*, 352–359. [CrossRef]
46. Schapire, R.E. The Boosting Approach to Machine Learning: An Overview. In *Nonlinear Estimation and Classification*; Denison, D.D., Hansen, M.H., Holmes, C.C., Mallick, B., Yu, B., Eds.; Springer: New York, NY, USA, 2003; pp. 149–171. ISBN 978-0-387-21579-2.
47. Mouton, A.; De Baets, B.; Goethals, P. Ecological relevance of performance criteria for species distribution models. *Ecol. Model.* **2010**, *221*, 1995–2002. [CrossRef]
48. James, G.; Witten, D.; Hastie, T.; Tibshirani, R. *An Introduction to Statistical Learning: With Applications in R*; Springer: Berlin/Heidelberg, Germany, 2014; ISBN 1-4614-7137-0.
49. Di, B.; Zhang, H.; Liu, Y.; Li, J.; Chen, N.; Stamatopoulos, C.A.; Luo, Y.; Zhan, Y. Assessing Susceptibility of Debris Flow in Southwest China Using Gradient Boosting Machine. *Sci. Rep.* **2019**, *9*, 12532. [CrossRef]
50. Chiu, D.; Wei, Y. *R for Data Science Cookbook*, 1st ed.; Packt Publishing: Birmingham, UK, 2016; ISBN B01ET5I38M.
51. Lee, S.; Ryu, J.-H.; Kim, I.-S. Landslide susceptibility analysis and its verification using likelihood ratio, logistic regression, and artificial neural network models: Case study of Youngin, Korea. *Landslides* **2007**, *4*, 327–338. [CrossRef]

52. Kang, Z.; Lee, C.-F.; Ma, A.; Luo, J. *Debris Flow Research in China*, 1st ed.; Science Press: Beijing, China, 2004; ISBN 978-7-03-013800-2.
53. Liu, C. Analysis on Genetic Model of Wenjiagou Debris Flows in Wenchuan Earthquake Area, Sichuan. *Geol. Rev.* **2012**, *58*, 709–716.
54. Stamatopoulos, C.A.; Di, B. Analytical and approximate expressions predicting post-failure landslide displacement using the multi-block model and energy methods. *Landslides* **2015**, *12*, 1207–1213. [CrossRef]
55. Chen, W.; Xie, X.; Wang, J.; Pradhan, B.; Hong, H.; Bui, D.T.; Duan, Z.; Ma, J. A comparative study of logistic model tree, random forest, and classification and regression tree models for spatial prediction of landslide susceptibility. *CATENA* **2017**, *151*, 147–160. [CrossRef]

© 2020 by the authors. Licensee MDPI, Basel, Switzerland. This article is an open access article distributed under the terms and conditions of the Creative Commons Attribution (CC BY) license (http://creativecommons.org/licenses/by/4.0/).

*Article*

# Evaluation of Satellite Precipitation Estimates over Australia

**Zhi-Weng Chua [1], Yuriy Kuleshov [1,2,*] and Andrew Watkins [1]**

1 Long-Range Forecasts, Bureau of Meteorology, Docklands 3008, Australia; zhi-weng.chua@bom.gov.au (Z.-W.C.); andrew.watkins@bom.gov.au (A.W.)
2 SPACE Research Centre, School of Science, Royal Melbourne Institute of Technology (RMIT) University, Melbourne 3000, Australia
\* Correspondence: yuriy.kuleshov@bom.gov.au or yuriy.kuleshov@rmit.edu.au; Tel.: +61-3-9669-4896

Received: 16 January 2020; Accepted: 14 February 2020; Published: 19 February 2020

**Abstract:** This study evaluates the U.S. National Oceanographic and Atmospheric Administration's (NOAA) Climate Prediction Center morphing technique (CMORPH) and the Japan Aerospace Exploration Agency's (JAXA) Global Satellite Mapping of Precipitation (GSMaP) satellite precipitation estimates over Australia across an 18 year period from 2001 to 2018. The evaluation was performed on a monthly time scale and used both point and gridded rain gauge data as the reference dataset. Overall statistics demonstrated that satellite precipitation estimates did exhibit skill over Australia and that gauge-blending yielded a notable increase in performance. Dependencies of performance on geography, season, and rainfall intensity were also investigated. The skill of satellite precipitation detection was reduced in areas of elevated topography and where cold frontal rainfall was the main precipitation source. Areas where rain gauge coverage was sparse also exhibited reduced skill. In terms of seasons, the performance was relatively similar across the year, with austral summer (DJF) exhibiting slightly better performance. The skill of the satellite precipitation estimates was highly dependent on rainfall intensity. The highest skill was obtained for moderate rainfall amounts (2–4 mm/day). There was an overestimation of low-end rainfall amounts and an underestimation in both the frequency and amount for high-end rainfall. Overall, CMORPH and GSMaP datasets were evaluated as useful sources of satellite precipitation estimates over Australia.

**Keywords:** satellite precipitation estimates; Australia; rain gauge precipitation measurements; satellite precipitation validation

## 1. Introduction

Precipitation is an essential climate variable and is one of the most important climate variables affecting human activities [1]. Variations in the intensity, duration, and frequency of precipitation directly impact water availability for many millions of people and industries. Measuring rainfall over broad areas enables efficient water management and disaster response and recovery.

The conventional method of using rain gauges to estimate spatial patterns of rainfall provides a direct measurement of surface rainfall but spatial density can be an issue across many parts of the world, including over the oceans, where the installation of an adequate rain gauge network is economically or physically unfeasible [2]. This greatly affects the ability to accurately assess rainfall across a region as it is a variable that exhibits a high degree of spatial variation and a point-based measurement may not provide an ideal representation of an area. Rain gauge estimates are subject to instrumental errors with many relying on manual sampling methods. Clock synchronization and mechanical faults are examples of potential issues [3]. Furthermore, they are also affected by localised effects including wind (precipitation can be prevented from entering the gauge), evaporation (some of the precipitation is evaporated before it can be recorded), wetting (some of the precipitation can be left behind in the gauge) and splashing effects (precipitation can incorrectly splash in and out of the gauge) [4,5]. For example, Groisman and Legates (1994) found biases due to wind-induced effects could be quite significant, especially around mountainous areas where the bias was as large as 40% [6].

Alternative physical-based precipitation datasets include those derived from ground-based radars and satellites. Radar estimates are derived from the detected reflectivity of hydrometeors but suffer from problems such as topography blockage, beam ducting, range-related and bright band effects [7,8]. Van De Beek et al. (2016) found that for a 3-day rainfall event in the Netherlands, the radar underestimated rainfall amount by over 50%, though after correction the difference was only 5-8% [9]. Correction to rain gauges is critical but this means radars are likely to perform poorly in areas that lack gauge coverage, hindering their ability to replace gauges. As a ground-based source, they are also affected by the same physical and economical limitations that are applicable to installation of rain gauge networks.

The use of meteorological satellites to monitor rainfall was introduced in the 1970s, providing a means to estimate rainfall across most of the globe. The first methods inferred precipitation intensity based on visible or infrared (IR) data by linking cloud-top temperature or reflectivity to rain rates through empirical relationships. Later methods used passive microwave (PMW) sensors that detect the radiation from hydrometeors and link this to rainfall rates, thereby providing a more direct interpretation of precipitation. However, the coverage of PMW satellites is much less than that of their IR counterparts.

Consequently, techniques have been developed to combine the increased accuracy of PMW estimates with the coverage provided by IR satellites. One method which can be referred to as the cloud-motion advection method involves using IR images to derive cloud-motion vectors and then using these vectors to advect PMW-based precipitation estimates to cover areas lacking in PMW coverage. The Climate Prediction Center morphing technique (CMORPH) developed by the National Oceanic and Atmospheric Administration (NOAA) was the first product of this kind and was followed by the Japan Aerospace Exploration Agency's (JAXA) Global Satellite Mapping of Precipitation (GSMaP) dataset [10,11]. CMORPH and GSMaP have undergone multiple advancements since their inception. The key improvements have been the introduction of the Kalman filter to modify the shape and intensity of the advected rainfall and the implementation of bias-correction using gauge data [10,12,13].

Many past verification studies have been performed with modern-day satellite technology, showing that, at least on a monthly basis, the technology can possess good potential [14]. However, few studies have been performed over Australia, with even fewer, if any, using the latest CMORPH and GSMaP datasets.

Continental studies in the past have indicated that performance varies greatly with rainfall type, amount, and season, with performance tending to be better for heavier rainfall regimes including those during summer and in the tropics [15,16]. Ebert et al. (2007) evaluated the performance of multiple satellite datasets, including CMORPH, over Australia for a two-year period and found the estimates were relatively unbiased over summer but the accuracy greatly deteriorated in winter [15]. Pipunic et al. (2015) examined Tropical Rainfall Measuring Mission (TRMM) 3B42RT satellite data over mainland Australia across a nine-year period and found a similar conclusion with detection of light rainfall (<3 mm/day) being unreliable while the most reliably detected regime was heavier rainfall associated with warm season convective systems, especially in the tropics [17]. Factors that make light rain detection difficult include subcloud evaporation and the poor recognition of clouds with warm cloud-top temperatures [18].

In general, CMORPH and GSMaP display an overestimation (underestimation) for low (high) rainfall rates. For example, Habib (2012) compared CMORPH data to seven rain gauges in southern Louisiana, USA from August 2004 to December 2006 and found a consistent positive (negative) bias for rainfall rates less (more) than 3 mm/h [19]. Ning et al. (2017) evaluated gauge-corrected GSMaP data against a gauge-based analysis (China daily Precipitation Analysis Product) over eastern China from

April 2014 to March 2016 and showed that GSMaP overestimated light precipitation (<16 mm/day) while underestimating heavier precipitation (>32 mm/day) [20]. Hit bias rather than false or missed event bias was noted as the major error with false event bias also being more significant than missed event bias. The introduction of a bias-correction scheme is largely able to correct a positive bias by scaling down the magnitudes, but the inability to correct missed events means there has been much less success in correcting the negative bias [13].

Previous studies have also indicated that a significant degradation of performance occurred over orography, with satellites underestimating rainfall over higher elevations [18,21]. The bias can be worse during winter where the poor detection of snowfall, as well as rainfall, over cold surfaces leads to both missed events and an underestimation of intensity [22]. Derin et al. (2016) performed an evaluation over the western Black Sea region of Turkey, an area featuring complex topography in the form of a mountain range, from 2007 to 2011, and found that CMORPH exhibited a bias of −54% for the windward side of the region during the warm season, increasing to −82% during the cold season [21].

Kubota et al. (2009) found that the greatest biases in GSMaP were over coastal areas with frequent orographic rainfall and that estimates were generally better over the ocean than over the land [23]. Coastal regions are likely to present difficulties as the retrieval algorithm struggles to account for both ocean and land surfaces in a single grid point.

This study aims to contribute to the validation of satellite rainfall data. It differs from earlier studies by evaluating satellite precipitation estimates over a relatively long period of record (18 years) with a focus on Australia, which has a relatively dense rain gauge network over a large area when compared to other world regions [15]. The use of a percentile-based verification statistic is an innovative feature of this study, while the use of both gridded and point gauge data as a reference adds additional insight compared to using just one. The CMORPH and GSMaP datasets were chosen due to their provision as part of the World Meteorological Organization (WMO) Space-based Weather and Climate Extremes Monitoring Demonstration Project (SEMDP) [24]. This project aims to introduce operational satellite rainfall monitoring products based on these two datasets, to East Asia and Western Pacific countries, of which many lack adequate rainfall monitoring capabilities due to the absence of an extensive and accurate rain gauge network. The verification of these datasets is thus an important step for the creation of these products. Moreover, the cloud-motion advection method used to blend PMW and IR data ranks amongst the best in terms of performance across various satellite methods used to estimate precipitation [25]. The variance of the errors in the satellite precipitation estimates with location, season, and rainfall intensity was investigated.

The paper is organised as follows. Section 2 describes the study area, datasets, and methods used in the study. Section 3 presents the results while Section 4 discuss the findings. Section 5 summarises the major findings and provides directions for future work.

## 2. Materials and Methods

### 2.1. Study Area

Australia has a land area of around 8.6 million km$^2$, making it the sixth-largest country in the world by land size. Its large geographical size means that it experiences a variety of climates, including temperate zones to the south east and south west, tropical zones to the north, and deserts or semi-arid areas across much of the interior [26]. The main orographic feature occurs in the form of the Great Dividing Range (GDR), a mountain range along the eastern side of the country that extends more than 3500 km from the north-eastern tip of Queensland, towards and along the coast of New South Wales, and into the eastern and central parts of Victoria. The width of the GDR ranges from about 160 to 300 km with a maximum elevation of 2228 m, though the typical elevation range for the highlands is from 300 to 1600 m [27]. In Figure 1, the domain of analysis is shown, with the stations used in the study also marked.

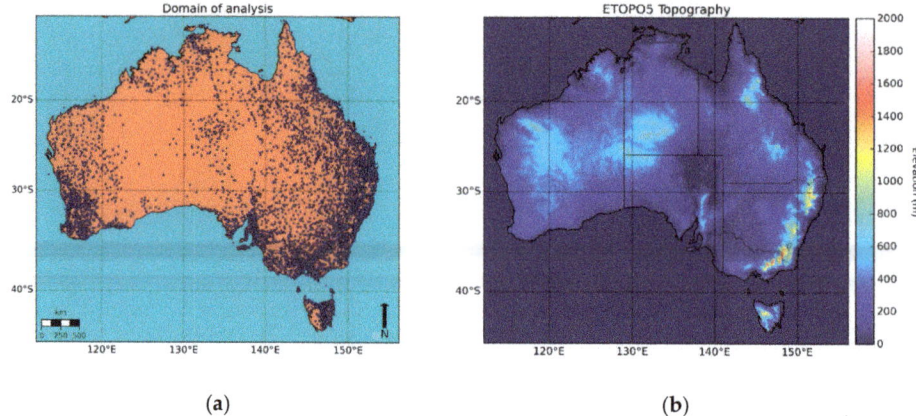

**Figure 1.** Domain of analysis. (**a**) Stations are marked as blue dots; (**b**) scale of topography.

*2.2. Datasets*

As part of the WMO SEMDP, access to GSMaP and CMORPH data were provided by JAXA and NOAA, respectively. Both datasets of satellite precipitation estimates employ the cloud-advection technique introduced in Section 1. The GSMaP version used was GSMaP Gauge-adjusted Near-Real-Time (GNRT) Version 6. To allow for a faster data latency, gauge adjustment over land was performed against the gauge-calibrated version (GSMaP gauge) from the past period, which, itself, is calibrated by matching daily satellite rainfall estimates to a global gauge analysis, CPC Unified Gauge-Based Analysis of Global Daily Precipitation (CPC Unified) [28]. Further details can be found in the GSMaP technical documentation [28].

Two versions of CMORPH were used. These were the bias-corrected CMORPH (CMORPH CRT) and the gauge-blended CMORPH (CMORPH BLD) datasets. Bias correction over land was also performed using the CPC Unified analysis but using a different algorithm that involves matching to probability distribution function (PDF) tables from the past 30 days. The gauge-blended version uses the bias-corrected version as a first guess and then incorporates the gauge data based on the density of the observations; further details can be found in [10,13].

Consequently, this study used two gauge-corrected sets (GSMaP and CMORPH CRT) and one that had been further processed by combining CMORPH CRT with gauge data (CMORPH BLD).

The reference datasets used were both based on the Bureau of Meteorology (BoM) rain gauges with the Australian Water Availability Project (AWAP) analysis being used as the reference dataset for the gridded comparison and the values from the stations themselves being used for the point comparison. The AWAP rainfall analysis is generated by decomposing the field into a climatology component and an anomaly component based on the ratio of the observed rainfall value to the climatology [29]. The Barnes successive-correction technique is applied to the anomaly component and added to the monthly climatological averages, which were derived using a three-dimensional smooth splice approach [29]. The climatological averages were generated from 30 years of monthly totals [29]. For the point comparison, only 'Series 0' stations were chosen as these stations are Bureau-maintained and conform to International Civil Aviation Organization (ICAO) standards. The minimum number of stations used across the period was 4764. As discussed earlier, even though rain gauge network measurements can be taken as 'truth', they still contain errors, which will artificially inflate the errors attributed to satellite measurements.

Details on the spatial and temporal resolutions of the gridded datasets along with their domains are shown in Table 1.

Table 1. Details about dataset used.

| Dataset | Resolution (°) | Start of Temporal Domain | Latitude Range (°) | Longitude Range (°) |
|---|---|---|---|---|
| CMORPH BLD | 0.25 | Jan 1998 | (−45, 40) | (50, 200) |
| CMORPH CRT | 0.25 | Jan 1998 | (−45, 40) | (50, 200) |
| GSMaP | 0.25 | Apr 2000 | (−45, 40) | (50, 200) |
| AWAP | 0.05 | Jan 1900 | (−44.525, −9.975) | (111.975, 156.275) |

The longest common period across the datasets using full years was chosen for the analysis (i.e., January 2001 to December 2018). A spatial domain of latitude from −44.625°N to −10.125°N and longitude from 112.125°E to 156.125°E was chosen as this domain centers on Australia. Ocean data were masked.

## 2.3. Method

The satellite datasets were compared against the gauge-based datasets. Both a gridded comparison and point comparison were performed. When performing the comparisons, all the datasets were linearly interpolated to the same spatial resolution. An interpolation to the coarsest resolution was chosen (i.e., 0.25°). Values at each grid box from these interpolated grids could then be compared against each other for the gridded comparison.

For the point comparison, values corresponding to the location of a station were linearly interpolated from each grid. These values could then be compared to the actual station value. Inclusion of the AWAP dataset was done to provide an additional reference. A complication arose from the fact that the gauge-based data values were 24 h accumulated values to 0900 local standard time (LST), while the satellite data values were values to 00 UTC. As this study is focused on monthly comparisons, the longer period greatly reduces the impact of this timing inconsistency. An elementary remedy would be to have shifted the gauge and AWAP values one day ahead of their satellite counterparts, reducing the inconsistency to two hours or less. Doing this adjustment resulted in improvements of less than 2% and so the unadjusted datasets were used for simplicity.

Both continuous and percentile-based statistics were calculated. The continuous statistics calculated were the mean bias (MB), root-mean-square error (RMSE), mean average error (MAE), and the Pearson correlation coefficient (R). The MB is the average difference between the estimated and observed values, which gives an indicator of the overall bias. The MAE measures the average magnitude of the error. To remove the effect of higher rainfall averages leading to larger errors, the MAE was also normalised through division by the average rainfall producing the normalised mean average error. The RMSE also measures the average error magnitude but is weighted towards larger errors. R is commonly known as the linear correlation coefficient as it measures the linear association between the estimated and observed datasets.

In addition to continuous verification statistics, a percentile-based verification can also be performed to measure how well the datasets reproduce the occurrence of low- and high-end values. This is a novel verification metric that the authors have deemed useful to assess because, even if the satellites performs poorly in terms of absolute values, they may still produce accurate values relative to their own climatology, meaning there is the potential to produce percentile-based products. Such products have already been produced (e.g., both NOAA and JAXA have generated satellite-derived versions of the Standardized Precipitation Index, as well as rainfall values expressed as high-end percentiles). The quintile for an observed month at a location could be derived by ranking that value against the same month but for different years across the verification period. The ranking can then be converted to a percentile through linear interpolation. If a bottom or top quintile was observed, the value from the satellite dataset was then investigated. If it was also registered in the same quintile, this was recorded as a success; otherwise, it was recorded as a failure. The number of successes was then converted to a hit rate. This hit rate was only calculated for the gridded comparison as the varying number of stations across the verification period made a point-based comparison more difficult. The use of quintiles provided greater differentiation of extreme values than terciles or quartiles, while the record length was considered too short for the use of deciles.

The equations for the metrics are summarised in Table 2 with $E_i$ representing the estimated value at a point or grid box $i$, $O_i$ being the observed value, and $N$ being the number of samples (across the whole domain and period) for the continuous metrics.

Table 2. Summary of metrics used.

| Metric | Equation | Range | Perfect Value | Unit |
|---|---|---|---|---|
| Mean bias (MB) | $\frac{1}{N}\sum_{i=1}^{N}(E_i - O_i)$ | $(-\infty, \infty)$ | 0 | mm/day |
| Mean average error (MAE) | $\frac{1}{N}\sum_{i=1}^{N}\|E_i - O_i\|$ | $[0, \infty)$ | 0 | mm/day |
| Normalised mean average error | $\frac{\frac{1}{N}\sum_{i=1}^{N}\|E_i - O_i\|}{\frac{1}{N}\sum_{i=1}^{N}E_i}$ | $[0, \infty)$ | 0 | |
| Root-mean-square error (RMSE) | $\sqrt{\frac{1}{N}\sum_{i=1}^{N}(E_i - O_i)^2}$ | $[0, \infty)$ | 0 | mm/day |
| Pearson correlation coefficient (R) | $\frac{\sum_{i=1}^{N}[(E_i-\overline{E})(O_i-\overline{O})]}{\sqrt{\sum_{i=1}^{N}(E_i-\overline{E})^2}\sqrt{\sum_{i=1}^{N}(O_i-\overline{O})^2}}$ | $[-1, 1]$ | 1 | |

## 3. Results

The results of the gridded continuous comparison against AWAP data are presented in Figure 2. The linear correlation of the satellite rainfall estimates ranges from 0.77 to 0.88, while the MAE ranges from 0.61 to 0.43 mm/day. The trend amongst all the metrics is the same with performance being the best for CMORPH BLD, then CMORPH CRT, and lastly GSMaP. CMORPH CRT and GSMaP display similar performances, while there is a clear increase in performance for CMORPH BLD.

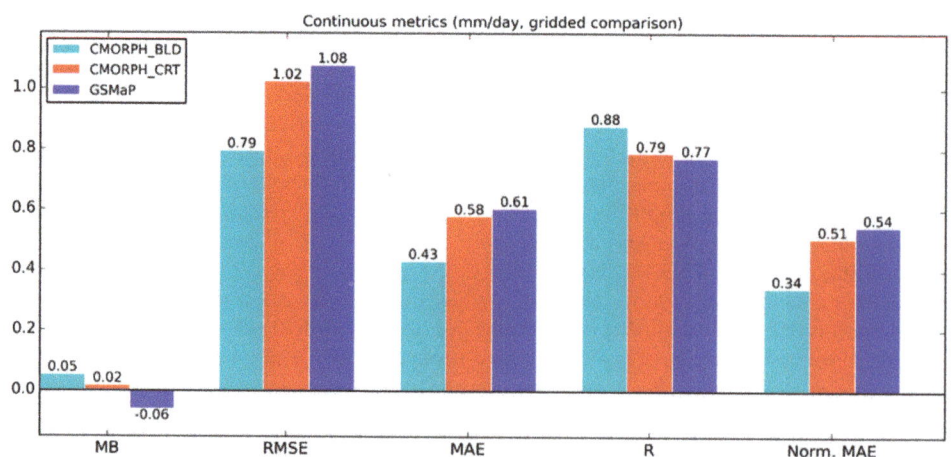

**Figure 2.** Gridded continuous comparison of satellite datasets against Australian Water Availability Project (AWAP) from January 2001 to December 2018. Mean bias (MB), root-mean-square error (RMSE), mean average error (MAE), Pearson correlation coefficient (R) and normalised mean average error (Norm. MAE) are displayed.

The gridded percentile-based comparison against AWAP data is shown in Figure 3. The satellite datasets obtain around a 70%–80% hit rate for the bottom quintile whilst scoring around 10% less for the top quintile. This suggests the rainfall values produced by the satellites are relatively accurate in terms of climatological occurrence, with better performance exhibited for low-end extremes. There appears to be potential in generating percentile-based products from satellite data.

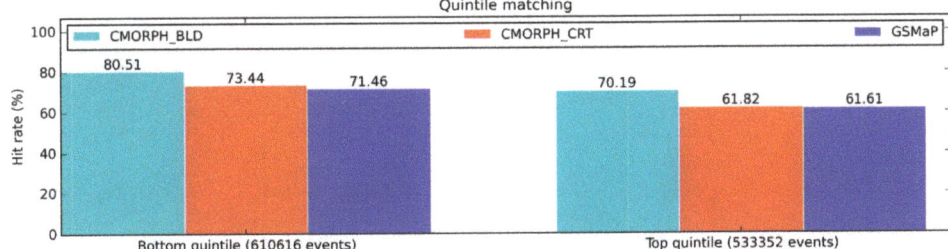

**Figure 3.** Gridded percentile-based comparison of satellite datasets against AWAP from January 2001 to December 2018. Bottom- and top-quintile hit rates are displayed.

Figure 4 displays the continuous statistics using point gauge data as truth. The comparison against point gauge data supports the gridded comparison with the CMORPH BLD error being about 50% larger than the AWAP error and CMORPH CRT and GSMaP being about 150% larger. The MB is negative for all the satellite datasets, indicating a slight tendency for underestimation of overall rainfall.

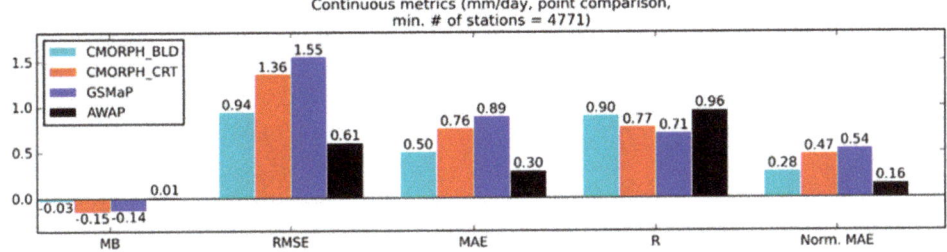

**Figure 4.** Comparison of satellite datasets against point gauge data from January 2001 to December 2018.

The ranking of performance between the satellite datasets remains the same for both the continuous and the percentile-based statistics. The benefit of blending in gauge data is again displayed, with CMORPH BLD showing significant improvement over the unblended datasets and skill comparable to AWAP. As the trend between MB, MAE, RMSE, and R is the same, future references to continuous statistics will refer to just MAE, normalised MAE and R for brevity.

The values and residuals of the datasets against point gauge data are shown in Figure 5. There appears to be a tendency towards an overestimation for low rainfall months and an underestimation for high rainfall months. AWAP and, to a lesser extent, CMORPH BLD were able to capture the high-end rainfall months more accurately, with observation of months where more than 40 mm/day was recorded, being distinctly better. All datasets appear to struggle with very high-end rainfall months (>60 mm/day). These gauge totals sit along the lower boundary, which indicates that the datasets observed little rainfall while the gauges observed a significant amount. The fact that even AWAP does not depict these totals well suggests that gridded datasets systematically struggle with these very high-end values. A likely reason is that the gridded datasets smooth down point values as part of their objective analysis process and so it is expected that high-end totals will be underrepresented by the grids. The impact from this effect would be worse if there were nearby gauges with low totals.

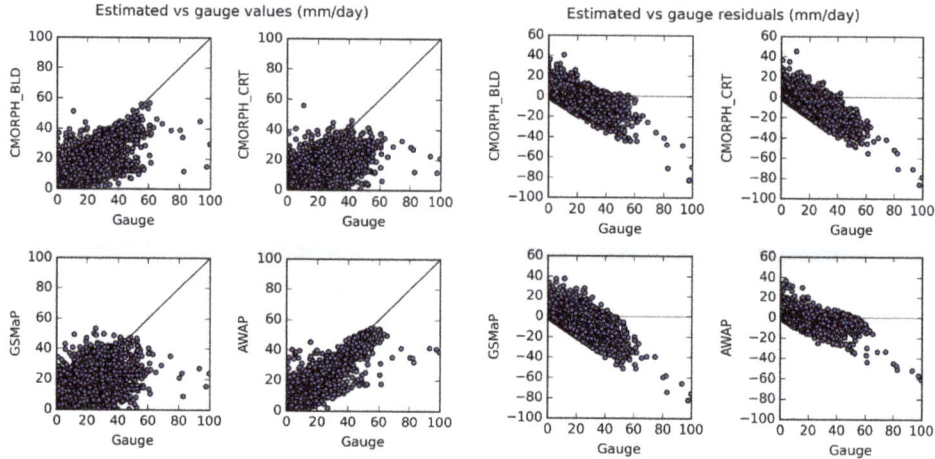

**Figure 5.** Scatterplot comparisons of satellite datasets against point gauge data from January 2001 to December 2018.

## 3.1. Variation with Geography

A gridded comparison was performed over the Australian domain with the geographical representations of the MB and MAE shown in Figure 6. The CMORPH CRT and CMORPH BLD datasets were chosen to allow an investigation into the effects of gauge correction. Generally, the satellite-derived data overestimate rainfall, except over western Tasmania where there is a significant underestimation.

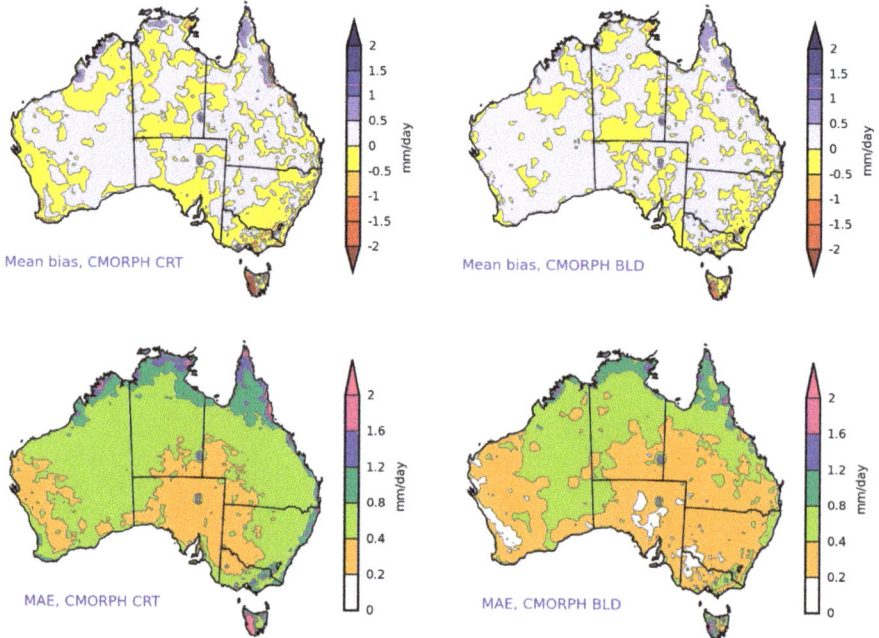

**Figure 6.** *Cont.*

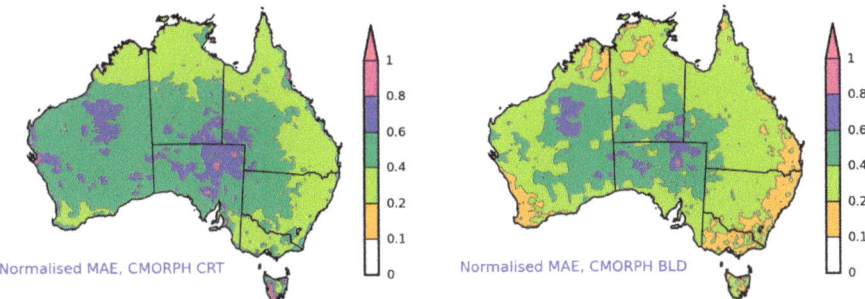

**Figure 6.** Mean bias, mean average error, and normalised mean average error for bias-corrected Climate Prediction Center morphing technique (CMORPH CRT) and gauge-blended CMORPH (CMORPH BLD) datasets from January 2001 to December 2018 using AWAP as truth.

The effects of normalisation are indicated along the northern coast of Australia and in western Tasmania where the unnormalised errors were previously the greatest but improve to about average after the adjustment, at least for the CMORPH BLD dataset.

The effect of gauge correction is especially evident around western Tasmania, as well as around western parts of Western Australia, the southern Australian coastline, the northern coastline of New South Wales, the Australian Alps, and the southwestern coast of Western Australia. In these areas, there are significant improvements in the normalised errors from the uncorrected dataset to the corrected one, indicating that there is a problem with satellite rainfall detection that cannot be accounted for by higher rainfall averages. Possible reasons will be discussed in the next section.

A point-based comparison using rain gauges categorised by states supported the gridded comparison with the results shown in Figure 7. The unnormalised MAE values suggest that performance is decreased in the tropical regions and in Tasmania, but after normalisation, the performance is much more even across the states. The performance is slightly worse in Queensland and South Australia, while gauge correction appears to have the greatest effect in Tasmania.

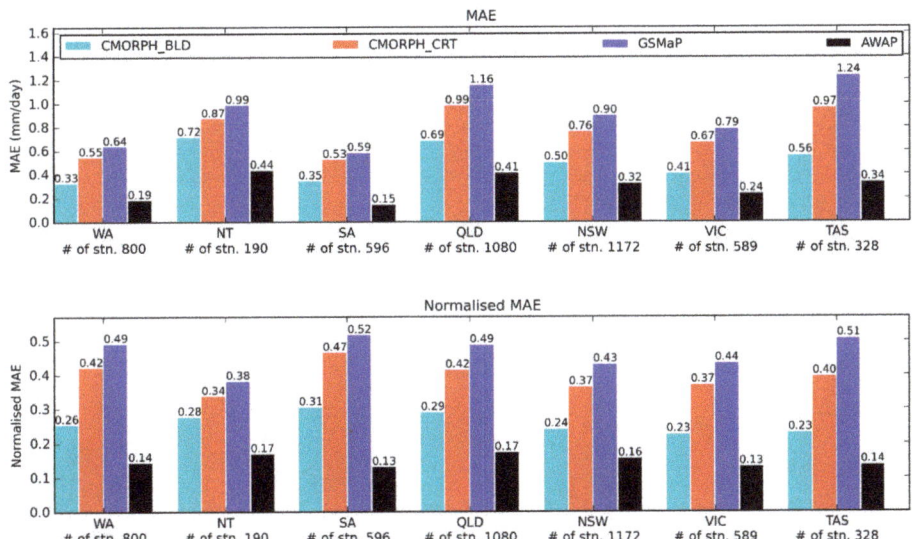

**Figure 7.** Point-based comparison categorised by states from January 2001 to December 2018 using station gauges as truth.

## 3.2. Variation with Seasons

A seasonal analysis was completed by categorising the data into four seasons with December, January, and February (DJF); March, April, and May (MAM); June, July, and August (JJA); and September, October, and November (SON) representing austral summer, autumn, winter, and spring respectively.

A gridded comparison showing the normalised MAE from the CMORPH BLD dataset is displayed in Figure 8. The greatest seasonal variation of the error is observed towards the interior and around the northern coastline with winter possessing the worst performance and summer having the best.

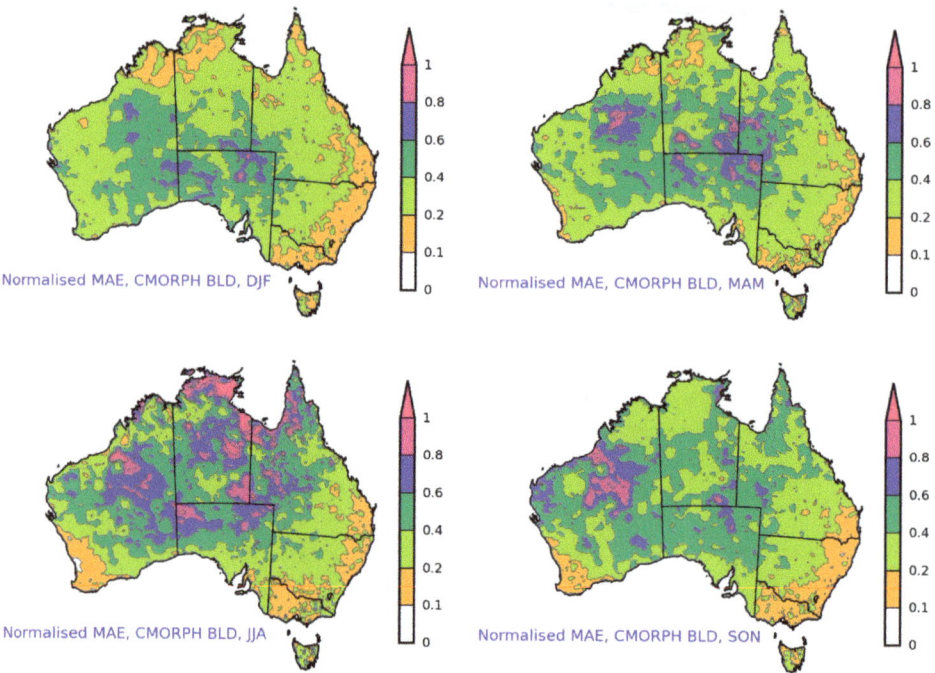

**Figure 8.** Gridded comparison categorised by seasons from January 2001 to December 2018 using AWAP as truth. Normalised mean average error from CMORPH BLD is displayed.

An analysis using point gauge data was also performed with the results shown in Figure 9. The MAE is the smallest in SON and largest in DJF where the error is approximately 50% greater. Normalisation of the errors results in the smallest relative error occurring in DJF and the largest in MAM and JJA, supporting the gridded comparison. The linear correlation coefficients across the seasons also suggest that DJF has the best performance across the seasons. The performance increase is more prominent in the non-gauge blended datasets, where the improvement is at least 10%.

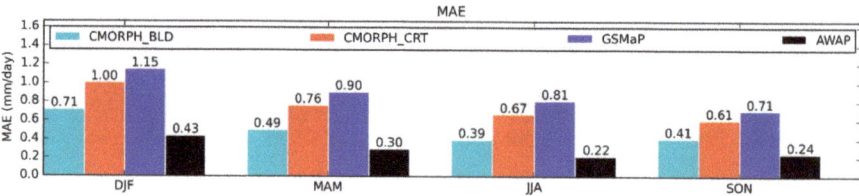

**Figure 9.** Cont.

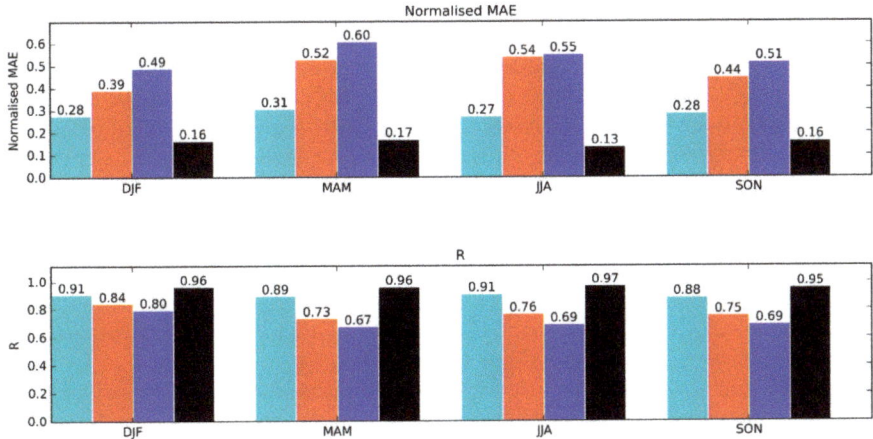

**Figure 9.** Point-based comparison categorised by seasons from January 2001 to December 2018 using station gauges as truth. Mean average error (MAE), normalised MAE, and R are displayed.

Overall, the performance appears to be relatively similar across the seasons with the exception of DJF, which shows a somewhat superior performance to the rest.

### 3.3. Variations with Rainfall Intensity

The effects of the intensity of the rainfall on the accuracy of the data were also analysed. The data were categorised into these bins: 0–0.2, 0.2–1, 1–2, 2–3, 3–4, 4–6, 6–9, and >9 mm/day. These rainfall ranges were chosen to ensure there were a reasonable amount of values in each bin with the values of 0.2 and 1 mm being specifically chosen as they correspond to the rainy-day threshold for BoM and a commonly used value in contingency statistics studies, respectively [15]. Continuous statistics for these bins were calculated along with a comparison of occurrence frequencies and cumulative volumes. These are shown in Figure 10.

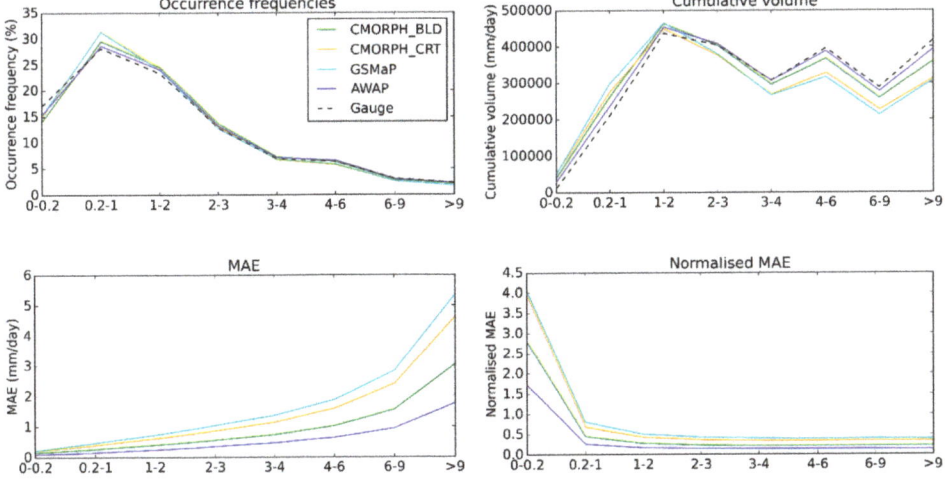

**Figure 10.** Point-based comparison categorised by rainfall intensities from January 2001 to December 2018 using station gauges as truth. Occurrence frequencies, cumulative volumes, mean average error (MAE), and normalised MAE are displayed.

The datasets appear to capture the correct frequency best for rainfall amounts between 3 and 6 mm/day. For higher amounts, the satellite-derived data underestimate the frequency, while for lower amounts, the frequency is underestimated for very low values (<0.2 mm/day) but overestimated between the range of 0.2 and 3 mm/day. The change in sign of the bias from 0–0.2 to 0.2–1 mm/day may indicate that very-low-rainfall events are being incorrectly attributed to the higher ranges. For values above 1 mm/day, the frequency matches the gauge data quite well.

Analysis of the cumulative volumes demonstrates that below 1–3 mm/day, the satellite-derived data overestimate the gauge amount, while above this range, they underestimate the amount. Combining this result with the frequency analysis suggests that although the frequency of very-low-rainfall events is underestimated, each event is an overestimation of reality.

The MAE suggests decreasing skill as the rainfall rate increases. The normalised MAE was calculated by normalising the MAE by the mean rainfall amount for each bin. It indicates that the relative error was the largest for very small values (<0.2 mm/day).

Overall, satellite-derived data appears to be most reliable for low-moderate rainfall totals (2–4 mm/day), with a significant underestimation of amounts occurring for high-end totals and an underestimation of frequency and overestimation of amounts occurring for very low totals.

## 4. Discussion

It is important to acknowledge the effect of the errors in the reference datasets. The errors in the quality-controlled gauge network used for the point comparison are minor; however, the same cannot be said for the AWAP dataset used for the gridded comparison. Jones et al. (2009) performed a cross-validation of AWAP against station observations and found the monthly rainfall mean bias, RMSE, MAE, and normalised MAE to be 0.016, 0.7, 0.38, and 0.21 mm/day, respectively [29]. The RMSE and MAE for the satellite datasets ranged between 0.79 to 1.08 and 0.43 to 0.61 mm/day respectively, indicating that the errors in AWAP are comparable to those in the satellite datasets using AWAP as truth. To gain a better idea of the true error of the datasets, the satellite datasets along with AWAP were compared to a climate reanalysis (ERA5). A climate reanalysis is a numerical representation of meteorological fields created by combining meteorological observations with climate models. A gridded comparison using ERA5 as the reference was completed and is presented in Figures 11 and 12. The results demonstrate comparable performance across the datasets. CMORPH BLD and AWAP displayed remarkably similar performances.

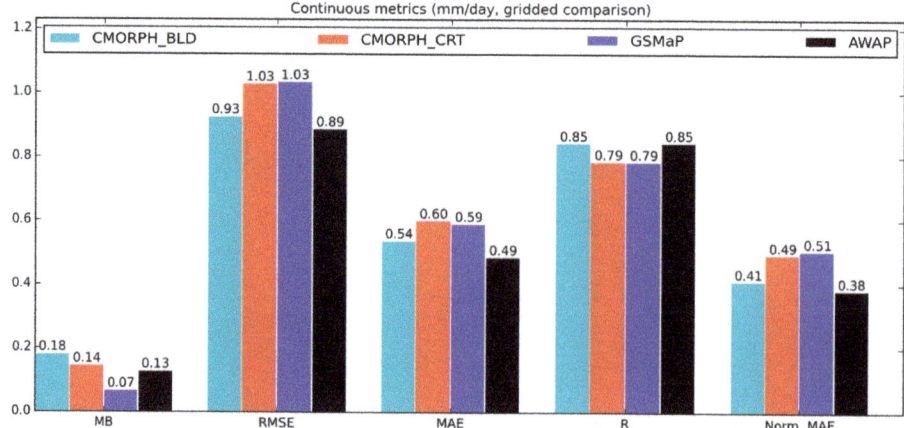

**Figure 11.** Gridded comparison of satellite datasets and AWAP against ERA5 reanalysis from January 2001 to December 2018. Mean bias (MB), root-mean-square error (RMSE), mean average error (MAE), R and normalised MAE are displayed.

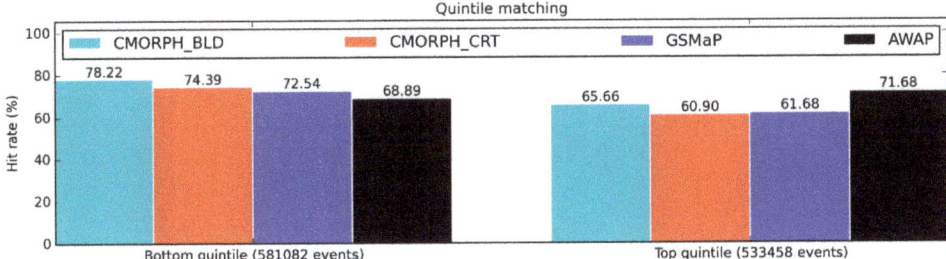

**Figure 12.** Gridded comparison of satellite datasets and AWAP against ERA5 reanalysis from January 2001 to December 2018. Quintile comparison is displayed.

The results of the error analysis of the gridded comparison are supported by the point-based comparison where both satellite datasets and AWAP were compared to station gauges with the errors in AWAP being smaller but still within the same order of magnitude as those from the satellite dataset. This highlights the caution needed in understanding that the gridded comparison results are unlikely to be a proper depiction of the true error of the satellite datasets.

There are certain regions where the performance of satellite rainfall detection is decreased. Past studies have indicated that the detection of cold frontal-based rainfall is poor [15,17]. The absence of ice crystals in the relatively low precipitating clouds typically associated with frontal rainfall hinders the ability of satellites to detect rainfall via scattering [15]. This is a likely factor behind the large errors over Tasmania, Western Australia, South Australia, and central Australia, areas where the prevalent rainfall generation mechanism is cold frontal systems. Errors are pronounced over the western half of Tasmania and the southwestern coast of Western Australia, areas of relatively high rainfall due to increased exposure to westerly flow and associated cold fronts.

Performance is also known to be decreased over topography [18,21]. Decreased performance is observed along the eastern coastline near the Great Dividing Range. The errors are greatest along the northern NSW coastline and the Australian Alps where the Great Dividing Range is at its highest elevations, leading to a strong orographic influence on rainfall.

A high-quality rain gauge network is extremely valuable for improving the accuracy of satellite-derived rainfall estimates as satellite estimates rely on gauges to calibrate or correct their raw values. The significantly greater number of gauges towards the coastline where most of Australia's population resides allows for a much greater improvement from gauge correction in contrast to the interior of the continent. Consequently, areas towards the coastlines that experience problematic regimes such as cold-frontal rainfall and orographically influenced rainfall greatly benefit from gauge correction, resulting in a performance similar to unproblematic regimes. However, the lack of rain gauges towards the interior means there are still large normalised errors in this region, even in the gauge-corrected dataset. This is compounded by the tendency of rainfall to be lighter towards the interior compared to the coast as light rainfall has been shown to be a problematic regime as well [17,18].

Low mean rainfall is another factor that would contribute to a large normalised MAE. Some areas of large normalised MAE around the interior of the continent can be seen to generally align with areas of low mean rainfall values, as seen in Figure 13, which depicts the seasonal mean rainfall across Australia. This is especially true during the austral winter 'dry' season for central Australia and northwards towards the Northern Territory coast.

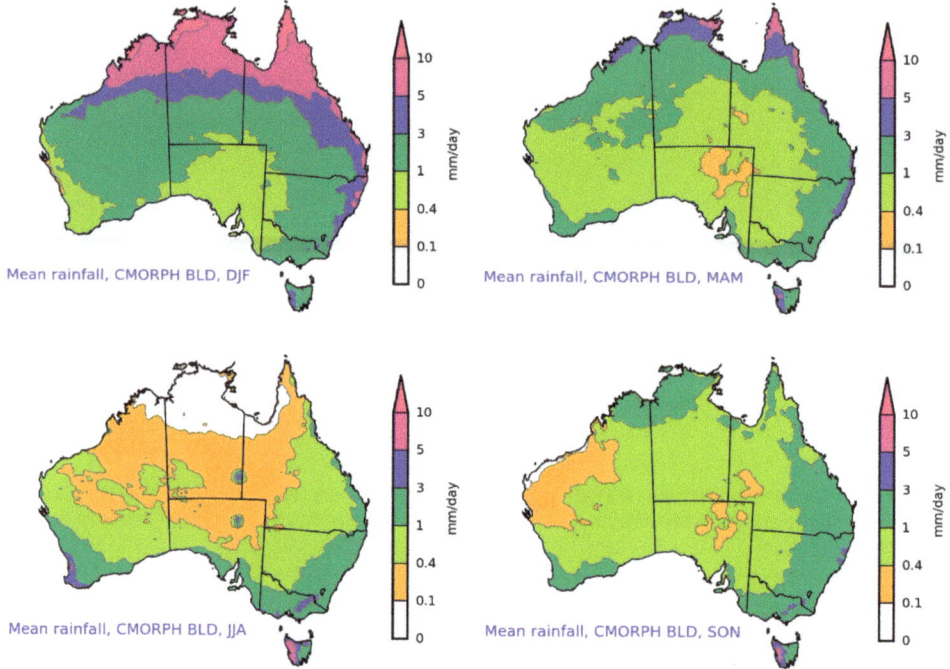

**Figure 13.** Mean CMORPH BLD rainfall by season from January 2001 to December 2018.

The importance of gauge correction is reduced for unproblematic regimes. For example, the normalised errors for the corrected and uncorrected datasets around the northern coastline of Australia are relatively similar, highlighting how the raw satellite algorithms exhibit decent performance in these areas, leading to gauge correction being less crucial. Tropical-based rainfall has been noted to be one of the better-performing regimes for satellite rainfall detection [15].

Satellite-derived precipitation estimates for austral winter demonstrate the worst performance, a result that agrees with past studies [15]. The difficulty of detecting cold-frontal rainfall, which is more frequent during winter, is most likely a key factor. The introduction of snow is another challenge for satellite detection of precipitation and is likely a contributing factor to the poor performance observed in western Tasmania and the Australian Alps. By contrast, the greater prevalence of convective-based rainfall in summer is a reason for this season performing the best [15,17].

An overestimation (underestimation) of low (high) rainfall rates was observed and is consistent with past literature [19,20].

It is natural to expect that CMORPH BLD (a gauge-blended dataset) should have at least equal performance to AWAP (a gauge-based analysis) as it relies on using gauges where the data exist whilst depending more heavily on satellites where there is little to no gauge data. However, the key assumption here is that satellite depiction of rainfall is superior to interpolation methods in areas with little to no data. This is not necessarily true, as, even though satellites are sourcing their data through a physical sensor, this process still relies heavily on calibration to rain gauge data. For locations where there is little to no gauge data, calibration and, subsequently, performance will be severely hindered. Furthermore, AWAP used a minimum number of stations exceeding 3000 while the satellite datasets are calibrated to the CPC Unified gauge analysis, which has a minimum number of stations across Australia at least an order of magnitude less than that of AWAP [29,30]. The ingestion of less data is likely to contribute to the discrepancies observed.

## 5. Conclusions

The high spatial variation of rainfall along with the issue of installing a sufficiently dense network of rain gauges in many areas around the world make satellites an attractive option in terms of their ability to provide a continuous estimate of near-surface rainfall. Numerous verifications of satellite-estimated rainfall have been performed in the past, but few studies have focused on Australia using a relatively long data record. This study aimed to fill that gap by performing a validation over Australia using monthly CMORPH (both the bias-corrected CMORPH CRT and the gauge-blended CMORPH BLD) and GSMaP (gauge-corrected) data across an 18 year period from 2001 to 2018.

Station data were used as a point of reference, both in the form of the AWAP analysis along with individual stations in order to enable both a gridded and point-based comparison, respectively. Both continuous statistics (MB, MAE, RMSE, and R) and percentile-based statistics (hit rate for bottom and top quintiles) were chosen. General performance along with the geographical, seasonal, and intensity dependencies were subsequently investigated.

Overall statistics showed that satellite performance was decent and, in the case of CMORPH BLD, somewhat comparable to the AWAP analysis used as truth. CMORPH BLD performed best followed by CMORPH CRT and then GSMaP. Linear correlations from 0.71 to 0.90 and a bottom quintile hit rate from 70% to 80% were especially encouraging.

A geographical analysis of the error dependency was completed by plotting the gridded errors over Australia, as well as by breaking down the point comparison into states. Western Australia, western Tasmania, central Australia, and the Australian Alps displayed large errors in the uncorrected datasets. Orographically influenced rainfall and cold frontal rainfall have been identified as problematic regimes by past studies and are applicable to these regions. The blending of gauge data was beneficial, especially for regions that had problematic rainfall regimes. However, a dense rain gauge network is also needed for accurate calibration, and it is likely that the lack of rain gauges towards the interior of the continent was probably the reason why little to no improvement was seen in the gauge-blended dataset over these areas.

Categorising the results by seasons demonstrated that the performance was relatively similar across the seasons, with satellite-derived precipitation estimates in austral summer performing best and those in austral winter performing worst. A categorisation by rainfall intensity suggested that the performance was best for moderate rainfall amounts (2–4 mm/day). The frequency of high-end rainfall was captured well but the amount was severely underestimated while low-end rainfall amounts were overestimated.

The main results from this study agree with past literature reconciling the performance of satellite-derived precipitation estimates over Australia with those seen around other regions in the world. The results obtained in this study are generally better than past studies. For example, Jiang et al. (2016) evaluated CMORPH CRT and CMORPH BLD over China on a monthly time scale from 2000 to 2012 and obtained slightly lower correlation coefficients of 0.72 and 0.83 respectively [16]. Possible reasons may be that satellite technology has continued to improve over the years, as well as Australia having a relatively high-quality and dense rain gauge network that allows for improved performance of gauge correction and blending.

The study supported the finding that orographically influenced rainfall and cold frontal rainfall are problematic regimes for satellite rainfall detection. Advancement in the detection of these regimes would be very beneficial. Gauge-blending was shown to be a worthy process; however, its performance is strongly tied to the availability of high-quality rain gauge network data, which do not exist in many regions. Considering that one of, if not, the most valuable use of satellite rainfall monitoring is in areas without rain gauges, an accuracy that is dependent on gauge-blending should not be relied on. The unblended datasets do demonstrate skilful performance, which would be useful for areas that lack a rain gauge network, but there is still a considerable amount of progress needed to bring unblended datasets to a level comparable to that of rain gauges.

To conclude, evaluation of satellite precipitation estimates (CMORPH and GSMaP) is an essential scientific contribution to WMO activities in assisting countries in Asia and the Pacific with improving precipitation monitoring (including accumulated heavy precipitation and drought monitoring) which WMO provides through its flagship initiatives such as the Space-based Weather and Climate Extremes Monitoring [24] and the Climate Risk and Early Warning Systems [31], among others.

**Author Contributions:** Conceptualisation, Z.-W.C., Y.K. and A.W.; methodology, Z.-W.C. and Y.K.; software, Z.-W.C.; validation, Z.-W.C.; formal analysis, Z.-W.C.; writing—original draft preparation, Z.-W.C.; writing—review and editing, Z.-W.C., Y.K., A.W.; visualisation, Z.-W.C.; supervision, Y.K. and A.W.; funding acquisition, Y.K. All authors have read and agreed to the published version of the manuscript.

**Funding:** This research was funded by the World Meteorological Organization as part of the Climate Risk and Early Warning Systems (CREWS) international initiative.

**Acknowledgments:** GSMaP data were provided by EORC, JAXA. CMORPH data were provided by CPC, NOAA. AWAP and station gauge data were provided by the Bureau of Meteorology. Contains modified Copernicus Climate Change Service Information [2019]. Neither the European Commission nor ECMWF is responsible for any use that may be made of the Copernicus Information or Data it contains. William Wang and Elizabeth Ebert provided useful comments that helped us to improve the quality of the initial manuscript. We are grateful to colleagues from the Climate Monitoring and Long-range Forecasts sections of the Bureau of Meteorology for their helpful advice and guidance.

**Conflicts of Interest:** The authors declare no conflict of interest. The funders had no role in the design of the study; in the collection, analyses, or interpretation of data; in the writing of the manuscript, or in the decision to publish the results.

## References

1. WMO. *Status of the Global Observing System for Climate*; WMO: Geneva, Switzerland, 2015; p. 54.
2. Kidd, C.; Becker, A.; Huffman, G.; Muller, C.; Joe, P.; Skofronick-Jackson, G.; Kirschbaum, D. So, how much of the Earth's surface is covered by rain gauges? *Bull. Amer. Meteor. Soc.* **2017**, *98*, 69–78. [CrossRef]
3. Michaelides, S.; Levizzani, V.; Anagnostou, E.; Bauer, P.; Kasparis, T.; Lane, J.E. Precipitation: Measurement, remote sensing, climatology and modeling. *Atmos. Res.* **2009**, *94*, 512–533. [CrossRef]
4. Michelson, D. Systematic correction of precipitation gauge observations using analyzed meteorological variables. *J. Hydrol.* **2004**, *290*, 161–177. [CrossRef]
5. Peterson, T.; Easterling, D.; Karl, T.; Groisman, P.; Nicholls, N.; Plummer, N.; Torok, S.; Auer, I.; Boehm, R.; Gullett, D.; et al. Homogeneity adjustments of in situ atmospheric climate data: A review. *Int. J. Climatol.* **1998**, *18*, 1493–1517. [CrossRef]
6. Groisman, P.; Legates, D. The accuracy of United States precipitation data. *Bull. Amer. Meteor. Soc.* **1994**, *75*, 215–227. [CrossRef]
7. Young, C.; Bradley, A.; Krajewski, W.; Kruger, A.; Morrissey, M. Evaluating NEXRAD Multisensor Precipitation Estimates for Operational Hydrologic Forecasting. *J. Hydrometeor.* **2000**, *1*, 241–254. [CrossRef]
8. Krajewski, W.; Smith, J. Radar hydrology: Rainfall estimation. *Adv. Water Resour.* **2002**, *25*, 1387–1394. [CrossRef]
9. Beek, C.; Leijnse, H.; Hazenberg, P.; Uijlenhoet, R. Close-range radar rainfall estimation and error analysis. *Atmos. Meas. Tech.* **2016**, *9*, 3837–3850. [CrossRef]
10. Joyce, R.; Janowiak, J.; Arkin, P.; Xie, P. CMORPH: A Method that Produces Global Precipitation Estimates from Passive Microwave and Infrared Data at High Spatial and Temporal Resolution. *J. Hydrometeor.* **2004**, *5*, 487–503. [CrossRef]
11. Okamoto, K.; Ushio, T.; Iguchi, T.; Takahashi, N.; Iwanami, K. The Global Satellite Mapping of Precipitation (GSMaP) project. In Proceedings of the International Geoscience and Remote Sensing Symposium (IGARSS), Seoul, Korea, 14 November 2005. [CrossRef]
12. Ushio, T.; Kachi, M. Kalman filtering applications for global satellite mapping of precipitation (GSMaP). In *Satellite Rainfall Applications for Surface Hydrology*; Springer: Dordrecht, The Netherlands, 2010; pp. 105–123. [CrossRef]
13. Xie, P.; Joyce, R.; Wu, S.; Yoo, S.; Yarosh, Y.; Sun, F.; Lin, R. Reprocessed, bias-corrected CMORPH global high-resolution precipitation estimates from 1998. *J. Hydrometeor.* **2017**, *18*, 1617–1641. [CrossRef]

14. Jiang, S.; Zhou, M.; Ren, L.; Cheng, X.; Zhang, P. Evaluation of latest TMPA and CMORPH satellite precipitation products over Yellow River Basin. *Water Sci. Eng.* **2016**, *9*, 87–96. [CrossRef]
15. Ebert, E.; Janowiak, J.; Kidd, C. Comparison of near-real-time precipitation estimates from satellite observations and numerical models. *Bull. Amer. Meteor. Soc.* **2007**, *88*, 47–64. [CrossRef]
16. Xu, R.; Tian, F.; Yang, L.; Hu, H.; Lu, H.; Hou, A. Ground validation of GPM IMERG and TRMM 3B42V7 rainfall products over Southern Tibetan plateau based on a high-density rain gauge network. *J. Geophys. Res.* **2017**, *122*, 910–924. [CrossRef]
17. Pipunic, R.; Ryu, D.; Costelloe, J.; Su, C. An evaluation and regional error modeling methodology for near-real-time satellite rainfall data over Australia. *J. Geophys. Res.* **2015**, *120*, 10767–10783. [CrossRef]
18. Dinku, T.; Ceccato, P.; Grover-Kopec, E.; Lemma, M.; Connor, S.; Ropelewski, C. Validation of satellite rainfall products over East Africa's complex topography. *Int. J. Remote Sens.* **2007**, *28*, 1503–1526. [CrossRef]
19. Habib, E.; Haile, A.; Tian, Y.; Joyce, R. Evaluation of the High-Resolution CMORPH Satellite Rainfall Product Using Dense Rain Gauge Observations and Radar-Based Estimates. *J. Hydrometeor.* **2012**, *13*, 1784–1798. [CrossRef]
20. Ning, S.; Song, F.; Udmale, P.; Jin, J.; Thapa, B.; Ishidaira, H. Error Analysis and Evaluation of the Latest GSMap and IMERG Precipitation Products over Eastern China. *Adv. Meteorol.* **2017**, *2017*, 1–16. [CrossRef]
21. Derin, Y.; Yilmaz, K. Evaluation of multiple satellite-based precipitation products over complex topography. *J. Hydrometeor.* **2014**, *15*, 1498–1516. [CrossRef]
22. Stampoulis, D.; Anagnostou, E. Evaluation of global satellite rainfall products over Continental Europe. *J. Hydrometeor.* **2012**, *13*, 588–603. [CrossRef]
23. Kubota, T.; Ushio, T.; Shige, S.; Kida, S.; Kachi, M.; Okamoto, K. Verification of high-resolution satellite-based rainfall estimates around japan using a gauge-calibrated ground-radar dataset. *J. Meteorol. Soc. Jpn.* **2009**, *87*, 203–222. [CrossRef]
24. Kuleshov, Y.; Kurino, T.; Kubota, T.; Tashima, T.; Xie, P. WMO Space-based Weather and Climate Extremes Monitoring Demonstration Project (SEMDP): First Outcomes of Regional Cooperation on Drought and Heavy Precipitation Monitoring for Australia and Southeast Asia. In *Rainfall—Extremes, Distribution and Properties*; IntechOpen: London, UK, 2019; pp. 51–57. [CrossRef]
25. Beck, H.; Vergopolan, N.; Pan, M.; Levizzani, V.; van Dijk, A.; Weedon, G.; Brocca, L.; Pappenberger, F.; Huffman, G.; Wood, E. Global-scale evaluation of 22 precipitation datasets using gauge observations and hydrological modeling. *Hydro. Earth Syst. Sci.* **2017**, *21*, 6201–6217. [CrossRef]
26. Beck, H.; Zimmermann, N.; McVicar, T.; Vergopolan, N.; Berg, A.; Wood, E. Present and future köppen-geiger climate classification maps at 1-km resolution. *Sci. Data* **2018**, *5*, 180214. [CrossRef] [PubMed]
27. Johnson, D. *The Geology of Australia*, 2nd ed.; Cambridge University Press: Cambridge, UK, 2009; p. 202. [CrossRef]
28. Global Satellite Mapping of Precipitation (GSMaP) for GPM Algorithm Theoretical Basis Document Version 6. Available online: https://eorc.jaxa.jp/GPM/doc/algorithm/GSMaPforGPM_20140902 (accessed on 14 September 2019).
29. Jones, D.; Wang, W.; Fawcett, R. High-quality spatial climate data-sets for Australia. *Aust. Meteorol. Ocean.* **2009**, *58*, 233–248. [CrossRef]
30. Chen, M.; Shi, W.; Xie, P.; Silva, V.; Kousky, V.; Higgins, R.; Janowiak, J. Assessing objective techniques for gauge-based analyses of global daily precipitation. *J. Geophys. Res.* **2008**, *113*, D4. [CrossRef]
31. Kuleshov, Y.; Inape, K.; Watkins, A.B.; Bear-Crozier, A.; Chua, Z.-W.; Xie, P.; Kubota, T.; Tashima, T.; Stefanski, R.; Kurino, T. Climate Risk and Early Warning Systems (CREWS) for Papua New Guinea. In *Drought–Detection and Solutions*; IntechOpen: London, UK, 2020; pp. 147–168. [CrossRef]

© 2020 by the authors. Licensee MDPI, Basel, Switzerland. This article is an open access article distributed under the terms and conditions of the Creative Commons Attribution (CC BY) license (http://creativecommons.org/licenses/by/4.0/).

*Article*

# Spatio–temporal Assessment of Drought in Ethiopia and the Impact of Recent Intense Droughts

Yuei-An Liou [1,2,*] and Getachew Mehabie Mulualem [2,3,4]

1. Center for Space and Remote Sensing Research, National Central University, No. 300, Jhongda Rd., Jhongli Dist., Taoyuan City 32001, Taiwan
2. Taiwan Group on Earth Observations, Zhubei City, Hsinchu County 30274, Taiwan
3. Taiwan International Graduate Program (TIGP), Earth System Science Program, Academia Sinica and National Central University, Taipei 112, Taiwan
4. College of Science, Bahir Dar University, P.O. Box 79, Bahir Dar, Ethiopia
* Correspondence: yueian@csrsr.ncu.edu.tw; Tel.: +886-3-4227151 (ext. 57631)

Received: 16 June 2019; Accepted: 30 July 2019; Published: 5 August 2019

**Abstract:** The recent droughts that have occurred in different parts of Ethiopia are generally linked to fluctuations in atmospheric and ocean circulations. Understanding these large-scale phenomena that play a crucial role in vegetation productivity in Ethiopia is important. In view of this, several techniques and datasets were analyzed to study the spatio–temporal variability of vegetation in response to a changing climate. In this study, 18 years (2001–2018) of Moderate Resolution Imaging Spectroscopy (MODIS) Terra/Aqua, normalized difference vegetation index (NDVI), land surface temperature (LST), Climate Hazards Group Infrared Precipitation with Stations (CHIRPS) daily precipitation, and the Famine Early Warning Systems Network (FEWS NET) Land Data Assimilation System (FLDAS) soil moisture datasets were processed. Pixel-based Mann–Kendall trend analysis and the Vegetation Condition Index (VCI) were used to assess the drought patterns during the cropping season. Results indicate that the central highlands and northwestern part of Ethiopia, which have land cover dominated by cropland, had experienced decreasing precipitation and NDVI trends. About 52.8% of the pixels showed a decreasing precipitation trend, of which the significant decreasing trends focused on the central and low land areas. Also, 41.67% of the pixels showed a decreasing NDVI trend, especially in major parts of the northwestern region of Ethiopia. Based on the trend test and VCI analysis, significant countrywide droughts occurred during the El Niño 2009 and 2015 years. Furthermore, the Pearson correlation coefficient analysis assures that the low NDVI was mainly attributed to the low precipitation and water availability in the soils. This study provides valuable information in identifying the locations with the potential concern of drought and planning for immediate action of relief measures. Furthermore, this paper presents the results of the first attempt to apply a recently developed index, the Normalized Difference Latent Heat Index (NDLI), to monitor drought conditions. The results show that the NDLI has a high correlation with NDVI ($r = 0.96$), precipitation ($r = 0.81$), soil moisture ($r = 0.73$), and LST ($r = -0.67$). NDLI successfully captures the historical droughts and shows a notable correlation with the climatic variables. The analysis shows that using the radiances of green, red, and short wave infrared (SWIR), a simplified crop monitoring model with satisfactory accuracy and easiness can be developed.

**Keywords:** drought; NDVI; NDLI; VCI; ENSO; time series analysis

## 1. Introduction

In the era of climate change, there is a continuous need to thoroughly assess vulnerabilities caused by complex environmental, ecological, and anthropogenic factors. Drought, as a natural phenomenon, creates numerous multidimensional effects on agriculture, human health, and disease prevalence [1]. Various drought management and vulnerability schemes were thus developed to mitigate the influences of natural and human-made disturbances at regional [2,3] and global scales [4,5]. Vulnerability assessment of natural disasters has become a necessity for policy-makers and practitioners in reducing the impacts associated with them [6,7].

Drought is dryness due to an acute shortage of water, which lasts for several months or years. Drought considerably endangers food and water security. As a complex natural event, it stems from a lack of precipitation over a prolonged period of time, and its effect can be only witnessed slowly over a period of time [8,9]. Besides the shortages of precipitation, droughts are associated with differences between actual and potential evapotranspiration, soil moisture deficits, and reduced groundwater or reservoir levels. These characteristics make the definition of drought complex and, thus, there is no single universally accepted definition. Owing to the lack of comprehensiveness of a single agreed definition, the identification and monitoring of key characteristics of drought is difficult.

Several studies have provided comprehensive reports on indices that are used to monitor the impacts of droughts [10–15]. Generally, a variety of drought indices were developed from climatic and satellite data. The most widely used indexes include the Palmer Drought Severity Index (PDSI), Standardized Precipitation Index (SPI) [16], normalized difference vegetation index (NDVI) [17], Normalized Difference Water Index (NDWI) [18], Vegetation Condition Index (VCI), and Temperature Condition Index (TCI) [13]. Remote sensing data-based indices have been widely used and compared with the other approaches for assessing drought, as they are among the best in detecting the onset of drought and measuring the intensity, duration, and impact of drought globally [19]. The remote-sensing based indices for quantifying the state of vegetation, namely the combination of visible and infrared bands, provide unique characterization for the vegetative area, including biomass, growth status, and leaf area coverage, and serve as a basis for the estimation of vegetation condition [20]. Surface temperature may serve as a basis for the estimation of vegetation condition and evapotranspiration [21]. The performance of drought indices generated based on MODIS reflectance and land surface temperature (LST), in association with the standardized precipitation index (SPI), were extensively investigated to assess drought conditions on a global scale to regional scale in the southern Great Plains, USA [22], China [23], in eastern Africa, and in southern and southeastern Africa [24–28].

Ethiopia faces drought conditions every eight–ten years [29]. The country has been facing drought at a growing incidence throughout the past many decades [30]. Among these, the 1984–1985 drought affected the lives of more than two hundred thousand people and millions of livestock [31]. The climate in Ethiopia is changing, even though significant trends are not clear [32]. OXFAM reports that according to the survey made questioning local people in Ethiopia, the climate is experiencing an increase in the rate of drought [33]. The farmers report that good harvests are less common due to an extended extreme dry season and strong rain in the wet season, followed by a prolonged absence of precipitation, which is likely due to a manifestation of global warming. Both the rise in temperature and the long absence of precipitation are major factors for causing droughts. The projected increase of weather events such as droughts due to climate change derails the availability of water and will lead to a cut in agricultural production.

Ethiopia's economy is essentially dependent on rain-fed agriculture, which is vulnerable to climate change [34]. 2015 was one of the driest years in large parts of Ethiopia [35]. The main rain season, locally called 'kiremt', was late and below normal conditions [36]. Consequently, the government called for emergency assistance for 10.2 million people [37]. The ultimate causes of this drought event originated from great distances, through atmospheric and oceanic circulations. The El Niño–Southern Oscillation (ENSO) phenomenon hugely impacts Ethiopian rainfall [38]. In particular, the warm-phase El Niño is closely linked with reserved rains during kiremt, over northern and central Ethiopia [39]. Under these circumstances, the evapotranspiration needs of plants were not met, leading to an intense reduction in vegetative production. Thus, the need to assess long-term vegetation trends and investigate the relationship between these changes and the variability in climatic conditions is increasingly important in Ethiopia.

The specific objectives of this research are: (i) to detect any long-term hydro-meteorological trends using the Mann–Kendall statistical test; (ii) to assess the drought patterns using the vegetation condition index; and (iii) to identify the main causes of NDVI change in relation to rainfall, soil moisture, LST, and ENSO. Additionally, this paper will be the first to attempt to incorporate the Normalized Difference Latent Heat Index (NDLI) as a proxy to evapotranspiration needs of the plant. NDLI, a combination of the green, red, and SWIR channels of the electromagnetic spectrum, has been found to be useful for the detection of plant water content [40]. It is highlighted that a better analysis of drought allows for the development and implementation of successful policies to better understand disruptive climate change in the region, to improve food security and strengthen climate resilience.

## 2. Study Area and Data

### 2.1. Study Area

The study area, Ethiopia, is located between 3°00′ to 15°00′N and 32°00′ to 48°00′E in the inner part of the Horn of Africa, as shown in Figure 1. The country has a total area of 1.1 million square kilometers, is landlocked, and has the second largest population in Africa, second to Nigeria. The elevation ranges from 194 to 4539 m above mean sea level. The highland, with an altitude of 1500 m or above, is located at the central and northern parts of the country and constitutes roughly 35% of the country [41]. In a traditional way, based on elevation, at least three climatic zones are identified—the tropical (lowland zone), which is below 1830 m in elevation and has mean annual temperatures of 20–28 °C; the subtropical zone, which includes the highland areas of 1830–2440 m in elevation and with mean annual temperatures of 16–20 °C; and the cool zone, which is above 2440 m in elevation and with mean annual temperatures of 6–16 °C [42].

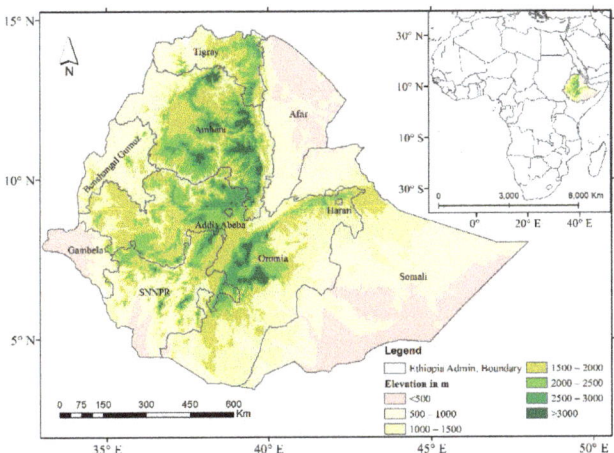

**Figure 1.** Location of the study area: the administrative boundary of Ethiopia, constituting the nine regional states, with a background showing an Advanced Spaceborne Thermal Emission and Reflection Radiometer digital elevation model of 30 m resolution.

Due to its complex topographical and geographical features, the climate of Ethiopia exhibits strong spatial variation and different rainfall regimes [43]. Thus, rainfall shows considerable spatial heterogeneity in Ethiopia [44]. Much of the region is generally bimodal, with long rains in JJAS (June–September) and short rains during OND (October–December). The meridional translation of the Intertropical Convergence Zone (ITCZ) across the equator is the main factor of the MAM (March–May) and OND seasons [45]. Topography also plays a role in affecting the annual cycle of precipitation. The highland areas receive an annual rainfall of about 1200 mm, with the least temperature variation, whereas the lowland areas (Afar and Somali regions) receive an annual rainfall of less than 500 mm with larger temperature variations [41]. The spatial distribution of Ethiopian drought indicates that most of the drought and food crises events are concentrated in the central and northern highlands, extending from North Shewa through Wollo to Tigray [46].

Based on the Climate Hazards Group Infrared Precipitation with Stations (CHIRPS) daily precipitation data obtained from the Climate Hazards Group at the University of California, Santa Barbara (UCSB) [35], the main rainfall season from June to September, locally called kiremt, accounts for 60–80% of the annual rainfall, with the remaining falling in the dry season, from October to May, Figure 2.

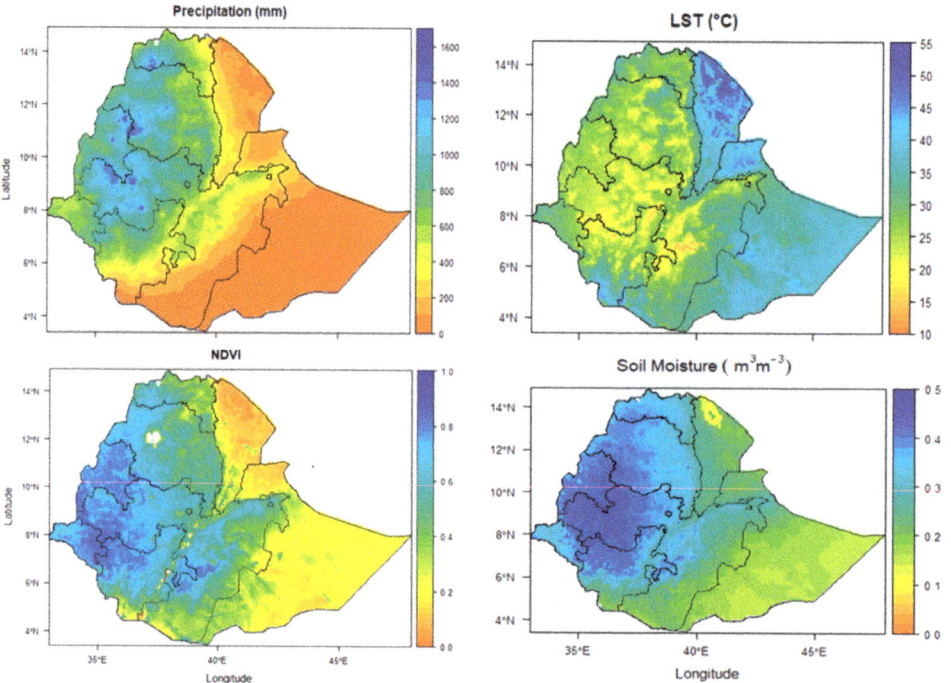

**Figure 2.** Long-term seasonal average of rainfall (mm), land surface temperature (LST, °C), normalized difference vegetation index (NDVI), and soil moisture ($m^3 m^{-3}$) for the period from 2001 to 2018.

The rainfall significantly varies between the northeastern and the western highlands of Ethiopia, where orographic rainfall is substantial. Figure 2 additionally depicts that an average seasonal LST of the land derived from the solar radiation (MOD11A2 Terra v.006 product) of Ethiopia is between 10 °C and 54 °C, with maximum temperatures concentrated on the lowland areas. Similarly, the soil moisture (derived from the FLDAS Noah Land Surface Model L4) distribution for the top 0–10 cm layer increases in the western and northern parts of the country. Moreover, NDVI distributions derived from the MOD13Q1 Terra v.006 product confirm healthy vegetation and forests, mainly located in the western parts of Ethiopia, which match with the rainfall, LST, and soil moisture patterns.

The land cover types for Ethiopia extracted from the European Space Agency (2016) Global Land Cover map are shown in Figure 3. The most dominant cover types are grassland, cropland, and shrubs, covering 29%, 26%, and 21% of the whole study area, respectively. The land cover in the highlands continually changes because of the persistent agricultural activities and higher population density as compared to the lowlands [47]. Large areas of agricultural farms, where people largely depend on the rain-fed farms, are of major concern due to recurrent drought incidents.

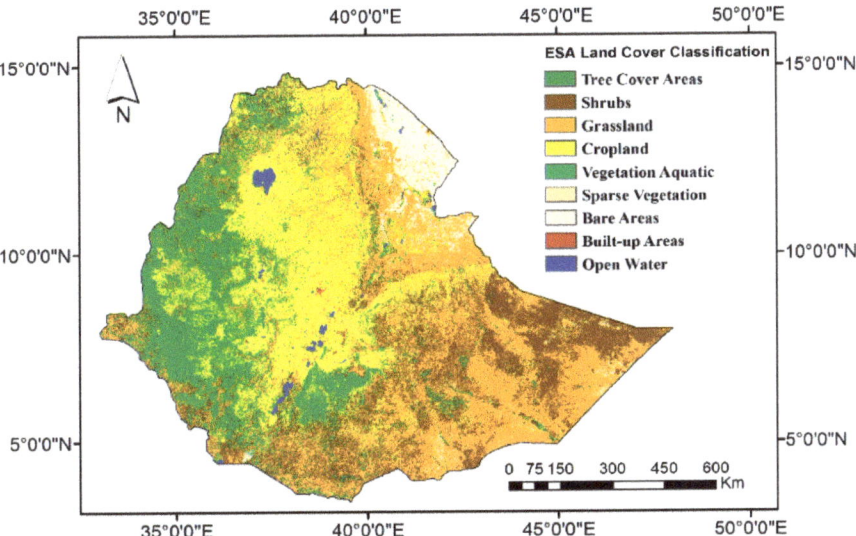

**Figure 3.** Land cover map of Ethiopia at 20 m spatial resolution during 2016, extracted from the European Space Agency.

*2.2. Datasets*

The data for this study were acquired from four sources. With extensively high temporal and spatial resolution as compared to the other satellites, the products of MODIS onboard NASA Terra and Aqua satellites were suited for this study because of their large geographic coverage. We used the monthly averaged MODIS Terra 16 day datasets for the period from 2001 to 2018 (18 years) that are archived in the Google Earth Engine (GEE) image collection. Time series NDVI and LST covering the whole study area at 250 m and 1 km spatial resolution were generated from MODIS/006/MOD13Q1 and MODIS/006/MOD11A1 version 6 surface reflectance composite, respectively. Similarly, surface reflectance products of MODIS/006/MOD09GA were generated for computing the NDWI and NDLI [40]. The data were extracted and processed using the JavaScript code editor in the GEE platform (https://earthengine.google.com/ Mountain View, CA, USA), which offers a parallel computing environment for processing large datasets. For monitoring the spatial and temporal conditions of drought, we chose NDVI, but we also included the other parameters that trigger dry conditions. NDVI, the most common index for remote sensing of vegetation, is known to be saturated over areas with high leaf area indexes. Numerous vegetation indexes using the same set of near-infrared and red channels have been developed, even though these indices do not enjoy the same popularity as NDVI, which is known for its capability to distinguish vegetation from other types of land cover, but is not really designed to sense the water content in the vegetation canopy. Nevertheless, remote sensing of the water content has important implications in agriculture and forestry. For the detection of plant water content, the near-infrared region (NIR) and shortwave infrared regions (SWIR) have been found to be useful. Thus, NDWI is defined in a similar way to NDVI but uses the near-infrared channel to monitor the water content of the vegetation canopy. Fluctuations in the vegetation canopy are indicators of drought stress [18]. Besides NIR and SWIR, two spectral regions of the electromagnetic spectrum have been found to be useful for the detection of plant water content: the green and red channels. Liou et al. [40] recently developed the NDLI, which uses the green, red, and SWIR channels. NDLI is sensitive to water availability for different land covers at the land–air interface and outperforms the different versions of NDWI indexes. The spectral indices are calculated using the following formulas:

$$NDVI = \frac{NIR - R}{NIR + R}$$

$$NDWI = \frac{NIR - SWIR}{NIR + SWIR}$$

$$NDLI = \frac{G - R}{G + R + SWIR}$$

where $G$, $R$, $NIR$, $SWIR$ are the spectral reflectance for MODIS band 4 (545–565 nm), band 1 (620–670 nm), band 2 (841–876 nm), and band 6 (1628–1652 nm), respectively.

The other data source used to generate time series rainfall data for the period from 2001 to 2018 was CHIRPS. CHIRPS is a 30+ year quasi-global rainfall dataset combining satellite observations from the Climate Prediction Center (CPC) and the National Climate Forecast System version 2 (CFSv2) and in situ precipitation observations [35,37]. It is widely used in Ethiopia for drought monitoring [38]. It is well demonstrated that CHIRPS can complement the sparse rain gauge network and provide high spatial and temporal resolution for trend analysis [48,49]. The CHIRPS data were accessed from the ftp server (ftp://ftp.chg.ucsb.edu/pub/org/chg/products/CHIRPS-2.0).

The monthly soil moisture (0–10 cm) was generated from the Famine Early Warning Systems Network (FEWS NET) Land Data Assimilation System (FLDAS) dataset, developed to assist food security assessments in data-sparse developing countries [50]. This is a natural tool to monitor drought conditions and was accessed from https://earlywarning.usgs.gov/fews/product/308.

In this study, we used the multivariate ENSO index (MEI) and the dipole mode index (DMI) to observe how vegetation responds to climatic conditions. The monthly mean MEI time series were retrieved from the National Oceanic and Atmospheric Administration (NOAA) website (https://www.esrl.noaa.gov/psd/enso/mei/, Washington, DC, USA). The MEI time series was calculated by taking the leading principal component time series of the empirical orthogonal function of the five variables, namely the sea level pressure, sea surface temperature, surface zonal winds, surface meridional winds, and Outgoing Longwave Radiation within the 30°S–30°N and 100°E–70°W region. Besides this, the DMI was calculated by taking the differences between the sea surface temperature anomalies in the western (50°E–70°E, 10°S–10°N) and eastern (90°E–110°E, 10°S–0°N) portions of the Indian Ocean [51]. The monthly DMI data was accessed from the Japan Agency for Marine-Earth Science and Technology (JAMSTEC) website (http://www.jamstec.go.jp/frcgc/research/d1/iod/iod/dipole_mode_index.html). The data used in these study are summarized on Table 1.

Table 1. Datasets characteristics and source.

| Data | Source | Characteristics |
|---|---|---|
| Precipitation | CHIRPS | Monthly precipitation at 0.05 × 0.05 from Jan. 2001 to Dec. 2018 |
| NDVI | MODIS | Monthly NDVI at 250 m from Jan. 2001 to Dec. 2018 |
| LST | MODIS | Monthly LST at 1 km from Jan. 2001 to Dec. 2018 |
| NDWI | MODIS | Estimated from surface reflectance at 500 m from Jan. 2001 to Dec. 2018 |
| NDLI | MODIS | Estimated from surface reflectance at 500 m from Jan. 2001 to Dec. 2018 |
| Soil Moisture | FLDAS Noah | Monthly soil moisture (0–10cm) at 0.10 × 0.10 from Jan. 2001 to Dec. 2018 |
| MEI | NOAA | Monthly MEI time series from Jan. 2001 to Dec. 2018 |
| DMI | JAMSTEC | Monthly DMI time series from Jan. 2001 to Dec. 2018 |

## 3. Methodology

### 3.1. Identification of Drought

A common way to calculate anomalies is to apply the Standardized Anomaly Index (SAI). SAI is a standardized departure from the long term mean and is calculated as:

$$SAI_i = \frac{x_i - \bar{x}}{\sigma}$$

where $x_i$ is the seasonal mean of variable $x$, $\bar{x}$ is the long term seasonal mean and $\sigma$ is the standard deviation of the seasonal mean of all data. The anomaly maps were created by subtracting the seasonal climatology mean from the seasonal values and then dividing this by the standard deviation. The resulting maps depict the intensity of how good or bad the current season is compared with the average situation. Seasonal anomaly maps of precipitation, NDVI, LST, and soil moisture for the years 2015–2018 from the 2001–2014 climatology were computed to identify and quantitatively measure which part of Ethiopia was severely affected in the year 2015 and its recovery to normal conditions.

Besides the common SAI, another method to compare the current NDVI with historical values is the Vegetation Condition Index [52]. The VCI has been extensively used to monitor vegetation conditions [53]. It normalizes NDVI on a pixel-by-pixel basis, scaling between the minimum and maximum values of NDVI:

$$VCI = 100 * \left( \frac{NDVI - NDVI_{min}}{NDVI_{max} - NDVI_{min}} \right)$$

where $NDVI$, $NDVI_{min}$, and $NDVI_{max}$ are the mean seasonal NDVI, and its absolute long-term minimum and maximum NDVI values, respectively, for each pixel. VCI varies in the range of 0 to 100 percent, reflecting relative changes in the vegetation condition from extremely low to high VCI [52]. As proposed by [52] and recently applied by [54], a threshold value of below 35% is used to indicate drought conditions as shown in Table 2.

Table 2. Drought categories derived from the Vegetation Condition Index (VCI).

| VCI Percentage | Drought Severity Level |
|---|---|
| >35 | No drought |
| 20–35 | Moderate drought |
| 10–20 | Severe drought |
| <10 | Extreme drought |

### 3.2. Mann–Kendall Trend Analysis

The Mann–Kendall method is a non-parametric rank-based test method, which is commonly used to identify a monotonic trend in climate, by using remote sensing and hydro-metrological data [55]. The usefulness of a non-parametric test relies on its resilience to outliers, non-normality, missing values, and seasonality and, hence, it is necessary for this study [56–58]. The univariate Mann–Kendall statistic S for time series data $(X_k, k = 1, 2, \ldots, n)$ is given as:

$$S = \sum_{j<1}^{n} sgn(X_i - X_j)$$

where $X_i$ and $X_j$ are the seasonal mean values in years $i$ and $j$, respectively, $i > j$, and $n$ is the length of the time series. The sign of all possible differences $X_i - X_j$ is computed as:

$$sgn(X_i - X_j) = \begin{cases} +1, & if\ X_i - X_j > 0 \\ 0, & if\ X_i - X_j = 0 \\ -1, & if\ X_i - X_j < 0 \end{cases}$$

When $n \geq 8$, the statistic S is approximately normally distributed with mean $E[S] = 0$, and variance $\sigma^2$ given by the following equation:

$$\sigma^2 = \left\{ n(n-1)(2n+5) - \sum_{j=1}^{p} t_j(t_j-1)(2t_j+5) \right\}/18$$

where $t_j$ is the number of data points in the jth tie group, and $p$ is the number of tie group in the time series. The test statistics $z$ is computed as:

$$z = \begin{cases} \frac{S-1}{\sigma}, & if\ S > 1 \\ 0, & if\ S = 0 \\ \frac{S+1}{\sigma}, & if\ S < 1 \end{cases}$$

Now, Z follows a standard normal distribution whereby its positive (negative) value indicates an upward (downward) trend. If Z is greater than $Z_{\alpha/2}$, where $\alpha$ represents the significance level, the trend is considered as significant. In this regard to the z-transformation, this study is considered a 9.5% confidence level, where the null hypothesis was no trend was rejected if $|z| > 1.96$, and the alternative hypothesis that increasing or decreasing monotonic trend exists in the series was accepted. The magnitude of the linear trend was then predicted by the Sen's slope estimator [59], i.e., the change per unit time of a trend was computed as:

$$Sen's\ slope\ =\ Median\ \{(x_i - x_j)/(i-j)\}, i > j,$$

where $x_i$ and $x_j$ are the changing values of the variable at time steps $i$ and $j$, respectively. A value close to zero means there is not much variation through time. A negative value of the slope depicts a negative trend, whereas a positive value indicates a positive trend. This method is recommended for remote sensing time series analysis and has been used for vegetation trend analysis [60]. The trend analysis described above was applied to the seasonal rainfall, NDVI, LST, and soil moisture values using the "spatialEco" package in R-project.

*3.3. Multiple Linear Regression*

Multiple linear regression is an extension of simple linear regression. It is used when to predict the value of a variable based on the value of two or more other variables. For instance, for analyzing a dependent variable (in this case NDVI) in light of related independent variables (precipitation, soil moisture, LST, NDWI, NDLI, MEI, DMI). It allows us to determine the overall fit of the model and the relative contribution of each of the predictors to the total variance explained. In this paper, we tried to quantify the susceptibility of NDVI to changes in climatic and hydro-metrological variables. Mathematically, a multiple linear regression model with $k$ predictor variables $x_1, x_2, \ldots, x_k$ and a response can be written as:

$$y\ =\ \beta_0 + \beta_1 x_1 + \beta_2 x_2 + \cdots + \beta_k x_k + \varepsilon\ where\ i = 1, 2, \ldots, k,$$

and $\varepsilon$ is the residual terms of the model, which tries to minimize, $y$ is the dependent variable in this case NDVI, $x_i$ represents the independent variables (precipitation, soil moisture, LST, NDWI, NDLI, MEI, DMI), $\beta_0$ is the intercept, and $\beta_1, \beta_2, \cdots, \beta_k$ are the coefficients of $x_i$. Before we chose to analyze our data using multiple regression, we made sure that assumptions required for multiple regression were met. We checked the existence of a linear relationship by inspecting the scatter and partial regression plots between NDVI and each of the independent variables. By using the variance inflation factor (VIF) values, we further checked whether the explanatory variables were highly correlated with each other or not. A VIF measures the extent to which multicollinearity has increased the variance of an estimated coefficient. It looks at the extent to which an explanatory variable can be explained by all the other explanatory variables in the equation.

## 4. Results and Discussion

The most recent ENSO, which was developed in 2014 and strengthened in the summer, has caused global impacts [61]. In Ethiopia, the dry kiremt seasons are closely linked to the significantly warmer Pacific sea surface temperatures [39]. Figure 4 depicts that the strongest kiremt precipitation anomalies derived from the CHIRPS datasets are located in the central and northwestern parts of Ethiopia, with maximum −4.6 standardized deviations anomalies around −460 mm/year in 2015. Vegetation in Ethiopia is sensitive to water availability and severely affected by low precipitation. Correspondingly, large area negative NDVI deviations are a result of water stress concentrated in the western, northern, and central parts of Ethiopia, with maximum NDVI departures by approximately −2.5 standardized anomalies below average. In the same way, the 10 cm soil moisture and LST follow the same patterns as those of the precipitation anomalies, by approximately −3 and 3.5 standardized deviations from their corresponding normal conditions, respectively. During 2016, due to the dry conditions linked with La Niña, the negative precipitations of the southern and eastern parts of Ethiopia persisted, with maximum −4.0 standardized deviations anomalies around −305 mm/year. The dry conditions evolved from the north and central regions to the south and east parts. Across the region, however, NDVI did not follow the same pattern and the vegetation productivity did not quickly decline. This may have been due to the extended availability of water stored in soils for growing crops [62]. Following the return of ENSO to neutral conditions in 2017, the central and northern regions of Ethiopia become more favorable for crop development. During this period, the cropland areas experienced enhanced precipitation and vegetation, which was also closely linked to the increase in soil moisture. The agricultural data obtained from the annual agricultural sample survey of the Central Statistics Agency indicated increments from 7.32 to 28.93 quintals per hectare for maize, from 5.05 to 26.76 quintals per hectare for Teff, and from 2.28 to 29.67 quintals per hectare for wheat [63].

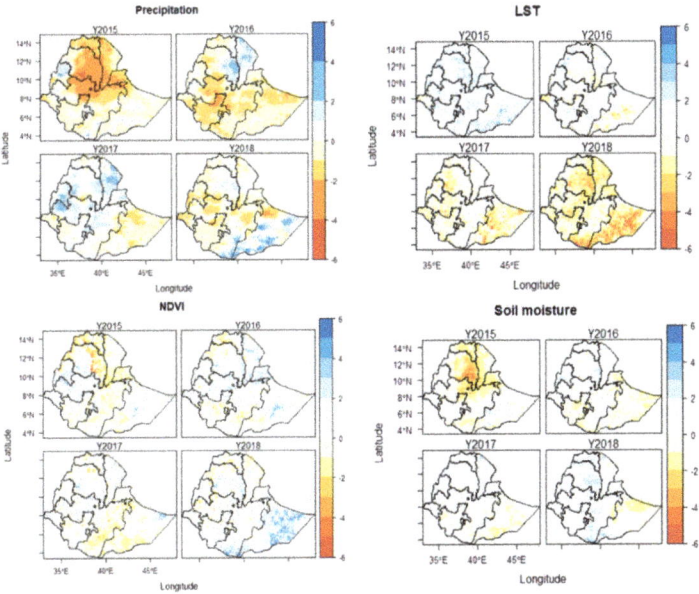

**Figure 4.** Standardized seasonal precipitation, LST, NDVI, and soil moisture anomalies from the 2001–2014 climatology averaged over June–September.

While in 2018 the precipitation showed negative anomalies, the maximum soil moisture and NDVI anomalies were about two standardized deviations above the average conditions. Similarly, the minimum LST departure was about −3 standardized deviations above the average conditions. It is worth mentioning that in 2018, compared to 2017, a higher precipitation in the southeast part of Ethiopia was observed, which was well-matched with increased NDVI.

## 4.1. Drought Patterns Based on VCI

Figure 5 depicts the spatio–temporal persistence of drought detected by VCI during the growing season in Ethiopia over the past two decades. It is shown that the growing season signifies the maximum vegetation growth, and demonstrates the suitability of VCI to detect drought and assist the measures of vegetation health.

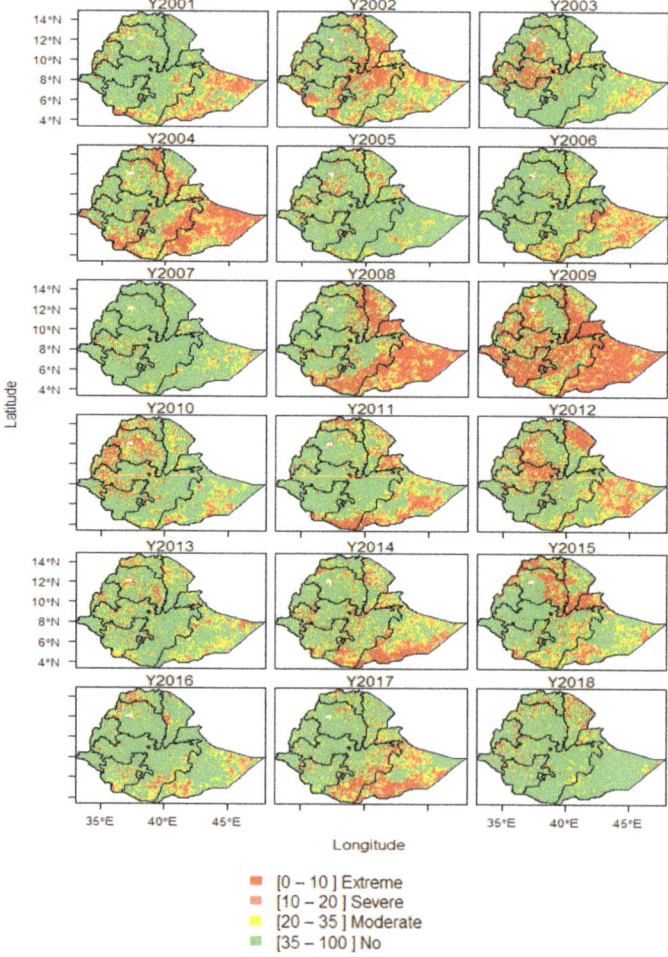

**Figure 5.** The spatio–temporal variability droughts detected by the NDVI-based vegetation condition. index for the growing season in Ethiopia for the period 2001 to 2018.

In this figure, regions which are greener indicate vegetation levels higher than the average conditions, whereas the red colors indicate poor conditions. Severe to extreme droughts were identified in the years 2002, 2003, 2004, 2009, 2010, 2012, and 2015 for the north, central, west, and southwest parts of the country, where the land is mainly covered by rain-fed agriculture. The results show a direct influence of ENSO on the vegetation of Ethiopia, especially during the El Niño years 2009–2010 and 2014–2015. During El Niño years, the NDVI values gradually declined and remained marginally below average. On the other hand, the years 2001, 2005, 2006, 2007, 2013, 2016, and 2018 reflect the near-normal NDVI throughout most of the rain-fed agriculture regions. In Ethiopia, an El Niño event would cause suppressed rainfall during the kiremt season, causing serious reductions in cereal yields and output [64]. On the other hand, when a La Niña event followed on from an El Niño, favorable and above average vegetation conditions were observed, for instance 2010–2011 and 2016–2017 La Niña events, which followed on from the 2009–2010 and 2014–2015 El Niño events, respectively.

## 4.2. Spatial and Temporal Trends

The spatial and temporal variability of the trends, together with the significance of the trends in precipitation, NDVI, soil moisture, and LST, are presented in Figures 6 and 7. The Mann–Kendall test was carried out to observe whether the mentioned variables changed over space during the 18 years period in the country. The areas in green (positive slope value) indicate an increasing monotonic trend in precipitation, NDVI, soil moisture, and LST, whereas areas in red (negative slope value) indicate a decreasing monotonic trend in precipitation, NDVI, soil moisture, and LST.

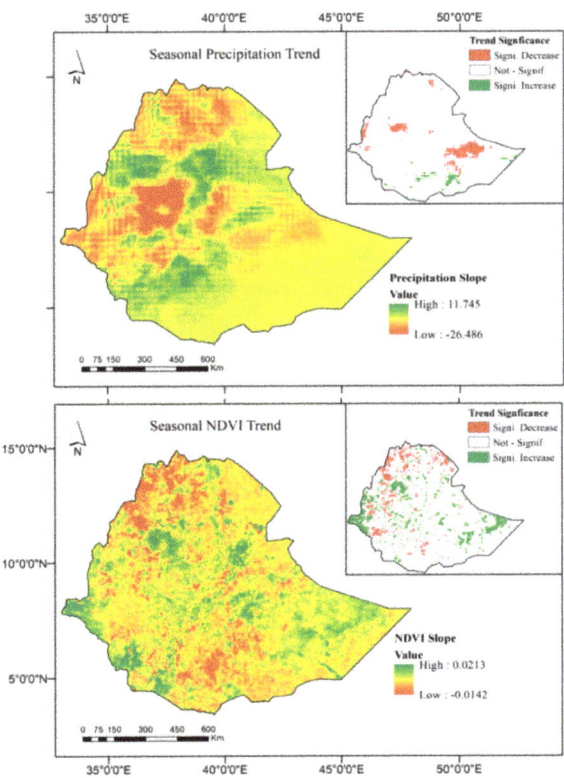

**Figure 6.** Spatial and temporal trends of seasonal precipitation, and NDVI in Ethiopia from 2001 to 2018. Positive slope values indicate an increasing monotonic trend, while negative slope values indicate a decreasing monotonic trend.

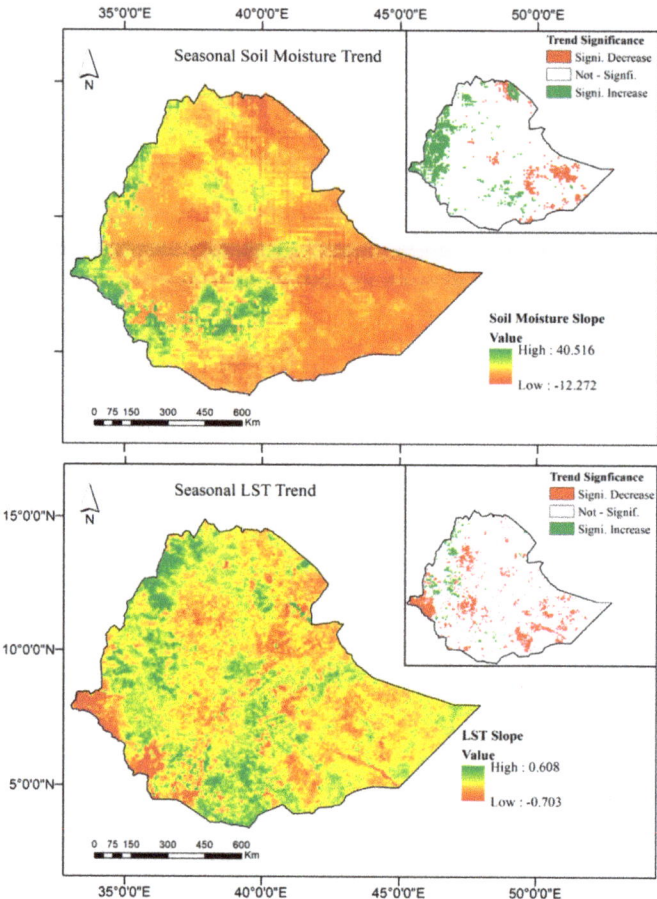

**Figure 7.** Spatial and temporal trends of seasonal LST and soil moisture in Ethiopia from 2001 to 2018. Positive slope values indicate an increasing monotonic trend, while negative slope values indicate a decreasing monotonic trend.

The pixel-based trend analysis shows the growing season trend values of precipitation range from −26 to 11 mm, with significant changes occurring in the central and northern parts of the country. Specifically, the northern, central, and rift valley regions of Ethiopia experienced a decreasing rainfall trend, whereas western Benshangul and the highlands of the central Amhara region show an increasing trend. On the other hand, the lowland pastoral regions of Somali region did not show a significant trend. Generally speaking, 52.8% of all pixels in the country show a decreasing trend and significant trends concentrate on the central and lowlands regions of the country.

With respect to the NDVI trend, the northern and northwestern areas of the Tigrai and Amhara region, as well as the southern region, showed a decreasing trend during the study period. The growing season NDVI values ranged from −0.0142 to 0.0213, and overall 41.67% of the country indicated a decreasing trend. The significant decreasing trends were located in the northwestern part. Similar pixel-based trend analysis for LST depicted in Figure 7 showed that LST increased for the northwestern, central highland, and southern parts of the country, whereas there was an estimated 11% significance decrease concentrated on the western parts of the Gambella region. These results are in agreement with the recent findings of Workie et al. [65], who used a linear regression approach to detect trends. Similar

### 4.3. Multi Linear Regression and Correlation Statistics

To facilitate relationships between NDVI and other parameters, a small box region (38E–39E, 9N–10N) which experienced significant decreasing trends, presented in Figure 4, Figure 6, and Figure 7 was extracted. Figure 8 shows the monthly anomalies time series plots for NDVI and soil moisture (Figure 8a), precipitation and LST (Figure 8b), and NDLI and NDWI (Figure 8c). Basically, the anomalies calculated by subtracting monthly climatology values from each month provide additional information about the variations present. The periods of severe droughts that resulted in countrywide drought conditions during the growing seasons are shaded with a box in Figure 8.

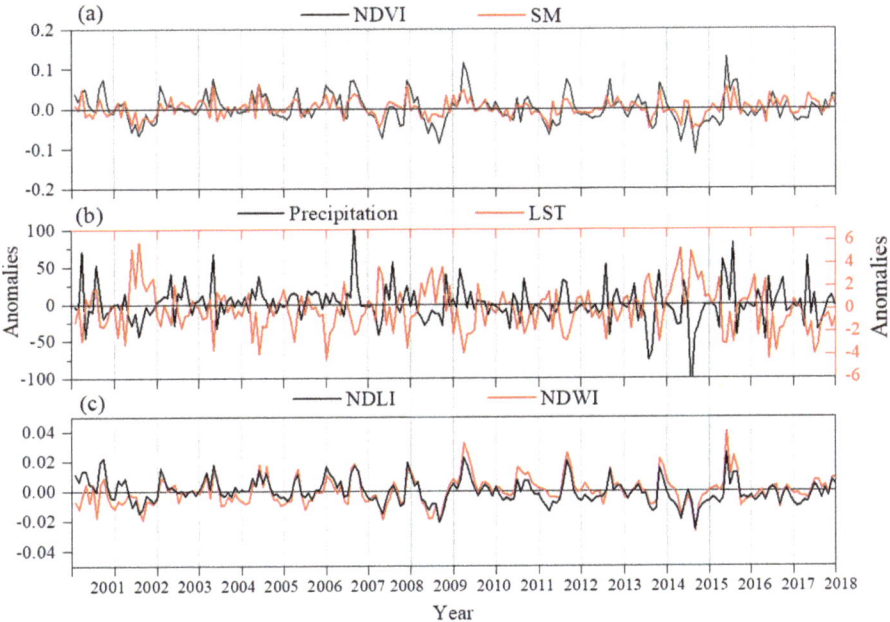

**Figure 8.** The monthly mean anomaly time series values of (**a**) NDVI and soil moisture, (**b**) precipitation and LST, and (**c**) NDLI and NDWI for 38E–39E, 9N–10N.

NDVI anomalies in this region were near normal for several years. In contrast, it showed slight green up in 2010 and late 2016, as conditions translate to weak La Niña. Maximum NDVI departures were observed in 2009 and 2015, where NDVI gradually decreased and remained slightly below average. In particular, the 2015 events were accompanied by higher precipitation anomalies of about −100 mm. There is an exact resemblance between the other parameters, with a clear identification of the drought and normal years. Considering the spatial drought patterns derived from VCI (Figure 5), the intense drought years certainly resulted in a decline in soil moisture and water availability. The water stress situations in the root zone were well captured by soil moisture values. The NDWI and NDLI indicate a similar pattern to that of NDVI, where they reached peaks in 2010 and 2016.

The Pearson correlation coefficients between NDVI and other factors (precipitation, soil moisture, LST, NDWI, NDLI, MEI, and DMI) on a seasonal time scale for the whole study record were computed to assess the relationship between them. The Pearson correlation coefficient was conducted using the statistics package in R. Figure 9 shows the heatmap, which summarizes the linear relationships between the parameters. There was a strong correlation between NDVI and precipitation ($r = 0.83$) soil moisture ($r = 0.83$), NDLI ($r = 0.96$), and NDWI ($r = 0.63$). The positive correlation between precipitation and NDVI implies that an enhanced precipitation supports vegetation growth and vice versa [66]. On the contrary, a significant negative correlation between NDVI and LST ($r = -0.76$) was observed. Furthermore, a less notable negative correlation of ($r = -0.43$, $r = -0.39$) was observed between NDVI and the two climatic indices MEI and DMI, respectively.

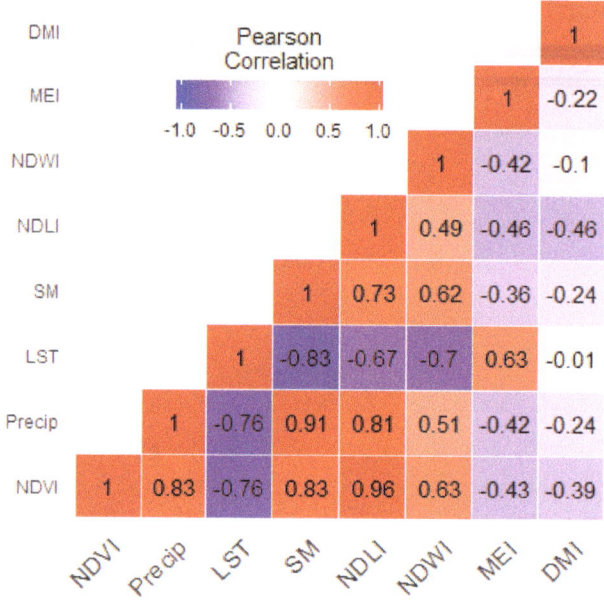

**Figure 9.** The heat map of Pearson correlation coefficients for NDVI, Precipitation, LST, soil moisture, NDLI, NDWI, MEI, and DMI.

Since there are substantial correlations among NDVI, Precipitation, LST, soil moisture, NDLI, NDWI, MEI, and DMI (Figure 9), the detection of multicollinearity is crucial before plugging data into a regression model. Multicollinearity denotes predictors that are correlated with other predictors. The most widely-used diagnostic for multicollinearity is the VIF. We can see from Table 3 that the VIFs are all down to satisfactory values; they are all less than 5. Even though there is some multicollinearity in our data, it is not severe enough to warrant further corrective measures.

The results in Table 3 reveal the statistically significant relationship between NDVI, NDLI, and NDWI and MEI, with $p$-values of $< 2.00 \times 10^{-16}, 2.86 \times 10^{-6}$, and 0.0576, respectively. The significant relationships between NDVI, and NDLI and NDWI make it clear that an increase in water availability causes an upward trend in NDVI, which implies a decline in drought [67]. The results indicate that water availability in the soil was the main influencing factor on the spatially averaged NDVI. The significantly negative correlation between MEI and NDVI reaffirms the claim that ENSO variability plays a major role in the climatic conditions and control vegetation growth conditions of central and northern parts of Ethiopia [68]. The overall multiple linear regression is significant, with a multiple R-squared value of 0.978 and adjusted R-squared value of 0.962. However, precipitation, soil moisture, and LST have insignificant regression coefficients due to their $p$-values, which are far greater than 0.05. This is due to the interaction (correlation) between the independent variables, and often since $p$-value is a function of sample size, as well as variance, there is no single rule for setting the "significance" threshold [69]. The insignificant association observed between precipitation and NDVI could also be due to the delayed response of vegetation to precipitation [70], where a time lag effect was not considered in this study. For future prediction, an optimal regression equation ($NDVI = -7.01 \times 10^{-5} + 3.75 \times NDLI + 0.518 \times NDWI - 1.386 \times 10^{-3} \times MEI$) was obtained via the backward elimination procedure in a stepwise regression analysis, which was achieved by dropping the least significant feature.

Table 3. The output of the multiple linear regression (MLR) model, in which NDVI was the dependent variable and precipitation, LST, soil moisture, NDWI, NDLI, MEI, and DMI were independent variables.

| Variable | Estimate | Std. Error | t-Value | p-Value | Sig | VIF |
|---|---|---|---|---|---|---|
| Precip. | $-2.215 \times 10^{-5}$ | $3.788 \times 10^{-5}$ | $-0.585$ | 0.5594 | | 2.387 |
| LST | $4.577 \times 10^{-4}$ | $6.140 \times 10^{-4}$ | 0.745 | 0.456 | | 3.433 |
| SM | $5.505 \times 10^{-2}$ | $6.104 \times 10^{-2}$ | 0.902 | 0.368 | | 4.622 |
| NDLI | 3.697 | $1.363 \times 10^{-1}$ | 27.125 | $<2.00 \times 10^{-16}$ | *** | 3.407 |
| NDWI | 0.528 | $1.097 \times 10^{-1}$ | 4.813 | $2.86 \times 10^{-6}$ | *** | 2.804 |
| MEI | $-1.470 \times 10^{-3}$ | $7.704 \times 10^{-4}$ | $-1.909$ | 0.0576 | . | 1.089 |
| DMI | $-4.552 \times 10^{-4}$ | $5.439 \times 10^{-3}$ | $-0.187$ | 0.852 | | 1.099 |

Significance. codes: 0 '***' 0.001 '**' 0.01 '*' 0.05 '.' 0.1 ' ' 1.

## 5. Conclusions

This study assessed the spatio–temporal variability of drought during the growing season in Ethiopia through VCI, anomaly maps, and trend analysis for the past two decades, from 2001 to 2018. The VCI results identified that severe to extreme countrywide droughts were identified in 2002, 2003, 2004, 2009, 2012, and 2015. On the other hand, the years 2001, 2005, 2006, 2007, 2013, 2016, and 2018 reflected near-normal NDVI throughout most of the rain-fed agriculture regions. These results are coherent with the findings of previous studies in indicating the onset, spatial, and temporal dynamics of agricultural drought in Ethiopia [18]. Pixel-based trend analysis showed that a significant precipitation decrease in the central areas is accompanied by a significant increase in LST. The increase in temperature in the growing season is of major concern, as it implies an increase in evapotranspiration and, thus, affects crop yields. Also, the browning in northwestern parts as estimated from NDVI trends was due to low rainfall and an increase in soil temperature. Furthermore, the anomaly maps for precipitation, soil moisture, and LST help us identify the locations and areas of potential concern regarding reduced crop harvest. We found that large areas of the central highland agricultural farms where people largely depend on rain-fed farms are of major concern due to recurrent drought incidents. Moreover, NDLI has a high correlation with NDVI, precipitation, LST, and soil moisture and successfully captured historical droughts (Figure 8). Additionally, the results of multilinear regression indicate that NDLI, NDWI, and MEI play a significant role in the variability of vegetation health. The analysis shows that using the radiances of green, red, and SWIR, a simplified crop monitoring model with satisfactory accuracy and easiness can be developed. Thus, NDLI can be a tool to help us better understand the vegetation vigor and moisture availability, and subsequently effectively assess large-scale temporal and spatial characteristics of drought.

This analysis can serve as an important input for food security studies and the planning of potential relief measures. However, this approach suffers from the low spatial and temporal resolution satellite images utilized, as this hugely impacts the quality of the trend analysis. Further research on detecting and assessing temporal and spatial trends is needed to offer essential information for planning agencies and government policies to monitor factors that trigger drought and to minimize their impact.

**Author Contributions:** G.M.M. and Y.-A.L. conceived the research, make helpful discussions during the conception of the research. G.M.M. conducted the research, performed analyses, and wrote the first manuscript draft. Y.-A.L. enhanced and finalized the manuscript for the communication with the journal.

**Funding:** This work was supported by the Ministry of Science and Technology under Grant MOST 105-2111-M-008-024-MY2 and Grant 105-2221-E-008-056-MY3.

**Acknowledgments:** The authors would like to thank Ravindra Babu Saginela for the discussion during the conception of the research.

**Conflicts of Interest:** The authors declare no conflict of interest.

## References

1. Singh, N.P.; Bantilan, C.; Byjesh, K. Vulnerability and policy relevance to drought in the semi-arid tropics of Asia—A retrospective analysis. *Weather Clim. Extrem.* **2014**, *3*, 54–61. [CrossRef]
2. Nguyen, A.K.; Liou, Y.-A.; Li, M.-H.; Tran, T.A. Zoning eco-environmental vulnerability for environmental management and protection. *Ecol. Indic.* **2016**, *69*, 100–117. [CrossRef]
3. Liou, Y.-A.; Nguyen, A.K.; Li, M.-H. Assessing spatiotemporal eco-environmental vulnerability by Landsat data. *Ecol. Indic.* **2017**, *80*, 52–65. [CrossRef]
4. Nguyen, K.A.; Liou, Y.A. Global mapping of eco-environmental vulnerability from human and nature disturbances. *Sci. Total Environ.* **2019**, *664*, 995–1004. [CrossRef] [PubMed]
5. Nguyen, K.-A.; Liou, Y.-A. Mapping global eco-environment vulnerability due to human and nature disturbances. *Methods X* **2019**, *6*, 862–875. [CrossRef] [PubMed]
6. Nguyen, K.-A.; Liou, Y.-A.; Terry, J.P. Vulnerability of Vietnam to typhoons: A spatial assessment based on hazards, exposure and adaptive capacity. *Sci. Total Environ.* **2019**, *682*, 31–46. [CrossRef] [PubMed]
7. Sitorus, E.; Nguyen, K.A.; Liou, Y.A. Forest fire impact on and vulnerability assessment of eco-environment in tropical rainforest: A case study of leuser ecosystem-Aceh, Indonesia. In *AGU Fall Meeting Abstracts*; 2018; Available online: http://adsabs.harvard.edu/abs/2018AGUFMGH23B1095S (accessed on 12 April 2019).
8. Cheng, C.-H.; Nnadi, F.; Liou, Y.-A. Energy budget on various land use areas using reanalysis data in Florida. *Adv. Meteorol.* **2014**, *2014*, 1–13. [CrossRef]
9. Hayes, M.; Svoboda, M.D.; Wardlow, B.D.; Anderson, M.; Kogan, F. Drought monitoring:Historical and current perspectives. In *Remote Sensing of Drought: Innovative Monitoring Approaches*; CRC Press: Boca Raton, FL, USA, 2012; pp. 1–19.
10. Kuri, F.; Murwira, A.; Murwira, K.S.; Masocha, M. Predicting maize yield in Zimbabwe using dry dekads derived from remotely sensed Vegetation Condition Index. *Int. J. Appl. Earth Obs. Geoinf.* **2014**, *33*, 39–46. [CrossRef]
11. Jiao, W.; Zhang, L.; Chang, Q.; Fu, D.; Cen, Y.; Tong, Q. Evaluating an enhanced vegetation condition index (VCI) based on VIUPD for drought monitoring in the continental United States. *Remote Sens.* **2016**, *8*, 224. [CrossRef]
12. Townshend, J.R.G.; Justice, C.O. Analysis of the dynamics of African vegetation using the normalized difference vegetation index. *Int. J. Remote Sens.* **1986**, *7*, 1435–1445. [CrossRef]
13. Kogan, F.N. Droughts of the late 1980s in the United States as derived from NOAA polar-orbiting satellite data. *Bull. Am. Meteorol. Soc.* **1995**, *76*, 655–668. [CrossRef]
14. Dorjsuren, M.; Liou, Y.-A.; Cheng, C.-H. Time series MODIS and in situ data analysis for Mongolia drought. *Remote Sens.* **2016**, *8*, 509. [CrossRef]
15. Tadesse, T.; Demisse, G.B.; Zaitchik, B.; Dinku, T. Satellite-based hybrid drought monitoring tool for prediction of vegetation condition in Eastern Africa: A case study for Ethiopia. *Water Resour. Res.* **2014**, *50*, 2176–2190. [CrossRef]
16. Mckee, T.B.; Doesken, N.J.; Kleist, J. The relationship of drought frequency and duration to time scales. In Proceedings of the 8th Conference on Applied Climatology, Anaheim, CA, USA, 17–22 January 1993; American Meteorological Society: Boston, MA, USA, 1993; pp. 179–184.
17. Rouse, J., Jr.; Haas, R.H.; Schell, J.A.; Deering, D.W. Monitoring vegetation systems in the Great Plains with ERTS, Third ERTS Symposium. *NASA* **1973**, *1*, 309–317.
18. Gebrehiwot, T.; Van der Veen, A.; Maathuis, B. Governing agricultural drought: Monitoring using the vegetation condition index. *Ethiop. J. Environ. Stud. Manag.* **2016**, *9*, 354. [CrossRef]
19. Cheng, C.-H.; Nnadi, F.; Liou, Y.-A. A Regional Land Use Drought Index for Florida. *Remote Sens.* **2015**, *7*, 17149–17167. [CrossRef]
20. Wu, D.; Qu, J.J.; Hao, X. Agricultural drought monitoring using MODIS-based drought indices over the USA Corn Belt. *Int. J. Remote Sens.* **2015**, *36*, 5403–5425. [CrossRef]

21. Liu, K.; Su, H.; Tian, J.; Li, X.; Wang, W.; Yang, L.; Liang, H. Assessing a scheme of spatial-temporal thermal remote-sensing sharpening for estimating regional evapotranspiration. *Int. J. Remote Sens.* **2018**, *39*, 3111–3137. [CrossRef]
22. Wan, Z.; Wang, P.; Li, X. Using MODIS Land Surface Temperature and Normalized Difference Vegetation Index products for monitoring drought in the southern Great Plains, USA. *Int. J. Remote Sens.* **2004**, *25*, 61–72. [CrossRef]
23. Du, L.; Tian, Q.; Yu, T.; Meng, Q.; Jancso, T.; Udvardy, P.; Huang, Y. A comprehensive drought monitoring method integrating MODIS and TRMM data. *Int. J. Appl. Earth Obs. Geoinf.* **2013**, *23*, 245–253. [CrossRef]
24. Mutowo, G.; Chikodzi, D. Remote sensing based drought monitoring in Zimbabwe. *Disaster Prev. Manag. An Int. J.* **2014**, *23*, 649–659. [CrossRef]
25. Rojas, O.; Vrieling, A.; Rembold, F. Assessing drought probability for agricultural areas in Africa with coarse resolution remote sensing imagery. *Remote Sens. Environ.* **2011**, *115*, 343–352. [CrossRef]
26. Anderson, W.B.; Zaitchik, B.F.; Hain, C.R.; Anderson, M.C.; Yilmaz, M.T.; Mecikalski, J.; Schultz, L. Towards an integrated soil moisture drought monitor for East Africa. *Hydrol. Earth Syst. Sci.* **2012**, *16*, 2893–2913. [CrossRef]
27. Masih, I.; Maskey, S.; Mussá, F.E.F.; Trambauer, P. A review of droughts on the African continent: A geospatial and long-term perspective. *Hydrol. Earth Syst. Sci.* **2014**, *18*, 3635–3649. [CrossRef]
28. Yang, S.; Meng, D.; Gong, H.; Li, X.; Wu, X. Soil drought and vegetation response during 2001–2015 in North China based on GLDAS and MODIS data. *Adv. Meteorol.* **2018**, *2018*, 1–14. [CrossRef]
29. Tsegay Wolde-Georgis. El Niño and Drought Early Warning in Ethiopia. *Internet J. Afr. Stud.* **1997**. Available online: https://papers.ssrn.com/sol3/papers.cfm?abstract_id=1589710 (accessed on 16 March 2019).
30. Edossa, D.C.; Babel, M.S.; Das Gupta, A. Drought analysis in the Awash River Basin, Ethiopia. *Water Resour. Manag.* **2010**, *24*, 1441–1460. [CrossRef]
31. Kumar, B.G. Ethiopian famines 1973–1985: A case-study. *Polit. Econ. Hunger* **1990**, *2*, 173–216.
32. Gore, T.; Hillier, D. Climate Change and Future Impacts on Food Security. *Oxfam Policy Pract. Agric. Food L.* **2011**, *11*, 57–62.
33. Ayalew, D.; Tesfaye, K.; Mamo, G.; Yitaferu, B.; Bayu, W. Variability of rainfall and its current trend in Amhara region, Ethiopia. *Afr. J. Agric. Res.* **2012**, *7*, 1475–1486.
34. Schmidt, W.; Peter Uhe, A.; Kimutai, J.; Otto, F.; Cullen, H. *Climate and Development Knowledge Network and World Weather Attribution Initiative Raising Risk Awareness*; Royal Netherlands Meteorological Institute: De Bilt, The Netherlands, 2017; pp. 2016–2017.
35. Funk, C.; Peterson, P.; Landsfeld, M.; Pedreros, D.; Verdin, J.; Shukla, S.; Husak, G.; Rowland, J.; Harrison, L.; Hoell, A. The climate hazards infrared precipitation with stations—A new environmental record for monitoring extremes. *Sci. Data* **2015**, *2*, 150066. [CrossRef] [PubMed]
36. Philip, S.; Kew, S.F.; Jan van Oldenborgh, G.; Otto, F.; O'Keefe, S.; Haustein, K.; King, A.; Zegeye, A.; Eshetu, Z.; Hailemariam, K.; et al. Attribution Analysis of the Ethiopian Drought of 2015. *J. Clim.* **2018**, *31*, 2465–2486. [CrossRef]
37. USAID. *El niño in Ethiopia, A Real-Time Review of Impacts and Responses 2015-2016*; USAID: Washington, DC, USA, 2016. Available online: https://www.agri-learning-ethiopia.org/wp-content/uploads/2016/06/AKLDP-El-Nino-Review-March-2016 (accessed on 28 May 2019).
38. Camberlin, P. Rainfall anomalies in the source region of the Nile and their connection with the Indian summer monsoon. *J. Clim.* **1997**, *10*, 1380–1392. [CrossRef]
39. Korecha, D.; Sorteberg, A. Validation of operational seasonal rainfall forecast in Ethiopia. *Water Resour. Res.* **2013**, *49*, 7681–7697. [CrossRef]
40. Liou, Y.-A.; Le, M.S.; Chien, H. Normalized difference latent heat index for remote sensing of land surface energy fluxes. *IEEE Trans. Geosci. Remote Sens.* **2019**, *57*, 1423–1433. [CrossRef]
41. Worqlul, A.W.; Jeong, J.; Dile, Y.T.; Osorio, J.; Schmitter, P.; Gerik, T.; Srinivasan, R.; Clark, N. Assessing potential land suitable for surface irrigation using groundwater in Ethiopia. *Appl. Geogr.* **2017**, *85*, 1–13. [CrossRef]
42. Viste, E.; Korecha, D.; Sorteberg, A. Recent drought and precipitation tendencies in Ethiopia. *Theor. Appl. Climatol.* **2013**, *112*, 535–551. [CrossRef]

43. Terefe, T.; Mengistu, G. Spatial and temporal variability of summer rainfall over Ethiopia from observations and a regional climate model experiment climate model experiments. *Theor. Appl. Climatol.* **2012**, *111*, 665–681.
44. Seleshi, Y.; Camberlin, P. Recent changes in dry spell and extreme rainfall events in Ethiopia. *Theor. Appl. Climatol.* **2006**, *83*, 181–191. [CrossRef]
45. Liebmann, B.; Hoerling, M.P.; Funk, C.; Bladé, I.; Dole, R.M.; Allured, D.; Quan, X.; Pegion, P.; Eischeid, J.K. Understanding recent eastern horn of africa rainfall variability and change. *J. Clim.* **2014**, *27*, 8630–8645. [CrossRef]
46. Gebrehiwot, T.; van der Veen, A.; Maathuis, B. Spatial and temporal assessment of drought in the Northern highlands of Ethiopia. *Int. J. Appl. Earth Obs. Geoinf.* **2011**, *13*, 309–321. [CrossRef]
47. Birhane, E.; Ashfare, H.; Fenta, A.A.; Hishe, H.; Gebremedhin, M.A.; Solomon, N. Land use land cover changes along topographic gradients in Hugumburda national forest priority area, Northern Ethiopia. *Remote Sens. Appl. Soc. Environ.* **2019**, *13*, 61–68. [CrossRef]
48. Larbi, I.; Hountondji, F.; Annor, T.; Agyare, W.; Mwangi Gathenya, J.; Amuzu, J.; Larbi, I.; Hountondji, F.C.C.; Annor, T.; Agyare, W.A.; et al. Spatio-temporal trend analysis of rainfall and temperature extremes in the Vea Catchment, Ghana. *Climate* **2018**, *6*, 87. [CrossRef]
49. Muthoni, F.K.; Odongo, V.O.; Ochieng, J.; Mugalavai, E.M.; Mourice, S.K.; Hoesche-Zeledon, I.; Mwila, M.; Bekunda, M. Long-term spatial-temporal trends and variability of rainfall over Eastern and Southern Africa. *Theor. Appl. Climatol.* **2019**, *137*, 1869–1882. [CrossRef]
50. McNally, A.; Arsenault, K.; Kumar, S.; Shukla, S.; Peterson, P.; Wang, S.; Funk, C.; Peters-Lidard, C.D.; Verdin, J.P. A land data assimilation system for sub-Saharan Africa food and water security applications. *Sci. Data* **2017**, *4*, 170012. [CrossRef]
51. Saji, N.H.; Goswami, B.N.; Vinayachandran, P.N.; Yamagata, T. A dipole mode in the tropical Indian Ocean. *Nature* **1999**, *401*, 360–363. [CrossRef]
52. LIU, W.T.; KOGAN, F.N. Monitoring regional drought using the Vegetation Condition Index. *Int. J. Remote Sens.* **1996**, *17*, 2761–2782. [CrossRef]
53. Winkler, L.; Gessner, U.; Hochschild, V. Identifying droughts affecting agrictlture in Africa based on remote sensing time series between 2000-2016: Rainfall anomalies and vegetation condition in the context of ENSO. *Remote Sens.* **2017**, *9*, 831. [CrossRef]
54. Measho, S.; Chen, B.; Trisurat, Y.; Pellikka, P.; Guo, L.; Arunyawat, S.; Tuankrua, V.; Ogbazghi, W.; Yemane, T. Spatio-temporal analysis of vegetation dynamics as a response to climate variability and drought patterns in the Semiarid Region, Eritrea. *Remote Sens.* **2019**, *11*, 724. [CrossRef]
55. Baniya, B.; Tang, Q.; Xu, X.; Haile, G.G.; Chhipi-Shrestha, G. Spatial and temporal variation of drought based on satellite derived vegetation condition index in Nepal from 1982. *Sensors* **2019**, *19*, 430. [CrossRef]
56. De Jong, R.; de Bruin, S.; de Wit, A.; Schaepman, M.E.; Dent, D.L. Analysis of monotonic greening and browning trends from global NDVI time-series. *Remote Sens. Environ.* **2011**, *115*, 692–702. [CrossRef]
57. Sobrino, J.A.; Julien, Y. Global trends in NDVI-derived parameters obtained from GIMMS data. *Int. J. Remote Sens.* **2011**, *32*, 4267–4279. [CrossRef]
58. Julien, Y.; Sobrino, J.A.; Mattar, C.; Ruescas, A.B.; Jiménez-Muñoz, J.C.; Sòria, G.; Hidalgo, V.; Atitar, M.; Franch, B.; Cuenca, J. Temporal analysis of normalized difference vegetation index (NDVI) and land surface temperature (LST) parameters to detect changes in the Iberian land cover between 1981 and 2001. *Int. J. Remote Sens.* **2011**, *32*, 2057–2068. [CrossRef]
59. Sen, P.K. Estimates of the regression coefficient based on Kendall's Tau. *J. Am. Stat. Assoc.* **1968**, *63*, 1379–1389. [CrossRef]
60. Tian, F.; Wang, Y.; Fensholt, R.; Wang, K.; Zhang, L.; Huang, Y. Remote sensing mapping and evaluation of NDVI trends from synthetic time series obtained by blending landsat and MODIS data around a coalfield on the Loess Plateau. *Remote Sens.* **2000**, *5*, 4255–4279. [CrossRef]
61. Ravindrababu, S.; Ratnam, M.; Basha, G.; Liou, Y.-A.; Reddy, N. Large anomalies in the tropical upper troposphere lower stratosphere (UTLS) trace gases observed during the Extreme 2015–2016 El Niño Event by using satellite measurements. *Remote Sens.* **2019**, *11*, 687. [CrossRef]
62. Anyamba, A.; Glennie, E.; Small, J.; Anyamba, A.; Glennie, E.; Small, J. Teleconnections and Interannual Transitions as Observed in African Vegetation: 2015. *Remote Sens.* **2018**, *10*, 1038. [CrossRef]

63. Cochrane, L.; Bekele, Y.W. Average crop yield (2001–2017) in Ethiopia: Trends at national, regional and zonal levels. *Data Br.* **2018**, *16*, 1025–1033. [CrossRef]
64. Korecha, D.; Barnston, A.G.; Korecha, D.; Barnston, A.G. Predictability of June–September rainfall in Ethiopia. *Mon. Weather Rev.* **2007**, *135*, 628–650. [CrossRef]
65. Workie, T.G.; Debella, H.J. Climate change and its effects on vegetation phenology across ecoregions of Ethiopia. *Glob. Ecol. Conserv.* **2018**, *13*, e00366. [CrossRef]
66. Yan, D.; Xu, T.; Girma, A.; Yuan, Z.; Weng, B.; Qin, T.; Do, P.; Yuan, Y.; Yan, D.; Xu, T.; et al. Regional Correlation between precipitation and vegetation in the Huang-Huai-Hai River Basin, China. *Water* **2017**, *9*, 557. [CrossRef]
67. Zhao, W.; Zhao, X.; Zhou, T.; Wu, D.; Tang, B.; Wei, H. Climatic factors driving vegetation declines in the 2005 and 2010 Amazon droughts. *PLoS ONE* **2017**, *12*, e0175379. [CrossRef]
68. Degefu, M.A.; Rowell, D.P.; Bewket, W. Teleconnections between Ethiopian rainfall variability and global SSTs: Observations and methods for model evaluation. *Meteorol. Atmos. Phys.* **2017**, *129*, 173–186. [CrossRef]
69. Greenland, S.; Senn, S.J.; Rothman, K.J.; Carlin, J.B.; Poole, C.; Goodman, S.N.; Altman, D.G. Statistical tests, P values, confidence intervals, and power: A guide to misinterpretations. *Eur. J. Epidemiol.* **2016**, *31*, 337–350. [CrossRef]
70. Wu, D.; Zhao, X.; Liang, S.; Zhou, T.; Huang, K.; Tang, B.; Zhao, W. Time-lag effects of global vegetation responses to climate change. *Glob. Chang. Biol.* **2015**, *21*, 3520–3531. [CrossRef]

© 2019 by the authors. Licensee MDPI, Basel, Switzerland. This article is an open access article distributed under the terms and conditions of the Creative Commons Attribution (CC BY) license (http://creativecommons.org/licenses/by/4.0/).

Article
# Multiple Kernel Feature Line Embedding for Hyperspectral Image Classification

Ying-Nong Chen [1,2]

1. Center for Space and Remote Sensing Research, National Central University, No. 300, Jhongda Rd., Jhongli Dist., Taoyuan City 32001, Taiwan; yingnong1218@csrsr.ncu.edu.tw; Tel.: +886-3-4227151 (ext. 57653)
2. Department of Computer Science and Information Engineering, National Central University, No. 300, Jhongda Rd., Jhongli Dist., Taoyuan City 32001, Taiwan

Received: 13 November 2019; Accepted: 3 December 2019; Published: 4 December 2019

**Abstract:** In this study, a novel multple kernel FLE (MKFLE) based on general nearest feature line embedding (FLE) transformation is proposed and applied to classify hyperspectral image (HSI) in which the advantage of multple kernel learning is considered. The FLE has successfully shown its discriminative capability in many applications. However, since the conventional linear-based principle component analysis (PCA) pre-processing method in FLE cannot effectively extract the nonlinear information, the multiple kernel PCA (MKPCA) based on the proposed multple kernel method was proposed to alleviate this problem. The proposed MKFLE dimension reduction framework was performed through two stages. In the first multple kernel PCA (MKPCA) stage, the multple kernel learning method based on between-class distance and support vector machine (SVM) was used to find the kernel weights. Based on these weights, a new weighted kernel function was constructed in a linear combination of some valid kernels. In the second FLE stage, the FLE method, which can preserve the nonlinear manifold structure, was applied for supervised dimension reduction using the kernel obtained in the first stage. The effectiveness of the proposed MKFLE algorithm was measured by comparing with various previous state-of-the-art works on three benchmark data sets. According to the experimental results: the performance of the proposed MKFLE is better than the other methods, and got the accuracy of 83.58%, 91.61%, and 97.68% in Indian Pines, Pavia University, and Pavia City datasets, respectively.

**Keywords:** manifold learning; hyperspectral image classification; feature line embedding; kernelization; multiple kernel learning

## 1. Introduction

In this big data era, deep learning has shown its convincing capabilities for providing effective solutions to the crucial areas such as hyperspectral image (HSI) classification [1], object detection [2], and face recognition [3]. Deep learning algorithms can extract substantial information and features from huge amount of data; however, if there is not a suitable dimensionality reduction (DR) algorithm to reduce the dimension of the training data effectively, the performance of deep learning algorithms could be seriously impacted [1]. Therefore, the DR algorithm has potential to improve the performance and explainability of deep learning algorithms.

Since most of the HSI are with high-dimensional spectral and abundant spectral bands, DR in HSI classification has been a critical issue. The major problem is that the spectral patterns of HSI are too similar to identify them clearly. Therefore, a powerful DR which can construct a high-dimensional discriminative space and preserve the manifold of discriminability in low-dimensional space is an essential step for HSI classification.

Recently, abundant DR schemes have been presented which could be grouped into three categories: global-based analysis, local-based analysis, and kernel-based analysis. In the global-based analysis category, those using subtracting the mean of population or mean of class from individual samples to obtain the scatter matrix, and try to extract a projection matrix to minimize or maximize the covariance matrix, include principal component analysis (PCA) [4], linear discriminant analysis (LDA) [5], and discriminant common vectors (DCV) [6]. In these methods, all the scatters of samples are demonstrated in the global Euclidean structure, which means that while samples are distributed in a Gaussian function or are linearly separated, these global-based analysis algorithms demonstrate superior capability in DR or classification. However, while the scatter of samples is distributed in a nonlinear structure, the performance of these global measurement algorithms would be seriously impacted since that in a space with high-dimension, samples' local structure is not apparent. In addition, the critical issue about global-based analysis methods is that, while the decision boundaries are predominantly nonlinear, the classification performance would decline sharply [7].

In the local-based analysis category, those using subtracting one sample from the other neighboring sample to obtain the scatter matrix, which is also termed manifold learning, can preserve the structure of locality of the samples. He et al. [8] presented the locality preserving projection (LPP) algorithm to keep the structure of locality of training data to identify faces. Because LPP applies the relationship between neighbors to reveal sample scatter, the local manifold of samples is kept and outperforms those in the category of global-based analysis methods. Tu et al. [9] proposed the Laplacian eigenmap (LE) algorithm, in which the polarimetric synthetic aperture radar data were applied to classify the land cover. The LE method preserves the manifold structure of polarimetric space with high-dimension into an intrinsic space with low-dimension. Wang and He [10] applied LPP as a data pre-processing step in classifying HSI. Kim et al. [11] proposed the locally linear embedding (LLE)-based method for DR in HSI. Li et al. [12,13] proposed the local Fisher discriminant analysis (LFDA) algorithm which considers the advantages of LPP and LDA simultaneously for reducing the dimension of HSI. Luo et al. [14] presented a neighborhood preserving embedding (NPE) algorithm, which was a supervised method for extracting salient features for classifying HSI data. Zhang et al. [15] presented a sparse low-rank approximation algorithm for manifold regularization, which takes the HSI as cube data for classification. These local-based analysis schemes all preserve the manifold of samples and outperform the conventional global-based analysis methods.

The kernel-based analysis category, in spite of the local-based analysis methods, has achieved a better result than those global-based ones. However, based on Boots and Gordon [16], the practical application of manifold learning was still constrained by noises due to that the manifold learning could not extract nonlinear information. Therefore, those using the idea of kernel tricks to generate nonlinear feature space and improve extracting the nonlinear information are kernel-based analysis methods. Since a suitable kernel function could improve a given method on performance [17]. Therefore, both categories of global-based and local-based analysis methods adopted the kernelization approaches to improve the performance of classifying HSI. Boots and Gordon [16] investigated the kernelization algorithm to mitigate the effect of noise to manifold learning. Scholkopf et al. [18] presented a kernelization PCA (KPCA) algorithm which can find a high-dimensional Hilbert space via kernel function and extract the salient non-linear features that PCA missed. In addition to single kernel, Lin et al. [19] proposed a multiple kernel learning algorithm for DR. The multiple kernel function was integrated, and the revealed multiple feature of data was shown in a low dimensional space. However, it tried to find suitable weights for kernels and DR simultaneously, which leads to a more complicated method. Therefore, Nazarpour and Adibi [20] proposed a kernel learning algorithm concentrating only on learning good kernel from some basic kernel. Although this method proposed an effective, simple idea for multiple kernel learning, it applied the global-based kernel discriminant analysis (KDA) method for classification, therefore, it could not preserve the manifold structure of high dimensional multiple kernel space; moreover, a combined kernel scheme, where multiple kernels were linearly assembled to extract both spatial and spectral information [21]. Chen et al. [22] proposed a kernel

method based on sparse representation to classify HSI data. A query sample was revealed by all
training data in a generated kernel space, and pixels in a neighboring area were also described by all
training samples in a linear combination. Resembling the multiple kernel method, Zhang et al. [23]
presented a multiple-features assembling algorithm for classifying HSI data, which integrated texture,
shape, and spectral information to improve the performance of HSI classification.

In previous works, the idea of nearest feature line embedding (FLE) was successfully applied in
reducing dimension on face recognition [24] and classifying HSI [25]. However, the abundant nonlinear
structures and information could not be efficiently extracted using only the linear transformation
and single kernel. Multple kernel learning is an effective tool for enhancing the nonlinear spaces
by integrating many kernels into a new consistent kernel. In this study, a general nearest FLE
transformation, termed multple kernel FLE (MKFLE), was proposed for feature extraction (FE) and DR
in which multiple kernel functions were simultaneously considered. In addition, the support vector
machine (SVM) was applied in the proposed multiple kernel learning strategy which uses only the
support vector set to determine the weight of each valid kernel function. Moreover, three benchmark
data sets were evaluated in the experimental analysis in this work. The performance of the proposed
algorithm was evaluated by comparison with state-of-the-art methods.

The rest of this study is organized as follows: The related works are reviewed in Section 2.
The proposed multple kernel learning method is introduced and incorporated into the FLE algorithm
in Section 3. Some experimental results and comparisons with some state-of-the-art algorithms for
classifying HSI are conducted to demonstrate the effectiveness of the proposed algorithm as introduced
in Section 4. Finally, in Section 5, conclusions are given.

## 2. Related Works

In this paper, FLE [24,25] and multple kernel learning were integrated to reduce the dimensions of
features for classifying the HSI data. A brief review of FLE, kernelization, and multple kernel learning
are introduced in the following before the proposed methods. Assume that $N$ $d$-dimensional training
data $X = [x_1, x_2, \ldots, x_N] \in R^{d \times N}$ consisting of $N_C$ land-cover classes $C_1, C_2, \ldots, C_{N_C}$. The projected
samples in low-dimensional space could be obtained by the linear projection $y_i = w^T x_i$, where $w$ is an
obtained linear transformation for dimension reduction.

### 2.1. Feature Line Embedding (FLE)

FLE is a local-based analysis for DR in which the sample data scatters could be shown in a form of
Laplacian matrix to preserve the locality by applying the strategy of point-to-line. The cost function of
FLE is minimized and defined as follows:

$$
\begin{aligned}
O &= \sum_i \left( \sum_{i \neq m \neq n} \|y_i - L_{m,n}(y_i)\|^2 l_{m,n}(y_i) \right) \\
O &= \sum_i \left( \sum_{i \neq m \neq n} \|y_i - L_{m,n}(y_i)\|^2 l_{m,n}(y_i) \right) \\
&= \sum_i \|y_i - \sum_j M_{i,j} y_j\|^2 \\
&\rightleftarrows tr\left(Y(I-M)^T(I-M)Y\right) = tr\left(w^T X(D-W)X^T w\right) \\
&= tr\left(w^T X L X^T w\right),
\end{aligned}
\quad (1)
$$

where point $L_{m,n}(y_i)$ is a projection sample on line $L_{m,n}$ for sample $y_i$, and weight $l_{m,n}(y_i)$ (being 0 or
1) describes the connection between point $y_i$ and the feature line $L_{m,n}$ which two samples $y_m$ and $y_n$
passes through. The projection sample $L_{m,n}(y_i)$ is described as a linear combination of points $y_m$ and
$y_n$: $L_{m,n}(y_i) = y_m + t_{m,n}(y_n - y_m)$, that $t_{m,n} = (y_i - y_m)^T(y_m - y_n)/(y_m - y_n)^T(y_m - y_n)$, and $i \neq m \neq n$.
Applying some simple operations of algebra, the discriminant vector from sample $y_i$ to the projection
sample $L_{m,n}(y_i)$. could be described as $y_i - \sum_j M_{i,j} y_j$, where two elements in the $i$th row in matrix
$M$ are viewed as $M_{i,m} = t_{n,m}$, $M_{i,n} = t_{m,n}$, and $t_{n,m} + t_{m,n} = 1$, while weight $l_{m,n}(y_i) = 1$. The other

elements in the $i$'th row are set as 0, if $j \neq m \neq n$. In Equation (1), the mean of squared distance for all training data samples to their nearest feature lines (NFLs) is then extracted as $tr(w^T X L X^T w)$, that $L = D - W$, and matrix $D$ expresses the column sums of the similarity matrix $W$. According to the summary of Yan et al. [26], matrix $W$ is expressed as $W_{i,j} = (M + M^T - M^T M)_{i,j}$, while $i \neq j$ is zero otherwise; $\sum_j M_{i,j} y_j = 1$. Matrix $L$ in Equation (1) could be expressed as a Laplacian form. More details could be referred to [24,25].

In supervised FLE, the label information is considered, and there are two parameters, $N_1$ and $N_2$, determined manually in obtaining the within-class matrix $S_{FLEw}$ and the between-class matrix $S_{FLEb}$, respectively:

$$S_{FLEw} = \sum_{c=1}^{N_C} \left( \sum_{x_i \in C_c} \sum_{L_{m,n} \in F_{N_1}(x_i, C_c)} (x_i - L_{m,n}(x_i))(x_i - L_{m,n}(x_i))^T \right), \text{ and} \quad (2)$$

$$S_{FLEb} = \sum_{c=1}^{N_C} \left( \sum_{x_i \in C_c} \sum_{l=1, l \neq c}^{N_C} \sum_{L_{m,n} \in F_{N_2}(x_i, C_l)} (x_i - L_{m,n}(x_i))(x_i - L_{m,n}(x_i))^T \right), \quad (3)$$

where $F_{N_1}(x_i, C_k)$ represents the set of $N_1$ NFLs within the same class, $C_c$, of point $x_i$, i.e., $l_{m,n}(y_i) = 1$, and $F_{K_2}(x_i, C_l)$ is a set of $N_2$ NFLs from different classes of point $x_i$. Then, the Fisher criterion $tr(w^T S_{FLEb} w / w^T S_{FLEw} w)$ is applied to be maximized and extract the transformation matrix $w$, which is constructed of the eigenvectors with the corresponding largest eigenvalues. Finally, a new sample in the low-dimensional space can be represented by the linear projection $y = w^T x$, and the nearest neighbor (one-NN) template matching rule is used for classification.

## 2.2. Kernelization

Kernelization is a function that maps a linear space $X$ to a nonlinear Hilbert space $H$, $\varphi : x \in X \to \varphi(x) \in H$, the conventional within-class and between-class matrix of LDA in space $H$ can be represented as:

$$S_{LDAw}^{\varphi} = \sum_{k=1}^{N_C} \left( \sum_{x_i \in C_k} (\varphi(x_i) - \overline{\varphi}_k)(\varphi(x_i) - \overline{\varphi}_k)^T \right), \text{ and} \quad (4)$$

$$S_{LDAb}^{\varphi} = \sum_{k=1}^{N_C} (\overline{\varphi}_k - \overline{\varphi})(\overline{\varphi}_k - \overline{\varphi})^T. \quad (5)$$

Here, $\overline{\varphi}_k = \frac{1}{n_k} \sum_{i=1}^{n_k} \varphi(x_i)$ and $\overline{\varphi} = \frac{1}{N} \sum_{i=1}^{N} \varphi(x_i)$ indicate mean of the class and mean of population in space $H$, respectively. In order to generalize the within-class and between-class scatters to the nonlinear version, the dot product kernel trick is used exclusively. The representation of dot product on the Hilbert space $H$ is given by the kernel function in the following: $k(x_i, x_j) = k_{i,j} = \varphi^T(x_i) \varphi(x_j)$. Considering the symmetric matrix $K$ of $N$ by $N$ be a matrix constructed by dot product in high dimensional feature space $H$, i.e., $K(x_i, x_j) = \langle \varphi(x_i) \cdot \varphi(x_j) \rangle = (k_{i,j})$ and, $i, j = 1, 2, \ldots, N$. Based on the kernel trick, the kernel operator $K$ makes the development of the linear separation function in space $H$ to be equivalent to that of the nonlinear separation function in space $X$. Kernelization can be applied in maximizing the between-class matrix and minimizing the within-class matrix, too, i.e., $max(w^T S_{LDAb}^{\varphi} w / w^T S_{LDAw}^{\varphi} w)$. This maximization is equal to the conventional eigenvector resolution: $\lambda S_{LDAw}^{\varphi} w = S_{LDAb}^{\varphi} w$, in which a set of eigenvalue $\alpha$ for $w = \sum_{i=1}^{N} \alpha_i \varphi(x_i)$ can be found so that the largest one obtains the maximum of the matrix quotient $\lambda = w^T S_{LDAb}^{\varphi} w / w^T S_{LDAw}^{\varphi} w$.

## 3. Multiple Kernel Feature Line Embedding (MKFLE)

Based on the analyses mentioned above, a suitable DR scheme can effectively generate discriminant non-linear space and preserve the discriminability of manifold structure into low dimensional feature space. Therefore, multiple kernel feature line embedding (MKFLE) was presented for classifying the HSI. The original idea of MKFLE is integrating the multiple kernel learning with the manifold learning method. The combination of multiple kernels not only effectively constructs the manifold of the original data from multiple views, but also increases the discriminability for DR. Then, the following manifold learning-based FLE scheme preserves the locality information of samples in the constructed Hilbert space. FLE has been applied in classifying HSI successfully. High-degree non-linear data geometry limits the effectiveness of locality preservation of manifold learning. Therefore, multiple kernel learning is applied as introduced in the following to mitigate the problem.

## 3.1. Multiple Kernel Principle Component Analysis (MKPCA)

In general, the multiple kernel learning is to transform the representation of samples in original feature space into the optimization of weights $\{\beta_m\}_{m=1}^{M}$ for a valid set of basic kernels $\{k_m\}_{m=1}^{M}$ based on their importance. The aim is to construct a new kernel $K$ via a linear combination of valid kernels as follows:

$$K = \sum_{m=1}^{M} \beta_m k_m, \ \beta_m \geq 0 \text{ and } \sum_{m=1}^{M} \beta_m = 1. \tag{6}$$

Then, a new constructed combined kernels function can be described as below:

$$k(x_i, x_j) = \sum_{m=1}^{M} \beta_m k_m(x_i, x_j), \ \beta_m \geq 0. \tag{7}$$

In this study, eight kernel functions all of the type Radial basis function (RBF), but with different distance functions and different parameters, are used as basic kernels. Therefore, there is no need to perform the kernel alignment or unify different kernels into the same dimension. While the optimization weights $\{\beta_m\}_{m=1}^{M}$ are determined, a new constructed kernel $K$ would be obtained. Let $\varphi : x \in X \rightarrow \varphi(x) \in H$ be a mapping function of kernel $K$ from feature space in low-dimension to Hilbert space $H$ in high-dimension. Denote $\Phi = [\varphi(x_1), \varphi(x_2), \ldots, \varphi(x_N)]$, and $\overline{\varphi} = \frac{1}{N}\sum_{i=1}^{N} \varphi(x_i)$. Without loss of generality, suppose that the training data are normalized in $H$, i.e., $\overline{\varphi} = 0$, then the total scatter matrix is described as $S_t^\varphi = \sum_{i=1}^{N}(\varphi_i - \overline{\varphi})(\varphi_i - \overline{\varphi})^T = \Phi\Phi^T$. In the proposed MKPCA, the criterion demonstrated in Equation (8) is applied to extract the optimal projective vector $v$:

$$J(v) = \sum_{i=1}^{N} \|v^T \varphi(x_i)\| = v^T S_t^\varphi v. \tag{8}$$

Then the solution for Equation (8) would be the eigenvalue problem: $\lambda v = S_t^\varphi v$ where $\lambda \geq 0$ and eigenvectors $v \in H$. Therefore, Equation (8) could be described as an equal problem:

$$\lambda q = Kq, \tag{9}$$

in which $K = \Phi^T \Phi$ is the kernel matrix. Assuming that $\{q_1, q_2, \ldots q_b\}$ are the corresponding $b$ eigenvalues to the Equation (9); then $v_i = \Phi q_i$ is the solution of Equation (8). Since the proposed MKPCA algorithm is a kind of modified KPCA, its kernel is a constructed ensemble of multiple kernels via a learned weighted combination. Therefore, the MKPCA based FE or DR needs only kernel functions in the input space instead of applying any nonlinear mapping $\varphi$ as kernel method. Furthermore, since each different data set has the nature of data itself, applying a fixed ensemble kernel for different applications would limit the performance. Therefore, an optimal weighted combination of all valid subkernels based on their separability is introduced in the following.

## 3.2. Multiple Kernel Learning based on Between-Class Distance and Support Vector Machine

In the proposed MKPCA, the new ensemble kernel function $K$ applied in Equation (6) is obtained through a linear combination of $M$ valid subkernel function, and $\beta_m$ is the weight of '$m$'th subkernel in the combination, which should be learned from the training data. Applying multiple kernels improves to extract the most suitable kernel function for the data of different applications. In this study, a new multiple kernel learning method is proposed to determine the kernel weight vector $\beta = [\beta_1, \beta_2, \ldots, \beta_M]$ based on the between-class distance and SVM.

Since the goal of the proposed MKFLE is for discrimination, our idea for optimization of the kernel weight vector $\beta$ is based on the maximizing between-class distance criterion as follows:

$$J_1(\beta) = tr(S_b^\varphi), \tag{10}$$

with

$$tr(S_b^\varphi) = tr\left(\sum_{i=1}^{N_c-1} \sum_{j=i+1}^{N_c} (\overline{\varphi}_i - \overline{\varphi}_j)(\overline{\varphi}_i - \overline{\varphi}_j)^T\right). \tag{11}$$

With $tr(AB) = tr(BA)$, Equation (11) could be described as follows:

$$\begin{aligned}
tr(S_b^\varphi) &= \sum_{i=1}^{N_c-1} \sum_{j=i+1}^{N_c} (\overline{\varphi}_i - \overline{\varphi}_j)^T(\overline{\varphi}_i - \overline{\varphi}_j) \\
&\rightleftarrows \sum_{i=1}^{N_c-1} \sum_{j=i+1}^{N_c} r_i r_j \left[1_i^T K_{i,i} 1_i - 2 1_i^T K_{i,j} 1_j + 1_j^T K_{j,j} 1_j\right] \\
&\rightleftarrows \sum_{m=1}^{M} \sum_{i=1}^{N_c-1} \sum_{j=i+1}^{N_c} r_i r_j \left[1_i^T K_{i,i}^m 1_i - 2 1_i^T K_{i,j}^m 1_j + 1_j^T K_{j,j}^m 1_j\right] \beta_m = B^T \beta,
\end{aligned} \tag{12}$$

in which $r_i = n_i/N$, $n_i$ is the amount of samples in $i$'th class, and $K_{i,j}$, $K_{i,j}^m$, $1_i$, are described as follows:

$$K_{i,j} = \sum_{m=1}^{M} \beta_m K_{i,j}^m, \ \beta_m \geq 0 \text{ and } \sum_{m=1}^{M} \beta_m = 1, \tag{13}$$

$$K_{i,j}^m = \begin{bmatrix} k_m(x_1^i, x_1^j) & \cdots & k_m(x_1^i, x_{n_j}^j) \\ \vdots & \ddots & \vdots \\ k_m(x_{n_i}^i, x_1^j) & \cdots & k_m(x_{n_i}^i, x_{n_j}^j) \end{bmatrix}, \tag{14}$$

$$1_i = \begin{bmatrix} 1/n_i \\ \vdots \\ 1/n_i \end{bmatrix}_{n_i \times 1}, \tag{15}$$

where $B$ is a $M \times 1$ vector, in which the elements are the traces of the between-class matrices of $M$ different kernels, and $\beta$ is a vector, in which the elements are the weights of subkernels.

The between-class distance is well for measurement of discrimination, however, while the $n_i$ and $n_j$ increase, the generalization of $\overline{\varphi}_i - \overline{\varphi}_j$ would decrease. To solve this problem, inspired from the SVM, support vectors between two classes are taken into consideration for computation of the between-class distance. In other words, since the support vectors are much more representative of the class for discrimination, only the support vectors between two classes are used for computation of the between-class distance to improve the generalization of $\overline{\varphi}_i - \overline{\varphi}_j$. Thus, based on the criterion in Equation (10), the integration of between-class distance and SVM is used as a criterion to find the optimal $\beta$, defined as follow:

$$J_2(\beta) = tr(S_b^{\varphi \ SV}), \tag{16}$$

where $S_b^{\varphi\ SV}$ is the between-scatter matrix formed by the support vectors between classes. Therefore, in a similar manner, the optimization problem in Equation (12) could be re-described as follows:

$$tr\left(S_b^{\varphi\ SV}\right) = \sum_{m=1}^{M} \sum_{i=1}^{N_c-1} \sum_{j=i+1}^{N_c} r_i^{SV} r_j^{SV} \left[ 1_i^{TSV} K_{i,i}^{mSV} 1_i^{SV} - 2 1_i^{TSV} K_{i,j}^{mSV} 1_j^{SV} + 1_j^{TSV} K_{j,j}^{mSV} 1_j^{SV} \right] \beta_m = B^T \beta, \quad (17)$$

where $r_i^{SV} = n_i^{SV}/N^{SV}$, $n_i^{SV}$ is the amount of support vectors of $i$'th class, and $N^{SV}$ is the amount of support vectors in all classes. The difference between criterion $J_2(\beta)$ and $J_1(\beta)$ is that the $J_2(\beta)$ applies only the support vectors between classes while the $J_1(\beta)$ uses all samples in the classes. Using Equation (17), the optimization problem is formulated as follows:

$$\max_{\beta} B^T \beta, \text{ subject to } \beta_m \geq 0 \text{ and } \sum_{m=1}^{M} \beta_m = 1. \quad (18)$$

In the optimization problem mentioned in Equation (18), each kernel is supposed to be a Mercer kernel. Therefore, the linear combination of these kernels is still a Mercer kernel. In addition, the sum of these weights is subject to be equal to one. Thus, the optimization problem of (18) is a linear programming (LP) problem which could be solved by a Lagrange optimization procedure. In this study, the proposed MKPCA applies $J_2(\beta)$ as multiple kernel learning criterion to find the optimal weights for subkernels.

In addition, radial basis function (RBF) kernel with Euclidian distance is applied as the kernel function of the method of single kernel function, such as Fuzzy Kernel Nearest Feature Line Embedding (FKNFLE), and KNFLE in [25]. In the proposed MKL scheme, eight kernel functions all of the RBF type [20] are applied with different distance measurements and different kernel parameters. The RBF kernel is defined as follows:

$$K_m(i,j) = k_m(x_i, x_j) = \exp\left( \frac{-d_m^2(x_i, x_j)}{\sigma_m^2} \right), \quad (19)$$

where $d(.,.)$ represents the distance function. There are four distance functions applied in the proposed MKL scheme, the first is the Euclidean distance function as follows:

$$d_m(x_i, x_j) = \sqrt{\left(x_i^1 - x_j^1\right)^2 + \cdots \left(x_i^d - x_j^d\right)^2}. \quad (20)$$

The second is the L1 distance function defined as follows:

$$d_m(x_i, x_j) = \sum_{l=1}^{d} \left| x_i^l - x_j^l \right|. \quad (21)$$

The third is the cosine distance function defined as follows:

$$d_m(x_i, x_j) = \cos(\theta) = \frac{x_i \cdot x_j}{\|x_i\| \|x_j\|}. \quad (22)$$

The fourth is the Chi-squared distance function defined as follows:

$$d_m(x_i, x_j) = \sum_{l=1}^{d} \frac{\left(x_i^1 - x_j^1\right)^2}{\left(x_i^1 + x_j^1\right)^2}. \quad (23)$$

In Equation (19), $\sigma_m$ is the kernel parameter, which could be obtained by the method in [20]. In this study, four kernels of Equations (20)–(23), their kernel parameter $\sigma_m$ are obtained by homoscedasticity method [20], and the other four kernels also apply the Equations (20)–(23) but with the mean of all distances as kernel parameters.

### 3.3. Kernelization of FLE

In the proposed MKFLE algorithm, the MKPCA is firstly performed to construct the new kernel via the proposed multiple kernel learning method. Then, all training points are projected into the Hilbert space $H$ based on the new ensemble kernel. After that, the FLE algorithm based on the manifold learning is performed to compute the mean of squared distance for total training samples to their nearest feature lines in high-dimensional Hilbert space, and which can be expressed as follows:

$$\begin{aligned}\sum_i \|\varphi(y_i) - L_{m,n}(\varphi(y_i))\|^{(2)} &= \sum_i \left\|\varphi(y_i) - \sum_j M_{i,j}\varphi(y_j)\right\|^2 \\ &= tr\left(\varphi^T(Y)(I-M)^T(I-M)\varphi(Y)\right) \\ &= tr\left(\varphi^T(Y)(D-W)\varphi(Y)\right) \\ &= tr\left(w^T\varphi(X)L\varphi^T(X)w\right).\end{aligned} \quad (24)$$

Then, the object function in Equation (24) could be described as a minimum problem and represented in a Laplacian form. The eigenvector problem of kernel FLE in the Hilbert space is represented as:

$$[\varphi(X)L\varphi^T(X)]w = \lambda[\varphi(X)D\varphi^T(X)]w. \quad (25)$$

To expand the applications of FLE algorithm to kernel FLE, the implicit feature vector, $\varphi(x)$, has no necessity to be calculated practically. The inner product representation of two data points in the Hilbert space is exclusively used with a kernel function as follows: $K(x_i, x_j) = \langle \varphi(x_i), \varphi(x_j) \rangle$. The eigenvectors of Equation (25) are described by the linear combinations of $\varphi(x_1), \varphi(x_2), \ldots, \varphi(x_N)$. The coefficient $\alpha_i$ is $w = \sum_{i=1}^{N} \alpha_i \varphi(x_i) = \varphi(X)\alpha$ where $\alpha = [\alpha_1, \alpha_2, \ldots, \alpha_N]^T \in R^N$. Then, the eigenvector problem is represented as follows:

$$KLK\alpha = \lambda KDK\alpha. \quad (26)$$

Assuming that the solutions of Equation (26) are the coefficient vectors, $\alpha^1, \alpha^2, \ldots, \alpha^N$ in a column format. Given a querying sample, $z$, and its projection on the eigenvectors, $w^k$, are computed by the following equation:

$$(w^k \cdot \varphi(z)) = \sum_{i=1}^{N} \alpha_i^k \langle \varphi(z), \varphi(x_i) \rangle = \sum_{i=1}^{N} \alpha_i^k K(z, x_i), \quad (27)$$

where $\alpha_i^k$ is the $i$th element of the coefficient vector, $\alpha^k$. The kernel function RBF (radial basis function) is used in this study. Thus, the within-class scatters and the between-class scatters in a kernel space are defined as follows:

$$S_{FLEw}^{\varphi} = \sum_{c=1}^{N_c} \left( \sum_{\varphi(x_i) \in C_c} \sum_{L_{m,n} \in F_{N_1}(\varphi(x_i), C_c)} (\varphi(x_i) - L_{m,n}(\varphi(x_i)))(\varphi(x_i) - L_{m,n}(\varphi(x_i)))^T \right), \text{ and} \quad (28)$$

$$S_{FLEb}^{\varphi} = \sum_{c=1}^{N_c} \left( \sum_{\varphi(x_i) \in C_c} \sum_{l=1, l \neq c}^{N_c} \sum_{L_{m,n} \in F_{N_2}(\varphi(x_i), C_l)} (\varphi(x_i) - L_{m,n}(\varphi(x_i)))(\varphi(x_i) - L_{m,n}(\varphi(x_i)))^T \right). \quad (29)$$

Since the Hilbert space $H$ constructed by the proposed MKPCA is an ensemble kernel space from multiple subkernels, there would be abundant useful non-linear information from different views for discrimination. Hence, applying the kernelized FLE to preserve those non-linear local structure in MKPCA would improve the performance of FE and DR. The pseudo-codes of the proposed MKFLE algorithm are tabulated in Table 1. In this study, it is proposed that a general form of the FLE method using the SVM-based multple kernel learning be used for FE and DR. The benefits of the proposed MKFLE are twofold: the multple kernel learning scheme based on the SVM strategy can generalize the optimal combination of weights; and the kernelized FLE algorithm based on the manifold learning can preserve the local structure information in high dimensional constructed multple kernel space as well as the manifold local structure in the dimension reduced space.

Table 1. The procedures of MKFLE (multple kernel feature line embedding) algorithm.

| | |
|---|---|
| Input: | A $d$-dimensional training data $X = [x_1, x_2, \ldots, x_N]$ consists of $Nc$ classes. |
| Output: | The projection transformation $w$. |
| Step 1: | Create $M$ kernels using Equation (19). |
| Step 2: | Apply the SVM algorithm to extract support vectors between classes for the criterion (16). |
| Step 3: | Determine the vector $\beta$ via solving the LP optimization problem of Equation (18). |
| Step 4: | Create a new kernel as linear combination of subkernels using Equation (7). |
| Step 5: | Project the $X$ into the new created kernel space $\varphi(X) = [\varphi(x_1), \varphi(x_2), \ldots, \varphi(x_N)]$. |
| Step 6: | PCA projection: Data points are projected from a space with high-dimension into a subspace with low-dimension by matrix $w_{PCA}$. |
| Step 7: | Calculation of the within-class matrix and between-class matrix using Equations (28) and (29), respectively. |
| Step 8: | Maximization of Fisher criterion: Fisher criterion $w^* = \arg\max S_{FLEb}^{\varphi}/S_{FLEw}^{\varphi}$ is maximized to extract the best projection matrix, which is composed of $\gamma$ eigenvectors with the largest eigenvalues. |
| Step 9: | Output the projection matrix: $w = w_{PCA} w^*$. |

## 4. Experimental Results

### 4.1. Data Sets Description

In this sub-section, in order to evaluate the effectiveness of the proposed MKFLE algorithm, some experimental results are conducted for classifying HSI. Three classic HSI benchmarks are applied for evaluation. The use-case of the three chosen images for evaluation are framed into the HSI analysis for land covers. The first data set was obtained from AVIRIS (Airborne Visible/Infrared Imaging Spectrometer), obtained by the Jet Propulsion Laboratory and NASA/Ames in 1992, and termed Indian Pines Site (IPS) image. This IPS image was collected from six miles in the western area of Northwest Tippecanoe County (NTC). A false color IR image of IPS dataset was shown in Figure 1a. The are 16 land-cover classes with 220 bands in the IPS dataset, e.g., Corn-notill(1428), Alfalfa(46), Corn(237), Corn-mintill(830), Grass-pasture(483), Grass-trees(730), Grass-pasture-mowed(28), Hay-windrowed(478), Oats(20), Soybeans-notill(972), Soybeans-mintill(2455), Soybeans-cleantill(593), Woods(1265), Wheat(205), Stone-Steel-Towers(93), and Bldg-Grass-Tree-Drives(386). The numbers shown in the parentheses were the pixel numbers in this IPS dataset. There were 10,249 pixels in the IPS dataset, and the ground truths for each pixel were labeled manually for testing and training. To evaluate the effectiveness of various algorithms, 10 classes with more than 300 samples were used in the experiments, e.g., a subset IPS-10 of 9620 pixels. There were nine hundred training samples in this IPS-10 subset chosen randomly from 9620 pixels, and the other remaining samples were applied for testing. In addition, all the tests were executed on a twelve-core intel i7-8700k CPU, Matlab (Mathworks) 2016b, Microsoft Win10, and 32gb ram.

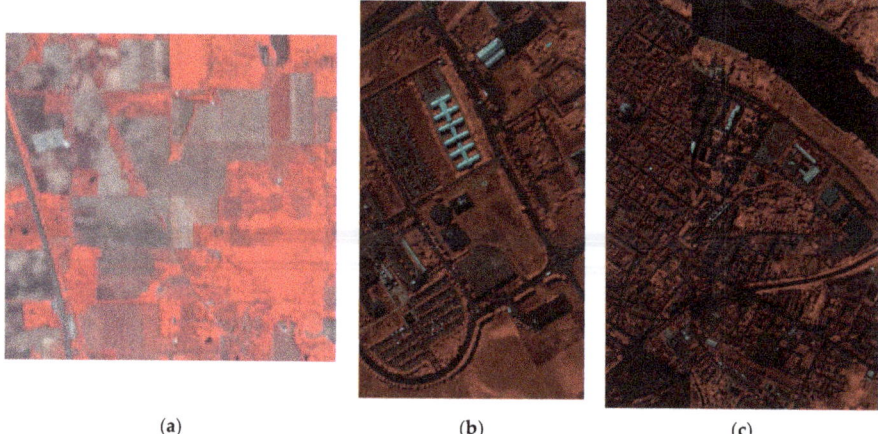

**Figure 1.** Datasets (**a**) Indian Pines Site (IPS); (**b**) Pavia University; and (**c**) Pavia City Center in false color of IR images.

The other two data sets, Pavia University, and Pavia City Center, were both the scenes covering the City of Pavia, Italy, obtained from the Reflective Optics System Imaging Spectrometer (ROSIS). They have 103 and 102 data bands, both with a spatial resolution of 1.3 m and a spectral coverage from 0.43 to 0.86 um. The dimension of these two images were 610 × 340 and 1096 × 715 pixels, respectively. The false color IR image of these two image were illustrated as Figure 1b,c. There were nine land-cover classes in each data set, and in each data set, the samples were divided into training and testing part, respectively. For example, in the Pavia University data set, there were 90 training samples in each class selected randomly for training, and the remaining 8046 samples were used for testing the performance. Based on the same manner, there were 810 training and 9529 testing samples used for the Pavia City Center data set, respectively.

*4.2. Classification Results*

The proposed MKFLE algorithm was compared with three state-of-the-art schemes, i.e., KNFLE, FNFLE, and FKNFLE [25]. The training samples were chosen randomly for computation of the transformation matrix, and the testing samples were applied to the nearest neighbor (NN) matching rule to matched with the training samples. The obtained average rates for each algorithm were run 30 times. In order to extract the suitable reduced dimensions of MKFLE, the obtained training samples were applied to measure the reduced dimensions versus the overall accuracy (OA) in the experimental datasets. As demonstration in Figure 2, the proposed MKFLE for datasets IPS-10, Pavia University, and Pavia City Center has the most suitable dimensions in 25, 65, and 65, respectively. From the classification results as shown in Figure 2, the proposed MKFLE outperforms all the other algorithms at the specific reduced dimensions on three datasets and with lower variant OA rates than the single kernel-based FKNFLE algorithm, which demonstrates the effectiveness of the proposed MKFLE. Based on observing Figure 2, a simple analysis was also done. When only fuzzy or single kernel method was applied in FLE, such as FNFLE and KNFLE, both of them obtained lower variant OA rates, and the kernel method was much more helpful than the fuzzy method since KNFLE outperformed FNFLE. Although FKNFLE outperformed the FNFLE and KNFLE, the variant OA of FKNFLE is large. Since the FKNFLE combined two different types of nonlinear and non-Euclidean information, it might cause the higher variant OA rates. In the meanwhile, multiple kernels applied in the proposed MKFLE used only nonlinear information with various parameters, which could improve the performance and obtain lower variant OA rates. In addition, since the MKL strategy was applied in the MKFLE training phases, which embedded different views of manifold structures from multiple kernel feature space, the reduced space obtained by the proposed MKFLE could be more general than FKNFLE, and obtained lower variant OA rates. From this analysis, the proposed MKFLE is much more superior than the FKNFLE, KNFLE, and FNFLE in HSI classification.

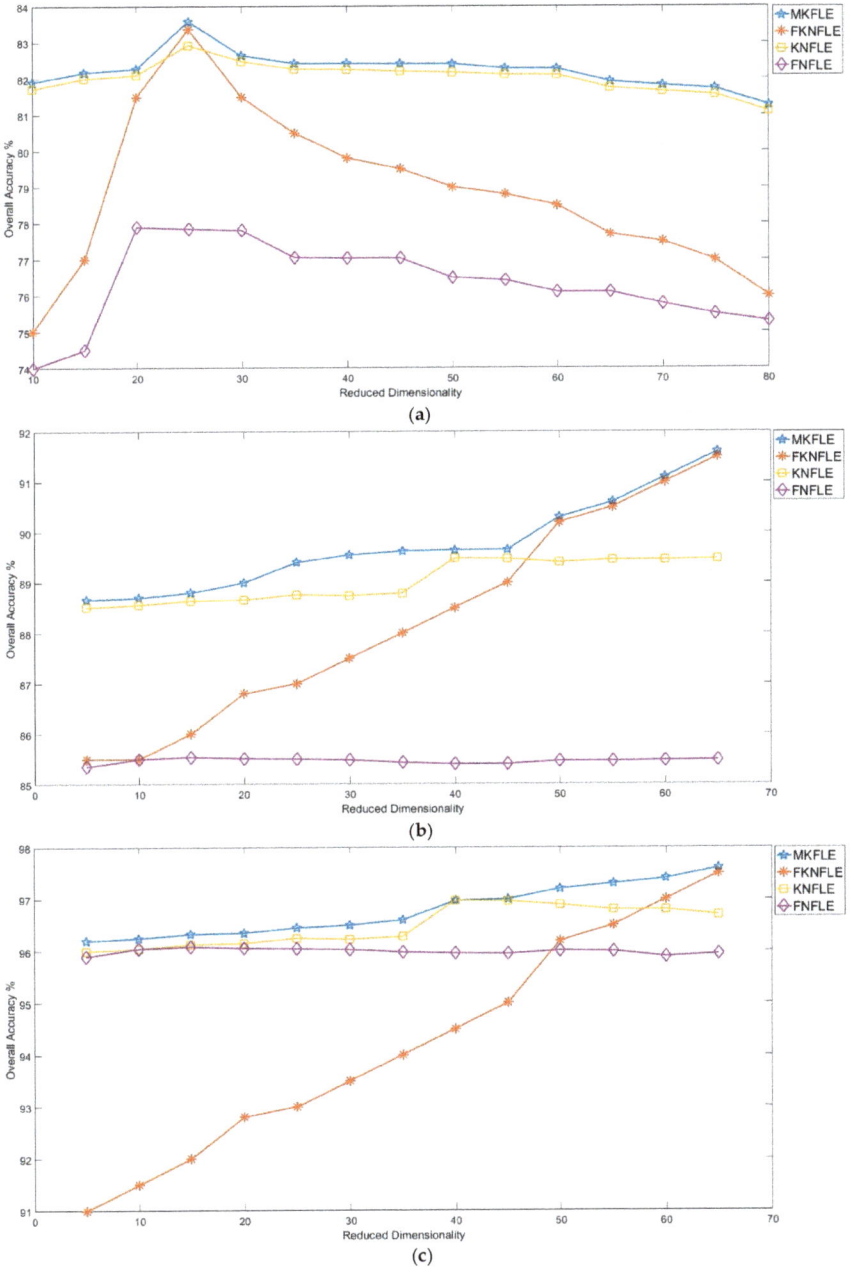

**Figure 2.** The reduced dimension versus the classification accuracy on three datasets applying various algorithms: (**a**) IPS-10; (**b**) Pavia University; (**c**) Pavia City Center.

Figure 3a shows the effect of changing the number of training samples on the average classification rates on dataset IPS-10; the proposed MKFLE algorithm has better performance than the other algorithms. The accuracy of MKFLE was 0.24% better than that of FKNFLE, which demonstrates that the proposed MKL strategy effectively enhanced the discriminative power of FLE. Figure 3b,c also shows the effect of changing the number of training samples on the overall accuracy on the benchmark datasets of Pavia University and Pavia City Center, respectively. Based on the overall accuracy in these two datasets, the proposed algorithm MKFLE outperforms the other algorithms. Next, Figure 4 demonstrates the classification results maps for the IPS-10 dataset. The various algorithms MKFLE, FKNFLE, KNFLE, and FNFLE are performed and obtained classification results on the maps of 145 × 145 pixels describing the ground truth. The proposed MKFLE obtained fewer speckle-like errors than those of the other algorithms. In the same manner, Figures 5 and 6 demonstrate the classification results maps for Pavia University and Pavia City Center datasets, respectively. Similarly, the proposed MKFLE obtained fewer speckle-like errors than in the case of the other algorithms.

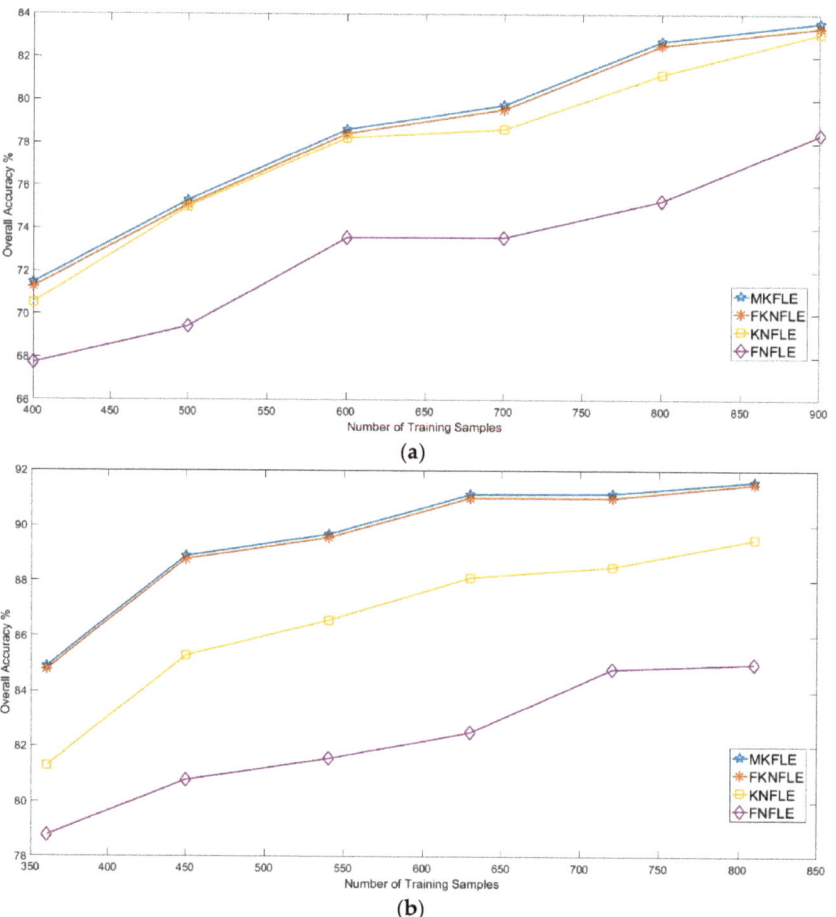

**Figure 3.** *Cont.*

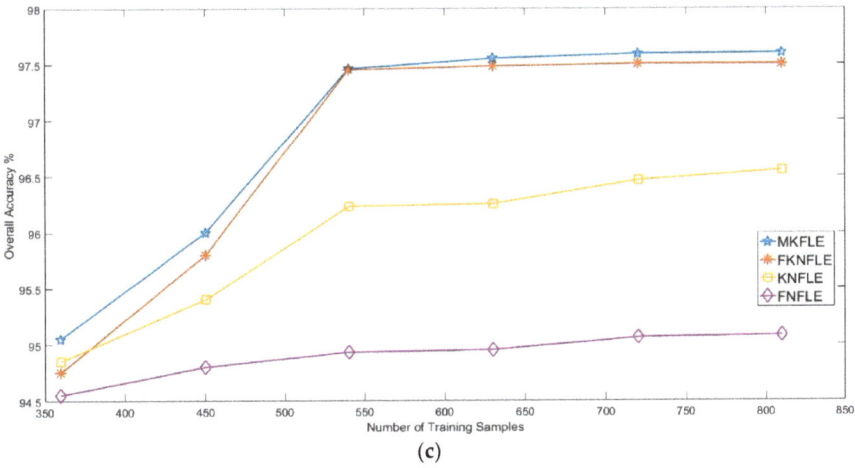

(c)

**Figure 3.** The number of training samples versus the accuracy rates for various datasets: (**a**) IPS-10; (**b**) Pavia University; and (**c**) Pavia City Center.

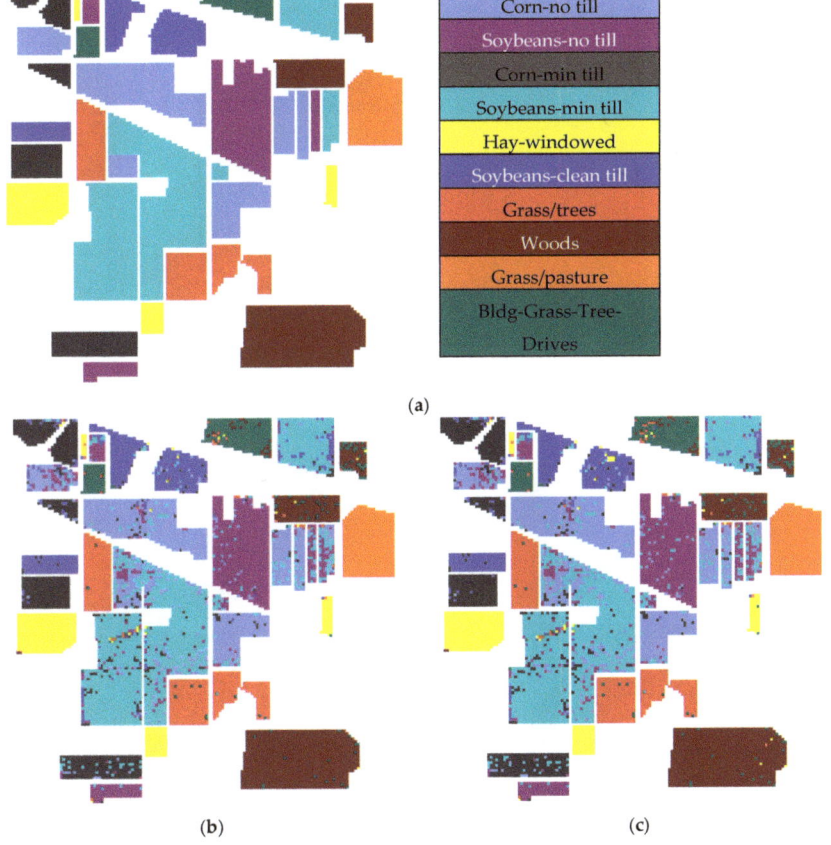

**Figure 4.** *Cont.*

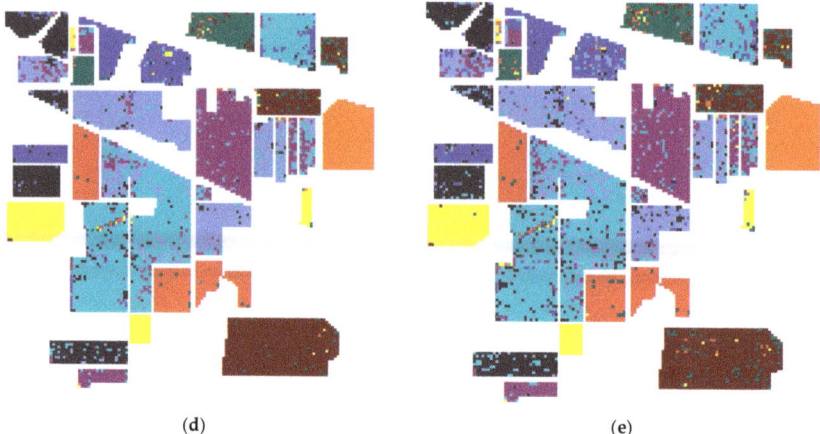

**Figure 4.** The maps of classification results on IPS dataset applying various algorithms: (**a**) The ground truth; (**b**) MKFLE; (**c**) FKNFLE; (**d**) KNFLE; (**e**) FNFLE.

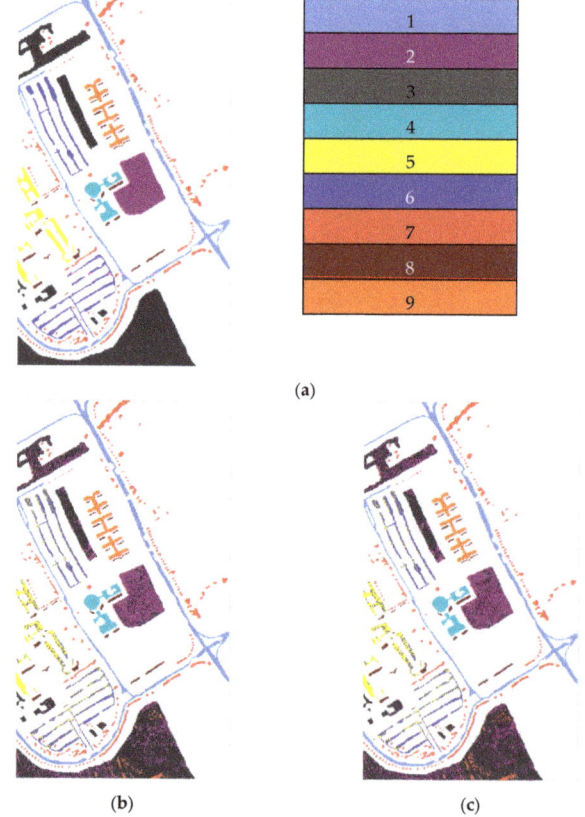

**Figure 5.** *Cont.*

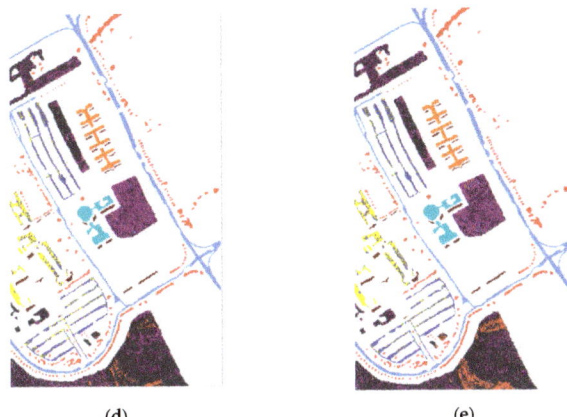

**Figure 5.** The classification results maps on Pavia University dataset applying various algorithms: (**a**) The ground truth; (**b**) MKFLE; (**c**) FKNFLE; (**d**) KNFLE; (**e**) FNFLE.

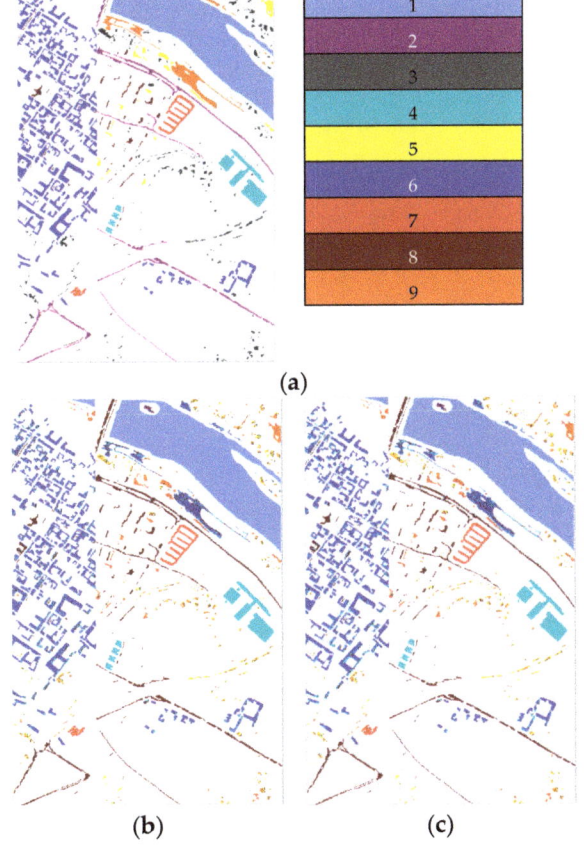

**Figure 6.** *Cont.*

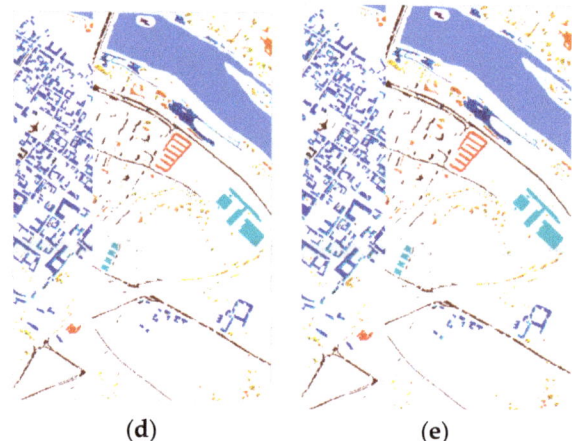

**Figure 6.** The maps of classification results on Pavia City Center dataset applying various algorithms: (**a**) the ground truth, (**b**) MKFLE; (**c**) FKNFLE; (**d**) KNFLE; (**e**) FNFLE.

Moreover, in order to evaluate the performance of the proposed MKFLE algorithm, the user's accuracy, producer's accuracy, kappa coefficients, and overall accuracy which were defined by the error matrices (or confusion matrices) [27] were tabulated in Tables 2–4. These four measures are briefly defined in the following. The user's accuracy and the producer's accuracy are two commonly applied measures for classification accuracy. The user's accuracy is the ratio of the number of pixels classified correctly in each class by the amount of pixel in the same class. The user's accuracy is a commission error, while the producer's accuracy measures the errors of omission and expresses the probability that some samples of a given class on the ground are actually identified as such. The kappa coefficient, which is also termed the kappa statistic, is defined as a measure of the difference between the changed agreement and the actual agreement. The proposed MKFLE algorithm achieved overall accuracies of 83.58% in IPS-10, 91.61% in Pavia University, and 97.68% in Pavia City Center with 0.829, 0.913, and 0.972 in kappa coefficients, respectively.

Furthermore, the main difference of computational complexity between MKFLE and FKNFLE is the SVM. Although the computational complexity of SVM is $O(N^2)$, which means that while the number of training samples increases, the training process of MKFLE would be time-consuming. However, since the training process is offline, the testing process is unaffected, and as aforementioned that if there is not a good DR algorithm to find a suitable lower dimensional representation for the training data, the performance of deep learning algorithms would be seriously impacted. Therefore, the proposed MKFLE is still competitive.

**Table 2.** The error matrix of classification for the IPS-10 dataset (in percentage).

| Classes | Reference Data | | | | | | | | | | User's Accuracy |
|---|---|---|---|---|---|---|---|---|---|---|---|
| | 1 | 2 | 3 | 4 | 5 | 6 | 7 | 8 | 9 | 10 | |
| 1 | 79.45 | 3.25 | 0.21 | 0.35 | 0 | 5.46 | 9.73 | 1.54 | 0 | 0 | 79.45 |
| 2 | 5.90 | 82.04 | 0 | 0.12 | 0 | 1.33 | 6.25 | 4.24 | 0 | 0.12 | 82.04 |
| 3 | 0 | 0 | 96.73 | 1.21 | 0.21 | 0.41 | 0 | 0.21 | 0.42 | 0.81 | 96.73 |
| 4 | 0 | 0 | 0.23 | 96.54 | 0 | 0 | 0 | 0 | 0 | 3.22 | 96.54 |
| 5 | 0 | 0 | 0.42 | 0 | 99.58 | 0 | 0 | 0 | 0 | 0 | 99.58 |
| 6 | 5.04 | 0.21 | 0.10 | 0.41 | 0 | 89.09 | 4.32 | 0.72 | 0 | 0.10 | 89.09 |
| 7 | 10.58 | 5.54 | 0.29 | 0.33 | 0.04 | 9.58 | 70.22 | 3.30 | 0 | 0.12 | 70.22 |
| 8 | 1.35 | 4.03 | 1.52 | 0.34 | 0 | 1.69 | 1.65 | 88.75 | 0 | 0.67 | 88.75 |
| 9 | 0 | 0 | 3.27 | 0.16 | 0 | 0 | 0 | 0 | 90.98 | 5.59 | 90.98 |
| 10 | 0 | 0 | 3.89 | 5.50 | 0 | 0 | 0 | 0.26 | 10.83 | 79.52 | 79.52 |
| Producer's Accuracy | 77.64 | 86.29 | 90.69 | 91.97 | 99.75 | 82.82 | 76.18 | 89.62 | 88.99 | 88.20 | |
| Kappa Coefficient: **0.829** | | | | | | Overall Accuracy: **83.58%** | | | | | |

Table 3. The error matrix of classification for the Pavia University dataset (in percentage).

| Classes | Reference Data | | | | | | | | | User's Accuracy |
|---|---|---|---|---|---|---|---|---|---|---|
| | 1 | 2 | 3 | 4 | 5 | 6 | 7 | 8 | 9 | |
| 1 | 90.18 | 3.15 | 0 | 0 | 0 | 3.24 | 1.35 | 1.26 | 0.81 | 90.18 |
| 2 | 2.31 | 92.82 | 0 | 2.31 | 0 | 1.55 | 0 | 1.01 | 0 | 92.82 |
| 3 | 0 | 0 | 90.39 | 2.38 | 1.38 | 0.99 | 2.88 | 0.99 | 0.99 | 90.39 |
| 4 | 0 | 1.23 | 2.84 | 90.55 | 1.42 | 1.42 | 1.31 | 1.23 | 0 | 90.55 |
| 5 | 0.63 | 1.13 | 0.75 | 1.26 | 92.22 | 0.63 | 1.44 | 0.81 | 1.13 | 92.22 |
| 6 | 1.01 | 1.09 | 1.28 | 1.56 | 1.19 | 92.86 | 0.55 | 0.46 | 0 | 92.86 |
| 7 | 0 | 1.12 | 0.51 | 0.61 | 2.09 | 0 | 93.58 | 1.07 | 1.02 | 93.58 |
| 8 | 0.47 | 1.42 | 0.95 | 1.33 | 2.18 | 1.90 | 0 | 91.09 | 0.66 | 91.09 |
| 9 | 1.14 | 0 | 2.06 | 2.01 | 0 | 2.09 | 0 | 2.15 | 90.55 | 90.55 |
| Producer's Accuracy | 94.19 | 91.03 | 91.50 | 88.76 | 91.77 | 88.73 | 92.55 | 91.02 | 95.15 | |
| Kappa Coefficient: 0.913 | | | | | | Overall Accuracy: 91.61% | | | | |

Table 4. The error matrix of classification for the Pavia City Center dataset (in percentage).

| Classes | Reference Data | | | | | | | | | User's Accuracy |
|---|---|---|---|---|---|---|---|---|---|---|
| | 1 | 2 | 3 | 4 | 5 | 6 | 7 | 8 | 9 | |
| 1 | 98.69 | 0.14 | 0.51 | 0.33 | 0.33 | 0 | 0 | 0 | 0 | 98.69 |
| 2 | 1.04 | 97.58 | 0.43 | 0 | 0 | 0.29 | 0.17 | 0.48 | 0 | 97.58 |
| 3 | 0.59 | 0.76 | 96.31 | 0.69 | 0.99 | 0 | 0 | 0 | 0.67 | 96.31 |
| 4 | 0 | 0.52 | 0.66 | 96.79 | 0.37 | 0.47 | 0.66 | 0.53 | 0 | 96.79 |
| 5 | 0 | 0 | 0.39 | 0.34 | 97.86 | 0.21 | 0.34 | 0.34 | 0.52 | 97.86 |
| 6 | 0.33 | 0.26 | 0.54 | 0 | 0 | 98.26 | 0 | 0.26 | 0.35 | 98.26 |
| 7 | 0.34 | 0.25 | 0 | 0.35 | 0 | 0.38 | 98.33 | 0.35 | 0 | 98.33 |
| 8 | 0 | 0 | 0.37 | 0.30 | 0.37 | 0.49 | 0.45 | 97.55 | 0.46 | 97.55 |
| 9 | 0.39 | 0.55 | 0.75 | 0.29 | 0.29 | 0 | 0 | 0 | 97.73 | 97.73 |
| Producer's Accuracy | 97.34 | 97.52 | 96.34 | 97.67 | 97.65 | 98.16 | 98.38 | 98.03 | 97.99 | |
| Kappa Coefficient: 0.972 | | | | | | Overall Accuracy: 97.68% | | | | |

## 5. Discussion

The proposed MKFLE dimension reduction algorithm was applied in HSI classification. Since the MKFLE algorithm applies the multiple kernel, which could extract much more useful nonlinear information, and accordingly obtained a better accuracy than that of FKNFLE to the value of 0.24%, 0.3%, and 0.09% for IPS-10, Pavia City Center, and Pavia University datasets, respectively. Although the improvements were few percentage, and increase the complexity of the algorithm. However, since the training process of MKFLE including the SVM process are offline, the testing process is unaffected. Besides, MKFLE is with lower variant OA rate. Therefore, the proposed MKFLE algorithm is suitable for dimension reduction.

## 6. Conclusions

In this study, a dimension reduction MKFLE algorithm based on general FLE transformation has been proposed and applied in HSI classification. The SVM-based multiple kernel learning strategy was considered to extract the multiple different non-linear manifold locality. The proposed MKFLE was compared with three previous state-of-the-art works, FKNFLE, KNFLE, and FNFLE. Three classic datasets, IPS-10, Pavia University, and Pavia City Center, were applied for evaluating the effectiveness of variant algorithms. Based on the experimental results, the proposed MKFLE had better performance than the other methods. More specifically, based on the 1-NN matching rule, the accuracy of MKFLE was better than that of FKNFLE to the value of 0.24%, 0.3%, and 0.09% for IPS-10, Pavia City Center, and Pavia University datasets, respectively. Moreover, the proposed MKFLE has higher accuracy and lower accuracy variant than FKNFLE. However, since SVM was applied in the training process of MKFLE, more training time than the FKNFLE was needed. Therefore, more efficient computational schemes for selecting the support vectors will be investigated in further research.

**Author Contributions:** Y.-N.C. conceived the project, conducted research, performed initial analyses and wrote the first manuscript draft. Y.-N.C. edited the first manuscript. Y.-N.C. finalized the manuscript for the communication with the journal.

**Funding:** This work was assisted by the Ministry of Science and Technology of Taiwan under Grant nos. MOST 108-2218-E-008-014 and MOST 108-2221-E-008 -073.

**Acknowledgments:** Constructive comments from anonymous reviewers helped the authors make significant improvements to the original manuscript.

**Conflicts of Interest:** The authors declare no conflict of interest.

## References

1. Zhu, L.; Chen, Y.; Ghamisi, P.; Benediktsson, J.A. Generative adversarial Networks for hyperspectral image classification. *IEEE Trans. Geosci. Remote Sens.* **2018**, *56*, 5046–5063. [CrossRef]
2. Redmon, J.; Divvala, S.K.; Girshick, R.B.; Farhadi, A. You only look once: Unified, real-time object detection. In Proceedings of the 2016 Proceedings CVPR '16. IEEE Computer Society Conference on Computer Vision and Pattern Recognition, Las Vegas, NV, USA, 27–30 June 2016; pp. 779–788.
3. Sun, Y.; Liang, D.; Wang, X.; Tang, X. Deepid3: Face recognition with very deep neural networks. *arXiv* **2015**, arXiv:1502.00873.
4. Turk, M.; Pentland, A.P. Face recognition using eigenfaces. In Proceedings of the 1991 Proceedings CVPR '91. IEEE Computer Society Conference on Computer Vision and Pattern Recognition, Maui, HI, USA, 3–6 June 1991; pp. 586–591.
5. Belhumeur, P.N.; Hespanha, J.P.; Kriegman, D.J. Eigenfaces vs. Fisherfaces: Recognition using class specific linear projection. *IEEE Trans. Pattern Anal. Mach. Intell.* **1997**, *19*, 711–720. [CrossRef]
6. Cevikalp, H.; Neamtu, M.; Wikes, M.; Barkana, A. Discriminative common vectors for face recognition. *IEEE Trans. Pattern Anal. Mach. Intell.* **2005**, *27*, 4–13. [CrossRef] [PubMed]
7. Prasad, S.; Mann Bruce, L. Information fusion in kernel-induced spaces for robust subpixel hyperspectral ATR. *IEEE Trans. Geosci. Remote Sens. Lett.* **2009**, *6*, 572–576. [CrossRef]
8. He, X.; Yan, S.; Ho, Y.; Niyogi, P.; Zhang, H.J. Face recognition using Laplacianfaces. *IEEE Trans. Pattern Anal. Mach. Intell.* **2005**, *27*, 328–340.
9. Tu, S.T.; Chen, J.Y.; Yang, W.; Sun, H. Laplacian eigenmaps-based polarimetric dimensionality reduction for SAR image classification. *IEEE Trans. Geosci. Remote Sens.* **2011**, *50*, 170–179. [CrossRef]
10. Wang, Z.; He, B. Locality preserving projections algorithm for hyperspectral image dimensionality reduction. In Proceedings of the 2011 19th International Conference on Geoinformatics, Shanghai, China, 24–26 June 2011; pp. 1–4.
11. Kim, D.H.; Finkel, L.H. Hyperspectral image processing using locally linear embedding. In Proceedings of the 1st International IEEE EMBS Conference on Neural Engineering, Capri Island, Italy, 20–22 March 2003; pp. 316–319.
12. Li, W.; Prasad, S.; Fowler, J.E.; Bruce, L.M. Locality-preserving discriminant analysis in kernel-induced feature spaces for hyperspectral image classification. *IEEE Geosci. Remote Sens. Lett.* **2011**, *8*, 894–898. [CrossRef]
13. Li, W.; Prasad, S.; Fowler, J.E.; Bruce, L.M. Locality-preserving dimensionality reduction and classification for hyperspectral image analysis. *IEEE Trans. Geosci. Remote Sens.* **2012**, *50*, 1185–1198. [CrossRef]
14. Luo, R.B.; Liao, W.Z.; Pi, Y.G. Discriminative supervised neighborhood preserving embedding feature extraction for hyperspectral-image classification. *Telkomnika* **2012**, *10*, 1051–1056. [CrossRef]
15. Zhang, L.; Zhang, Q.; Zhang, L.; Tao, D.; Huang, X.; Du, B. Ensemble manifold regularized sparse low-rank approximation for multi-view feature embedding. *Pattern Recognit.* **2015**, *48*, 3102–3112. [CrossRef]
16. Boots, B.; Gordon, G.J. Two-manifold problems with applications to nonlinear system Identification. In Proceedings of the 29th International Conference on Machine Learning, Edinburgh, UK, 26 June–1 July 2012.
17. Odone, F.; Barla, A.; Verri, A. Building kernels from binary strings for image matching. *IEEE Trans. Image Process.* **2005**, *14*, 169–180. [CrossRef] [PubMed]
18. Scholkopf, B.; Smola, A.; Muller, K.R. Nonlinear component analysis as a kernel eigenvalue problem. *Neural Comput.* **1998**, *10*, 1299–1319. [CrossRef]
19. Lin, Y.Y.; Liu, T.L.; Fuh, C.S. Multiple kernel learning for dimensionality reduction. *IEEE Trans. Pattern Anal. Mach. Intell.* **2011**, *33*, 1147–1160. [CrossRef]

20. Nazarpour, A.; Adibi, P. Two-stage multiple kernel learning for supervised dimensionality reduction. *Pattern Recognit.* **2015**, *48*, 1854–1862. [CrossRef]
21. Li, J.; Marpu, P.R.; Plaza, A.; Bioucas-Dias, J.M.; Benediktsson, J.A. Generalized composite kernel framework for hyperspectral image classification. *IEEE Trans. Geosci. Remote Sens.* **2013**, *51*, 4816–4829. [CrossRef]
22. Chen, Y.; Nasrabadi, N.M.; Tran, T.D. Hyperspectral image classification via kernel sparse representation. *IEEE Trans. Geosci. Remote Sens.* **2013**, *51*, 217–231. [CrossRef]
23. Zhang, L.; Zhang, L.; Tao, D.; Huang, X. On combining multiple features for hyperspectral remote sensing image classification. *IEEE Trans. Geosci. Remote Sens.* **2012**, *50*, 879–893. [CrossRef]
24. Chen, Y.N.; Han, C.C.; Wang, C.T.; Fan, K.C. Face recognition using nearest feature space embedding. *IEEE Trans. Pattern Anal. Mach. Intell.* **2011**, *33*, 1073–1086. [CrossRef]
25. Chen, Y.N.; Hsieh, C.T.; Wen, M.G.; Han, C.C.; Fan, K.C. A dimension reduction framework for HIS classification using fuzzy and kernel NFLE transformation. *Remote Sens.* **2015**, *7*, 14292–14326. [CrossRef]
26. Yan, S.; Xu, D.; Zhang, B.; Zhang, H.J.; Yang, Q.; Lin, S. Graph embedding and extensions: A framework for dimensionality reduction. *IEEE Trans. Pattern Anal. Mach. Intell.* **2007**, *29*, 40–51. [CrossRef] [PubMed]
27. Lillesand, T.M.; Kiefer, R.W. *Remote Sensing and Image Interpretation*; Wiley: New York, NY, USA, 2000.

© 2019 by the author. Licensee MDPI, Basel, Switzerland. This article is an open access article distributed under the terms and conditions of the Creative Commons Attribution (CC BY) license (http://creativecommons.org/licenses/by/4.0/).

*Article*

# Temporal Variation and Spatial Structure of the Kuroshio-Induced Submesoscale Island Vortices Observed from GCOM-C and Himawari-8 Data

**Po-Chun Hsu, Chia-Ying Ho, Hung-Jen Lee, Ching-Yuan Lu and Chung-Ru Ho \***

Department of Marine Environmental Informatics, National Taiwan Ocean University, Keelung 20224, Taiwan; hpochun@ntou.edu.tw (P.-C.H.); 20781001@ntou.edu.tw (C.-Y.H.); lecgyver@mail.ntou.edu.tw (H.-J.L.); 20681002@ntou.edu.tw (C.-Y.L.)
\* Correspondence: b0211@mail.ntou.edu.tw

Received: 13 February 2020; Accepted: 8 March 2020; Published: 9 March 2020

**Abstract:** Dynamics of ocean current-induced island wake has been an important issue in global oceanography. Green Island, a small island located off southeast of Taiwan on the Kuroshio path was selected as the study area to more understand the spatial structure and temporal variation of well-organized vortices formed by the interaction between the Kuroshio and the island. Sea surface temperature (SST) and chlorophyll-a (Chl-a) concentration data derived from the Himawari-8 satellite and the second generation global imager (SGLI) of global change observation mission (GCOM-C) were used in this study. The spatial SST and Chl-a variations in designed observation lines and the cooling zone transitions on the left and right sides of the vortices were investigated using 250 m spatial resolution GCOM-C data. The Massachusetts Institute of Technology general circulation model (MITgcm) simulation confirmed that the positive and negative vortices were sequentially detached from each other in a few hours. In addition, totals of 101 vortexes from July 2015 to December 2019 were calculated from the 1-h temporal resolution Himawari-8 imagery. The average vortex propagation speed was 0.95 m/s. Totals of 38 cases of two continuous vortices suggested that the average vortex shedding period is 14.8 h with 1.15 m/s of the average incoming surface current speed of Green Island, and the results agreed to the ideal Strouhal-Reynolds number fitting curve relation. Combined with the satellite observation and numerical model simulation, this study demonstrates the structure of the wake area could change quickly, and the water may mix in different vorticity states for each observation station.

**Keywords:** island wake; vortex; sea surface temperature; chlorophyll-a; Himawari-8; GCOM-C

## 1. Introduction

*1.1. Background*

Island wakes have been studied for many years in the global oceans. Different driving forces cause this phenomenon and these wakes have different characteristics according to the island's scale and water depth [1–4]. Green Island, a small island located off southeast of Taiwan, is on the path of the Kuroshio to the East China Sea (Figure 1). Green Island is an obstacle on the "ocean highway". Kuroshio passes Green Island and causes a change in flow fields, such as the Von Kármán vortex street [5–9]. In the lee of the island, a high chlorophyll-a (Chl-a) concentration and low sea surface temperature (SST) wake region can be formed [8]. This recirculation area contained a cyclonic/anticyclonic vorticity pair accompanied by a density overturn and water upwelling [9]. The increased surface Chl-a concentration and decreased SST are induced by upwelling from deep-layer waters in the island wakes [5,9]. Like a supply depot for marine life, the SST fronts between the wake area and the Kuroshio could provide favorable conditions for marine fishing grounds. The island wake induced upwelling, downstream mixing and eddies could result in phytoplankton biomass and chlorophyll concentrations increased near the island, which is called the island mass effect [10]. Enhanced phytoplankton can elicit a biological response in fishes, carbon cycles and food web. There are more than 600 species of fish around Green Island [11]. It is also an important habitat for a coral reef [12]. Therefore, studying the wake of the island is an important issue.

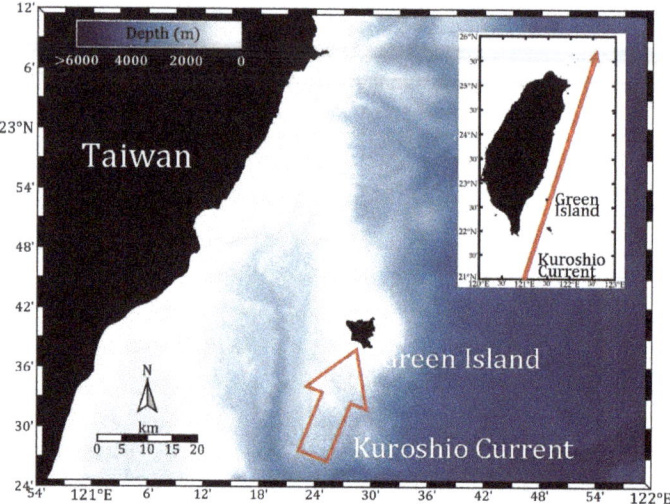

**Figure 1.** The bottom topography around Green Island and the path of the Kuroshio (red line and arrow).

Previous researches have used high-resolution satellite imagery, acoustic Doppler current profilers, cruise observations, and numerical models to analyze the Green Island wakes [5,8,9]. An island-induced ocean vortex train (IOVT) [7] is formed leeward of Green Island by the incoming Kuroshio [8]. When an ocean vortex forms, observations from the moored acoustic Doppler current profiler (ADCP) suggest that the velocity on the western side of Green Island increases [8]. In addition, the IOVT shows variability under wind forcing [8,13]. Southerly wind helps the wake expand because it lies in the same direction as the incoming current, while the northerly wind compresses the development of the wake [8,13]. The island wakes influence the area of SST drops, which could extend 35 km downstream of Green Island, and vortex shedding could propagate as far as 80 km downstream of the island [9]. Research on underwater island wakes suggests that a depth range with a high inverted Reynolds number is supposed to have density overturns and a high Thorpe scale with a turbulent kinetic energy dissipation rate of $O(10^{-6}$–$10^{-5})$W/kg, which corresponds to an eddy diffusivity of $O(10^{-2}$–$10^{-1})$m$^2$/s [9]. Numerical model research has suggested that the vortex street features are adapted by inertial and barotropic instabilities [14], and the shedding period of the vortex is synchronized to a tidal period [15].

*1.2. Objectives*

Although there are many studies on the Green Island wakes, the dynamics of these wakes remain difficult to be determined. Cruise and instrument observations can measure detailed data on the underwater wakes, but the advantages of remote sensing could be used to obtain accurate and simultaneous data for sea surface wakes. Before the next measurement, it is necessary to clarify the spatial scale and dynamic processes of the wakes on the island surface using satellite data. Only by determining the feedback of the surface seawater to the wake can a researcher accurately design the observation line and measurement frequency by cruise.

The purpose of this study is to use the satellite data to obtain novel results of the Green Island wake in order to make up for the shortcomings of in-situ measurements and the inability for the model to be measured. They include (1) designing the imaginary lines to describe the Chl-a and SST response to the wake from the second generation global imager (SGLI) data on the global change observation mission (GCOM-C), (2) using the Himawari-8 SST data to calculate the wake vortex detachment propagating speed and using the maximum cross-correlation (MCC) method to estimate the incoming velocity of the wake, (3) calculating and discussing the Strouhal number–Reynolds number relationship of wake dynamics.

Understanding the spatial structure of surface wakes can help determine the biological hotspots in the wake area and improve knowledge of oceanic wakes. The evolution process of a wake can be separated into four types, including a wake occurring alone, a wake occurring with a tail stretching downstream, an S-shaped meandering wake, and a wake with a small cyclonic/anticyclonic vortex pair downstream [9]. However, the detailed spatial structures of each type of wake remain unknown. The SGLI observations with 250 m spatial resolution will improve our understanding of ocean change mechanisms through long-term monitoring. The use of imaginary design observation lines helps to understand the sea surface characteristics of the wake region at the same time but different locations.

Using high spatial resolution sun-synchronous satellite data could successfully detect the SST drop in the cold wake area but could not further track the process of vortex movement in continuous an hour interval due to the lack of a high temporal resolution [9]. To study the dynamics of island wakes, the most important factor is to solve the setting of the two major parameters, the Strouhal number ($St$) and the Reynolds number ($Re$), as well as the propagation velocity of the cold vortex ($U_e$). However, it is very difficult to measure these two parameters with a cruise observation in the ocean. For the Green Island area, there must be a certain number of research surveys to measure the current velocity and SST variations in front of and downstream of the island. Previous studies applied parameters such as the estimated distance between two consecutive vortices and the radius of vortices from Synthetic Aperture Radar (SAR) images to the empirical equation to obtain the incoming current velocity and the propagation velocity of the vortex [16]. However, SAR imagery is a snapshot image, which does not have a high temporal resolution or a sufficient image, making it difficult to engage in a long-term observation of wake dynamics. With advances in satellite imagery, the Himawari-8 geostationary satellite data provide a useful solution to the small cold vortex detachment process and detect the vortex trajectory in wake dynamics. Understanding the relationship between $St$ and $Re$ and the speed and period of cold vortex detachment can be useful for improving numerical simulations.

Recognizing the Green Island wake not only requires us to physical discuss wake dynamics but also to investigate the development of ocean environmental sustainability. The results could help marine fisheries and chemists better understand the status of this area.

## 2. Materials and Methods

The process of this study was divided into three steps. Firstly, the in-situ observations were used to present the underwater structure in the wake region. Secondly, the detailed structure of the surface wake was analyzed using the SGLI data from the GCOM project. Finally, the Himawari-8 imagery was used to analyze the vortices movement process, and the MITgcm (Massachusetts Institute of Technology General Circulation Model) numerical model was used to simulate the wake variation.

The MCC method was applied to hourly Himawari-8 SST images to calculate the incoming surface current velocity for Green Island (22.5°–22.7°N, 122.3°–122.5°E). The principle of the MCC method is to find the maximum correlation between a template image of SST patterns and the selected search image in the next time interval. In this way, we can determine how far the center point of the template image has moved in an hour and estimate the flow velocity. The MCC method needs to adjust the parameter selection settings for different research scopes [17–19]. In this study, the pixel resolution of 2 km was increased to 1 km by interpolation. Considering the complex flow field variations and spatial scale of the SST variation in southeast Taiwan, the four template images ($7 \times 7$, $9 \times 9$, $11 \times 11$, and

13 × 13 pixels) and the sides of the search image (7 km to each direction from the center of the template image) were chosen. In this study, the acceptance criteria for the cross-correlation coefficient should be larger than 0.9. With this parameter set, the minimum and the maximum speeds in the estimate were about 0.28 m/s and 1.94 m/s, respectively. The maximum flow velocity per template image was selected as the flow velocity in this area. Finally, the average maximum speed of the four template images during each vortex shedding case was calculated.

*2.1. Sea Surface Temperature and Chlorophyll-a Concentration*

The GCOM-C, carrying an SGLI conducts surface and atmospheric measurements, such as measurements of clouds, aerosols, ocean color, vegetation, and snow and ice. The SGLI is an optical sensor capable of multi-channel observations at wavelengths from near-UV to thermal infrared and obtains global observation data once every 2 or 3 days. The SGLI data are provided by the Japan Aerospace Exploration Agency/National Aeronautics and Space Administration. The data are available from January 2018, with a 250 m spatial resolution. The standard product and algorithm handbooks can be found on https://suzaku.eorc.jaxa.jp/GCOM_C.

Himawari-8 is the 8th of the Himawari geostationary Japanese weather satellites operated by the Japan Meteorological Agency. Himawari-8 carries the Advanced Himawari Imager with a wide spectral range and very high spatial and 10-minute temporal resolutions. The research products of SST and Chl-a used in this paper were supplied by the P-Tree System, Japan Aerospace Exploration Agency (JAXA). The standard product can be found on https://www.eorc.jaxa.jp/ptree/index.html. The SST data (2 km spatial resolution) and Chl-a (5 km spatial resolution) are available from 7 July 2015, with a 1-h temporal resolution. The Himawari-8 data have good accuracy and have been applied to calculate short-term sea surface currents [17,20].

*2.2. Ocean Currents*

OSCAR (Ocean Surface Current Analysis Real-time) contains near-surface ocean current estimates derived using quasi-linear and steady flow momentum equations and combining geostrophic, Ekman, and Stommel shear dynamics, and a complementary term from the surface buoyancy gradient, detailed calculations can be referred to the User's Handbook [21]. These data were collected from various satellites and in situ instruments and directly estimated from the sea surface height, surface vector wind, and SST. These data were generated by the Earth Space Research (ESR) and are available from October 1992 to present, with a 1/3 degree spatial resolution and a 5-day temporal resolution. The OSCAR third-degree resolution ocean surface currents (OSCAR_L4_OC_third-deg) data which can be accessed through the https://podaac.jpl.nasa.gov/dataset/OSCAR_L4_OC_third-deg from NASA PODAAC (Physical Oceanography Distributed Active Archive Center). The OSCAR data are highly associated and accurate with the global tropical moored buoy array, and it has also been successfully used to explore the island wake around Palau in the western tropical North Pacific Ocean [1,22,23].

*2.3. Numerical Model*

MITgcm is a numerical model designed for the study of the atmosphere, ocean, and climate. The source code for this model can be download from http://mitgcm.org/. We used this model to solve the Navier–Stokes equations under hydrostatic and Boussinesq approximations using the finite volume method [24]. The south, north, and east directions of the model domain are open boundaries that use the Orlanski radiation condition [25]. The HYbrid Coordinate Ocean Model and the Navy Coupled Ocean Data Assimilation (HYCOM + NCODA) Global 1/12 degree analysis (GLBu0.08) was used for the initial background data and the boundary driving force in the model. In this study, the case of the summer season was based on the average of June to August from 2013 to 2018. These data can be download from https://www.hycom.org/. The study area in the model set is from 120.85°E to 122.5°E and from 22.25°N to 23.5°N. The horizontal resolution is 500 m. The model uses z-coordinates. There are 80 levels in the vertical direction. The top layer's thickness is 5 m, and the other thicknesses increase at a 5.26% increase rate for each layer. The maximum depth is 5513 m. The horizontal eddy viscosity is 40 m$^2$/s. The Laplacian diffusion of heat and salt laterally is 4 m$^2$/s. The vertical diffusion of temperature and salt depends on the buoyancy frequency [26].

### 2.4. In-Situ Observation

In-situ observations were conducted in November 2012 using R/V Ocean Researcher I (OR1) in the lee of Green Island (the datasets are available from http://www.odb.ntu.edu.tw/en/). Seven stations (A1–A7) were selected to detect the vertical layer of the island wake from the southeast coast of Taiwan across the Kuroshio and the wake area north of Green Island (see Figure 2a). The shipboard acoustic Doppler current profiler (Sb-ADCP) and Conductivity Temperature Depth (CTD) surveys were also collected during the cruise. The AquaTracka III fluorometer is used for in-situ detection of Chl-a, dissolved compounds re-emit a fraction of this energy as fluorescence at longer wavelengths when they absorb light and the intensity of fluorescence is directly proportional to the concentration. The measurement range of the 75 kHz Sb-ADCP on board R/V OR1 was 16.56–650 m, with a bin size of 8 m. It recorded the current velocity with 2 min ensembles averaged over 30 pings.

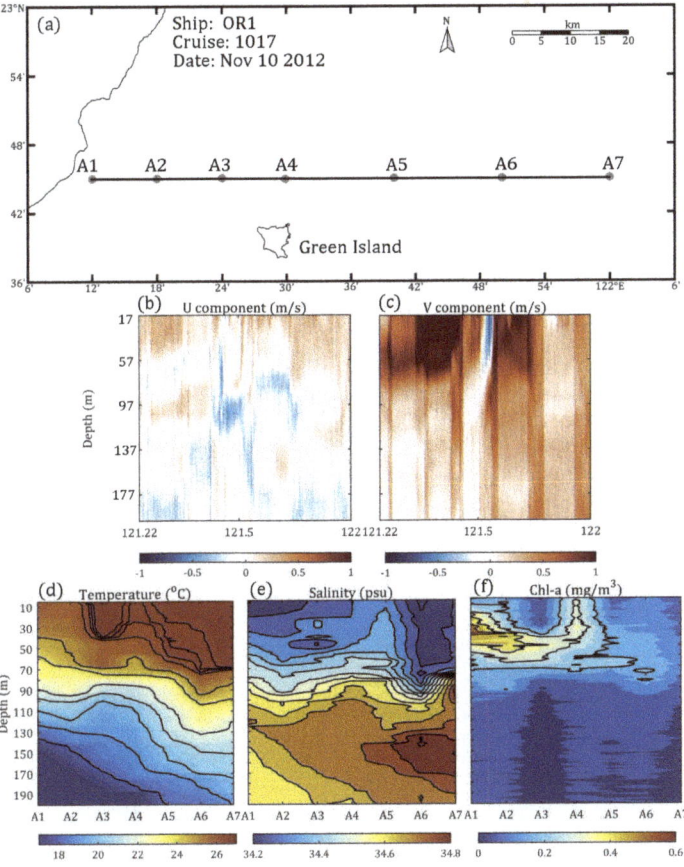

**Figure 2.** (a) The cruise experiment results with (b,c) velocity (U component positive in the east, V component positive in the north), (d) temperature, (e) salinity, and (f) Chl-a from stations A1 to A7 on 10 November 2012.

## 3. Results

### 3.1. Field Experiment

In this field experiment, the area between A4 and A5 was significantly affected by the island wake (Figure 2b,c), the isotherms and isohalines were uplifted from a depth of 100–120 m to the surface and the high Chl-a concentration water also upwelled from a depth of 70 m to the surface (Figure 2d–f). The wake water has lower SST, higher salinity, and higher Chl-a than the Kuroshio water. Moreover, the cruise observations could not measure the whole area of the wake at the same time, and the hydrological structure may be changed due to the dynamic nature of the island wakes. Therefore, remote sensing data with high temporal and spatial resolutions are recommended to discuss the surface structures of wakes.

### 3.2. Spatial Structure of Island Wake

In this section, GCOM-C SGLI data with a 250 m grid resolution are used to clarify these properties. Notably, these high-resolution cloud-free images are used to observe the spatial changes of SST and Chl-a at different locations throughout the wake region at the same time, and we designed imaginary observation lines for each case. In case 1, the classic and most common wake type with ten east-west observation lines is used to discuss the wake from the lee of Green Island to downstream (Figures 3 and 4). In case 2, an S-shaped meandering wake with three lines along the wake and one imaginary line are used to discuss the spatial–structural differences between two cold vortices (Figure 5). Case 3 offers a good comparative example to show the change of vorticity in the wake region during the vortex transition (Figure 6). A numerical model simulation is used to illustrate this phenomenon (Figure 7). Case 4 presents a typical theoretical von Kármán vortex street, but such phenomena rarely occur in an SST response (Figure 8).

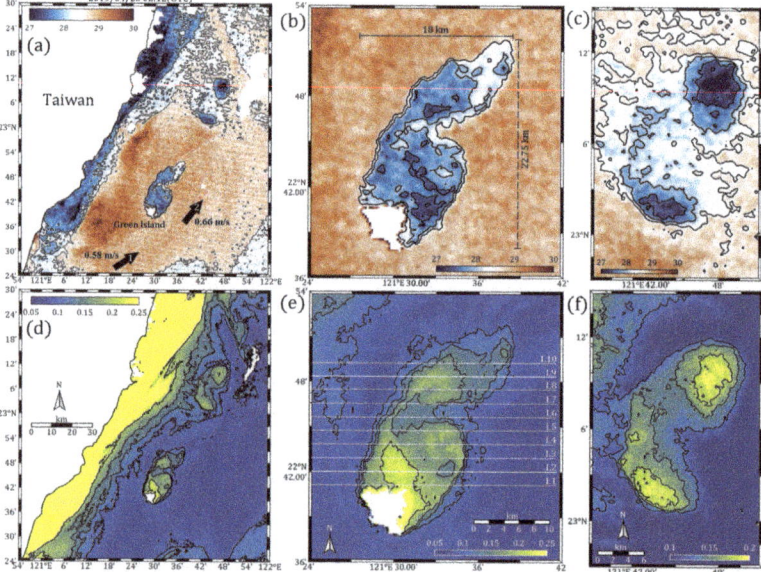

**Figure 3.** Case of the island vortex obtained from the global change observation mission (GCOM-C) second-generation global imager (SGLI) data taken at 02:12 (UTC), 25 April 2019. (**a**) Sea surface temperature (SST) (°C), (**b**,**c**) zoom in on vortices of (**a**,**d**) Chl-a (mg/m$^3$), (**e**,**f**) zoom in on vortices of (**d**). The black arrow in (**a**) is the current velocity from the OSCAR data. The first arrow (22.33°N, 121.33°E) has a speed of 0.58 m/s, and the second arrow (22.67°N, 121.67°E) has a speed of 0.66 m/s.

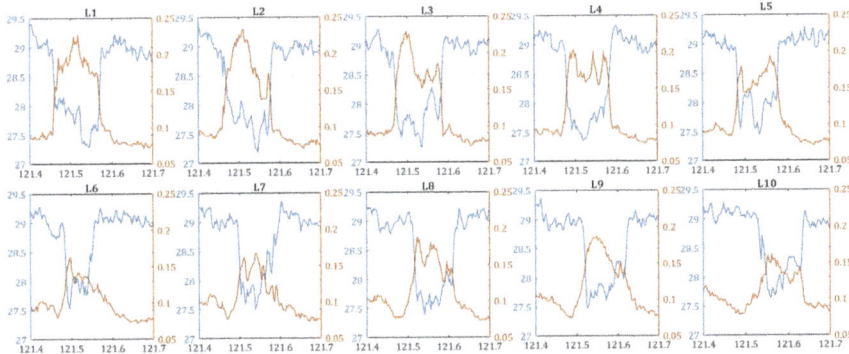

**Figure 4.** SST and chlorophyll-a (Chl-a) of L1 to L10 in Figure 3e.

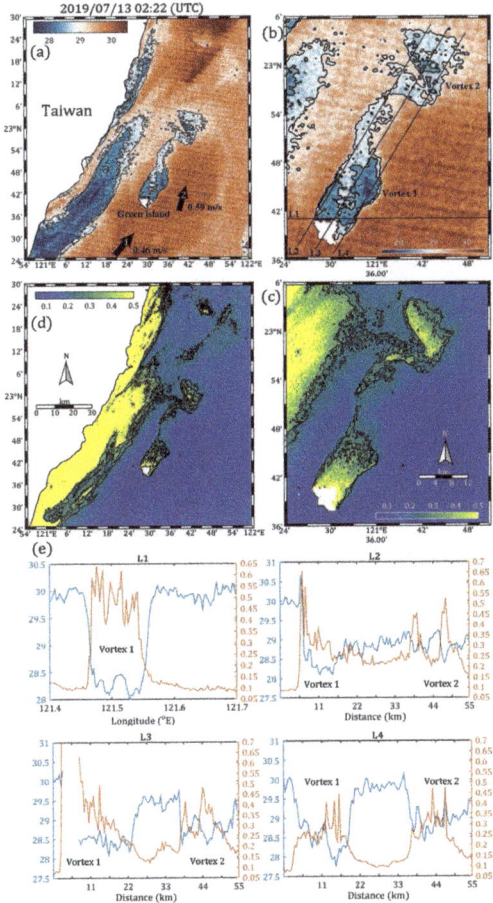

**Figure 5.** Case of the island vortex obtained from GCOM-C SGLI image taken at 02:22 (UTC), July 13, 2019. (**a**,**b**) SST (°C), (**c**,**d**) Chl-a (mg/m$^3$), (**e**) SST, and Chl-a values of L1 to L4 in (b). The black arrow in (**a**) is the current velocity from the OSCAR data. The first arrow (22.33°N, 121.33°E) has a speed of 0.46 m/s, and the second arrow (22.67°N, 121.67°E) has a speed of 0.59 m/s.

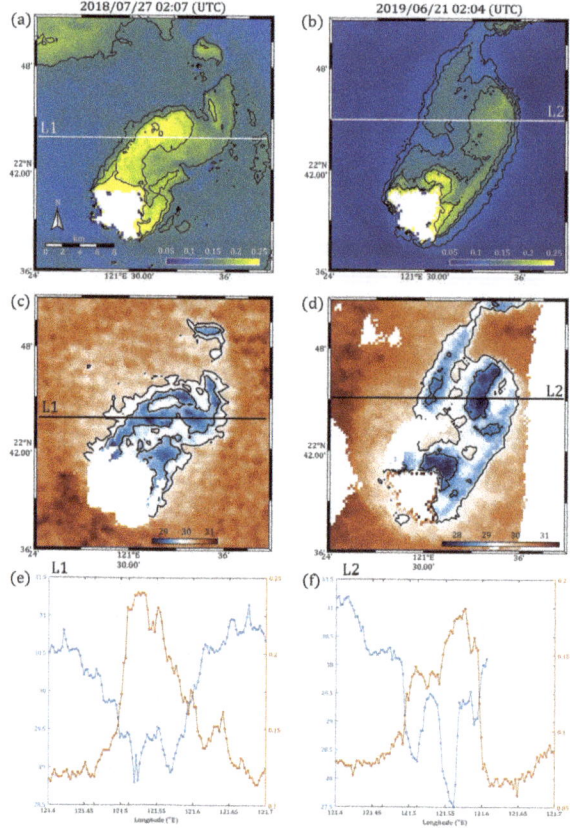

**Figure 6.** Two different spatially distributed vortices with (**a**,**b**) Chl-a (mg/m$^3$) and (**c**,**d**) SST (°C). Two cases obtained from GCOM-C SGLI data taken at 02:07 (UTC) 27 July 2018 (left) and at 02:04 (UTC) 21 June 2019 (right), (**e**,**f**) SST and Chl-a values of L1 and L2.

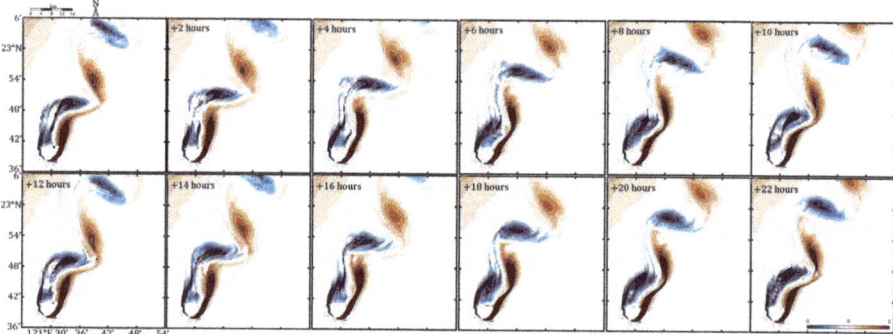

**Figure 7.** Results of the MITgcm numerical mode lasting one day. The background is the dimensionless parameter *Ro*.

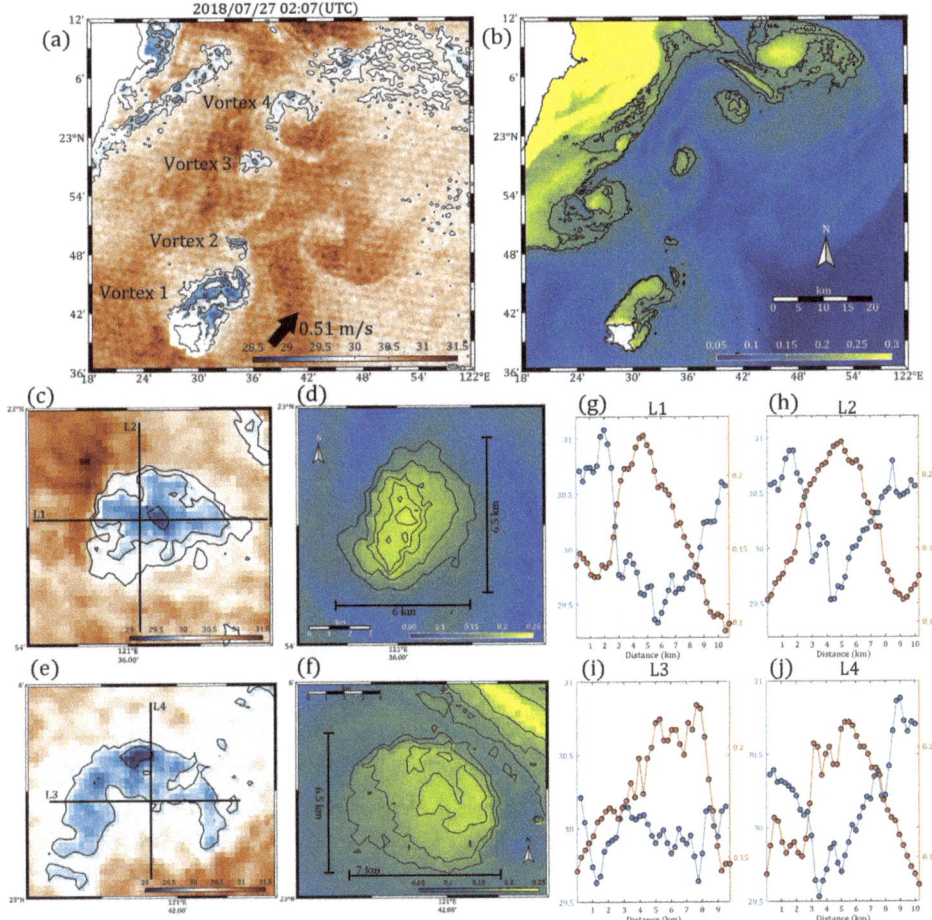

**Figure 8.** Case of the island vortex train with (**a**) SST (°C) and (**b**) Chl-a (mg/m$^3$) obtained from GCOM-C SGLI data taken at 02:07 (UTC), 27 July 2018 (UTC). (**c**,**d**) are zoom in on SST and Chl-a of vortex 3. (**e**,**f**) same as (**c**,**d**), but for vortex 4. (**g**–**j**) are SST and Chl-a values of L1 to L4. The black arrow (22.67°N, 121.67°E) in (**a**) is the current velocity from the OSCAR data with a speed of 0.51 m/s.

The typical Green Island wake (Figure 3) presents with a cold wake extending about 25 km behind the island. We assumed that the 10 lines cross the wake (L1 to L10 in Figure 3e). If a cruise survey is performed, only one of these lines could be observed in one hour due to the speed limit of the voyage. The lowest SST and the highest Chl-a concentration in the wake region were found at the imaginary lines of L1 to L4. L1 is about 1 km behind the island, and the SST of the wake area (27.3 °C) can reach 2.1 °C lower than that of the Kuroshio water (29.4 °C) as seen in Figure 4. The Chl-a concentration reached 0.23 mg/m$^3$, which was three times that of the Kuroshio water (0.07 mg/m$^3$). We can use this case to quantify the magnitude of the horizontal gradient of SST and Chl-a. From L1 to L4, the SST gradient is about 0.5 °C/km, and the Chl-a concentration is about 0.025 mgm$^{-3}$/km. Notably, the dynamic process of a wake can change rapidly, and the seawater structure of the entire wake area will likely change drastically within one hour. It can be seen here that from L7 to L10, there is a small vortex that may detach soon and propagate downstream. Around 23°–23.2°N, 121.7°–121.8°E, two small vortices were captured (Figure 3c,f). However, we are unsure whether this was caused by the shedding of the vortex train or by the coastal current.

Unlike the low-SST wake in Figure 3, which was concentrated only in a certain range behind the island, Figure 5 shows a case of a vortex train, which the low SST and high Chl-a concentration continue to extend downstream. The imagery line L1 in Figure 5b has the same purpose as L1 in Figure 3e. In this case, a Chl-a concentration of 0.66 mg/m$^3$ was observed, which was eight times the Kuroshio water, and the difference in SST was 2 °C. The distance between the two vortices is about 30 km (L2 to L4 in Figure 5e). L3 and L4 lines in Figure 5b passed through two vortices. The data sequence shows the situation of a low–high–low SST and high–low–high Chl-a concentration. Since the downstream vortex is affected by seawater mixing, the SST is usually higher than that of the wake close to the island.

A classic state in wake dynamics is illustrated in Figure 6, which was not captured in previous studies on Green Island. With high-resolution satellite data, we can observe the cooling zone transitions on the left and right side of the island wake region (usually corresponding to the magnitude of the vorticity). From the data in Figure 6e, it can be seen that there are two peaks in the low-SST wake region, 28.8 °C and 28.9 °C, respectively, and the corresponding Chl-a concentrations are 0.24 mg/m$^3$ and 0.20 mg/m$^3$. The left side of the wake region has a lower SST and a higher Chl-a concentration. From the data in Figure 6f, two peaks in the low-SST wake region can be found (28.3 °C and 27.5 °C), and the corresponding Chl-a concentrations are 0.14 mg/m$^3$ and 0.16 mg/m$^3$. The right side of the wake region has lower SST and higher Chl-a concentration. To understand the wake evolution more clearly, we used the MITgcm model to present the variations in Rossby number ($Ro$) during the wake's dynamic process (Figure 7):

$$Ro = \frac{U}{fL} \sim \frac{\zeta}{f} \tag{1}$$

$$\zeta = \frac{\partial v}{\partial x} - \frac{\partial u}{\partial y} \tag{2}$$

where $U$ and $L$ are the velocity and length scales of the phenomenon, $f$ is the Coriolis frequency, and $\zeta$ is the relative vorticity. The quantified vorticity using $Ro$ helps to interpret the eddy strength. A very small $Ro$ is typical to general circulation. A very high $Ro$ can be seen in rapidly rotating eddies. The simulation results show that positive and negative vorticity are sequentially detached from each other in a few hours. This clearly illustrates the results we captured from the GCOM-C images (Figure 6). This also demonstrates the importance of being careful when making cruise measurements, because each wake can change quickly, and the water may mix in different vorticity states for each observation station.

An ideal von Carmen vortex street can be seen in Figure 8a,b. Four sub-mesoscale vortices detached from the island to downstream. Vortex 3 (Figure 8c,d) and vortex 4 (Figure 8e,f) were selected for observations, and two imaginary lines that crossed the highest Chl-a position of the vortex were designed. The size of the vortex's core is about 5 km, which is close to the size of Green Island. The maximum Chl-a concentration in both vortices was 0.22 mg/m$^3$. An interesting phenomenon was also seen here. In the vortex core, the Chl-a concentration did not exactly match that of the SST. As a rotating vortex, the SST may increase during the propagation process due to the mixing of the Kuroshio. The SST variation of the vortex in the model experiment is difficult to simulate because it is not known which factors alongside seawater mixing could affect the vortex when it detaches downstream. In the satellite observations, it was often found that vortex street rarely appeared, which was very different from the model simulations [9]. In the next section, the geosynchronous Himawari-8 satellite data were used to analyze and discuss the dynamic process of the vortex train.

### 3.3. Temporal Variation and Vortex Trajectory

The previous section presented the spatial structure of biological hotspots in the wake area, and the dynamic process of the vortex is mentioned in this section. To study the dynamics of island wakes, the most important factor is to solve the setting of the two major parameters, the Strouhal number ($St$) and the Reynolds number ($Re$), as well as the propagation velocity of the vortex ($U_e$). The $St$ is a dimensionless number describing the oscillating flow mechanisms as:

$$St = \frac{L}{TU_0} \tag{3}$$

where $T$ is the vortex shedding period, $U_0$ is the incoming current speed, and $L$ is the characteristic length. The $Re$ is an important dimensionless quantity in fluid mechanics used to help predict the flow patterns in different fluid flow situations as:

$$Re = \frac{U_0 L}{v_h} \tag{4}$$

where $v_h$ is the horizontal eddy viscosity.

With continuous cloud-free data, the trajectory and velocity of the vortex could be tracked and calculated. A vortex has the characteristics of a closed isotherm of SST, and the lowest SST in the eddy can represent its center position [9]. In this way, the trajectory of the vortex can be tracked. An example of two consecutive vortices within 24 h of SST images is shown in Figure 9, with the temporal ranging from 21:00 (UTC) on 12 July 2016 to 20:00 (UTC) on 13 July 2016. We can see how the wake vortex detached and passed downstream. The first vortex was formed at 23:00 (UTC) on 12 July 2016 (the red star in Figure 9), and the second vortex was formed at 11:00 (UTC) on 13 July 2016 (the red dot in Figure 9), the shedding time interval was 12 h. The vortex detaching happens when the instability disturbances by background flow. The average speeds of the two vortices were 1.21 m/s and 1.45 m/s, respectively. To understand the variations of the sub-mesoscale Green Island vortices, a total of 101 vortex cases from July 2015 to December 2019 were calculated. A total of 78 cases occurred in the summer (June to August). In 101 cases, the average vortex propagation velocity was 0.95 ± 0.28 m/s, and the maximum and minimum values were 1.82 m/s and 0.31 m/s, respectively. A total of 101 vortex trajectories and their probability distributions are shown in Figure 10a,b. Figure 10c presents a histogram of property velocity statistics for 101 vortex cases. As mentioned in previous studies [8], the angle of the vortex distribution and the angle of the incoming velocity in front of the island are highly correlated. Therefore, based on the trajectory distribution in Figure 10a, the angle of the incoming current field in front of the island can be estimated. More than half of the vortex cases have propagation velocities between 0.8 and 1.2 m/s. The probability distribution of the trajectory is shown in Figure 10b. Most of the vortices with low SST and high Chl-a concentrations are concentrated within 30 km northeast of Green Island. The biological hotspots in the wake area mainly appear in this area of 900 km² and propagate downstream with the vortex stream. Based on the MCC method, the calculation results show that the average incoming flow velocity $U_0$ that caused the vortex shedding cases in Figure 10a is 1.15 ± 0.22 m/s.

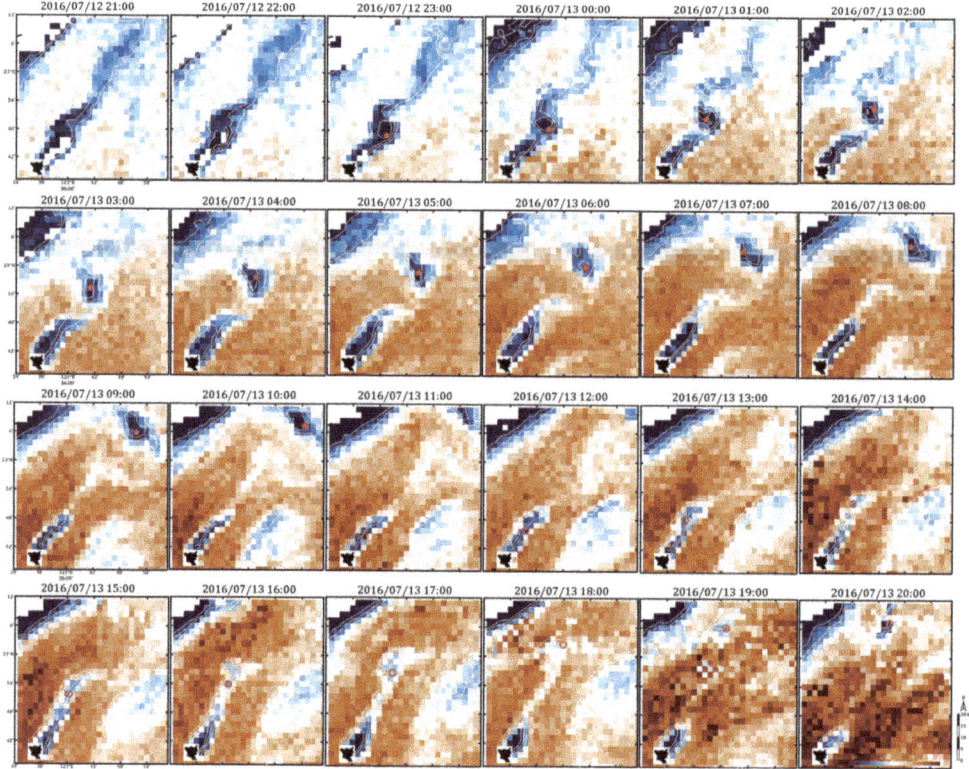

**Figure 9.** The 24-h continuous Himawari-8 SST images from 21:00 UTC on 12 July 2016 to 20:00 UTC on 13 July 2016. Red stars and red dots represent the center positions of the two vortex cases.

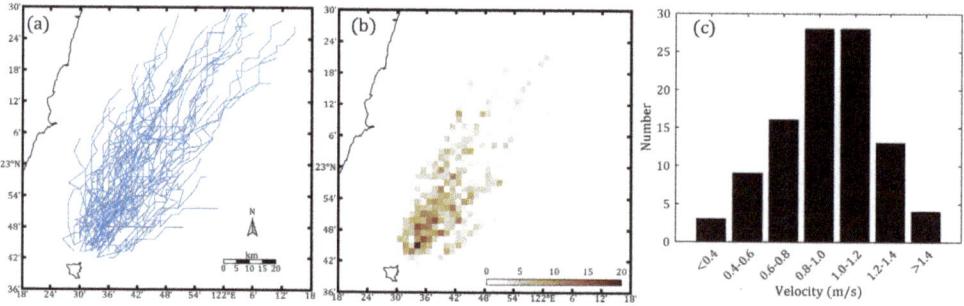

**Figure 10.** (**a**) trajectory of 101 vortex cases, (**b**) the distribution probability (%) of the vortices for 101 cases, and (**c**) a histogram of the property speed statistics for 101 vortex cases.

Next, we use the case of vortex shedding to calculate $St$, and we must determine the time interval between two continuous vortices. Therefore, a total of 38 cases were provided in this study for calculations. The results show that the average vortex shedding period is 14.8 h. Based on the settings of $L = 5500$ m and $v_h = 100$ m$^2$/s, the average $St$ and $Re$ are 0.114 and 64, respectively, which are consistent with the fitted curve values [27] as shown in Figure 11.

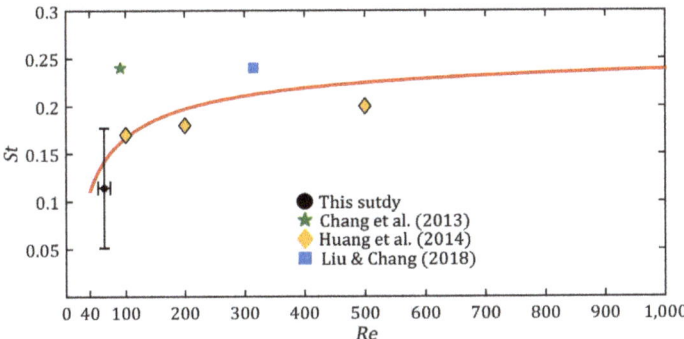

**Figure 11.** The Strouhal number (*St*) versus the Reynolds number (*Re*) diagram. The point for this study is expressed as the mean value with one standard deviation.

## 4. Discussion

### 4.1. The Relationship Between Re and St

The relationship between *Re* and *St* plays a major role in determining island wake patterns, evolution, and vortex shedding. The fitting curve relation between *Re* and *St* for cylinder wakes were formulated as [27]:

$$St = 0.273 - 1.11Re^{-\frac{1}{2}} + 0.482Re^{-1} \quad (5)$$

In theory, the *Re* value depends only on the size of $U_0$ when focusing on the same obstacle. However, Green Island is an irregular island, and there were differences in the choice of the diameter (*L*) of the obstacle. According to the measurements from the National Land Surveying and Mapping Center, Taiwan, the size of Green Island is 15.09 km$^2$, 4.6 km wide, 5.2 km long, and the longest side is about 6 km. Therefore, it is appropriate to choose *L* = 5–6 km for research on Green Island. However, some researchers have instead chosen *L* = 7 km, which indicates a width of 100 m for the island in the water [5,14,15]. It is worth noting that the choice of *L* will significantly affect the values of *St* and *Re*. In addition, the value selection of horizontal eddy viscosity ($v_h$) when calculating the *Re* value is a difficult problem to be solved. Since this value is not easy to measure with observations, previous studies have used estimated values. In the open ocean, the $v_h$ varies from 10$^2$ to 10$^5$ m$^2$/s [28]. For the mesoscale or sub-mesoscale ocean processes, the $v_h$ is generally chosen as 10$^3$ to 10$^4$ m$^2$/s [16]. Next, we sort the data of various researchers and recommend what types of values are most suitable for studying Green Island. Table 1 shows the corresponding data for each point in Figure 11. In this study, *L* and $v_h$ were selected as 5500 m and 100 m$^2$/s, respectively. The results of 38 cases showed that the cases fit the equation of the fitted curve. With the average $U_0$= 1.15 m/s, the average *Re* is 64, and the average *St* is 0. 114. We recommend that L be selected from 5–6 km. If a larger value of *L* is selected, the *St* may be too large (>0.3). A $v_h$ value of 100 m$^2$/s is suitable, as a $v_h$ value that is too small may cause the *Re* value to be too large. Although previous studies were based on observational data, there was only one sample. Take the case of this study as an example. If the $v_h$ value was chosen to be 15 m$^2$/s [14] or even 0.2–7 m$^2$/s [15], the points of the case could be far from the fitted curve equation.

Table 1. Corresponding values for each point in Figure 10.

| L (m) | $U_0$ (m/s) | $v_h$ (m²/s) | $Re$ | $St$ | Reference |
|---|---|---|---|---|---|
| 7000 | 1.3 | 100 | 91 | 0.24 | [5] |
| 5000 | 1 | 50 | 100<br>200<br>500 | 0.17<br>0.18<br>0.20 | [6] |
| 7000 | 0.675 | 15 | 315 | 0.24 | [14] |
| 5500 | $\overline{U_0} = 1.15$ | 100 | $\overline{Re} = 64$ | $\overline{St} = 0.114$ | This study |

Next, we will discuss the cases in our sample with $Re > 40$ (vortex street occurrence) but $St < 0.1095$ (the minimum value of the fitted curve equation). According to our parameter settings, if $Re$ is greater than 40, $U_0$ must be larger than 0.73 m/s. The minimum $U_0$ in 38 cases is 0.85 m/s. To make $St > 0.1095$, when $U_0 = 1$ m/s, $T < 13.95$ h, when $U_0 = 1.2$ m/s, $T < 11.62$ h, when $U_0 = 1.4$ m/s, $T < 9.96$ h. According to the field observation data [14], the vortex shedding period is 12 h. However, there are many cases in this study that showed a longer period for vortex shedding, which results in a lower $St$. According to our observations, a low SST vortex formed over a long period of time, but it did not detach for a long time. This verifies the results of previous studies that have used about 15 years of MODIS SST data to determine that about 87% of cases feature wakes occurring alone [9]. However, the cause of this situation is still unknown. It may be due to local wind, or it may be due to the seawater mixing at lower depths.

Another interesting feature is the vortex propagation velocity. According to numerical simulations [6], $U_e$ is between 0.3 and 0.51 m/s. Based on a near-shore observation [15], $U_e$ is 0.347 m/s. These two results indicate that $U_e$ is only 1/3 to 1/2 of $U_0$. However, this study indicates that the average $U_e$ is 0.95 m/s, which is about 83% of the $U_0$ value. In the island vortex of the Luzon Strait, a previous study calculated that $U_e$ is about 89% of $U_0$ [16]. We believe that the calculations in this study are reasonable. The previous results may be greatly underestimated because the complete period of the vortex movement is not measured.

### 4.2. Seasonal Changes in $U_0$ and Chlorophyll-a Concentrations

The velocity data for OSCAR based on the satellite altimeter, temperature gradients, and wind fields, it can explore seasonal variations. In addition to the Kuroshio being possibly affected by mesoscale eddy invasions [29], the seasonal variation of velocity also significantly affects wake evolution. Figure 12a–c show the distribution of the flow field of the Kuroshio annually and in summer and winter, respectively. Figure 12d shows the velocity changes each month. Table 2 summarizes the incoming flow speeds from 2010 to 2019. The fastest incoming flow speed was in July, which was 2.4 times the slowest in November. The Kuroshio is variable in different seasons. In addition to analyzing the changes in SST when observing the development of wakes [9], it is also important to observe changes in Chl-a concentration in the wake region, which may affect the aggregation of phytoplankton and fish populations. Based on the average Chl-a concentration (0.15 mg/m³) in the wake region, we analyzed the percentage of data that exceeds this value in different seasons (Figure 13). Since the surface water is relatively cold, and the wind is relatively strong in winter, more nutritious seawater in the deeper layer was mixed into the euphotic zone near the surface of the ocean, which nourishes the growth of phytoplankton. Therefore, such a broadly high Chl-a concentration is not related to the wake development. In summer, due to the vigorous development of wakes, there is a good chance that high Chl-a concentration could be produced within 15 km of the lee of the island.

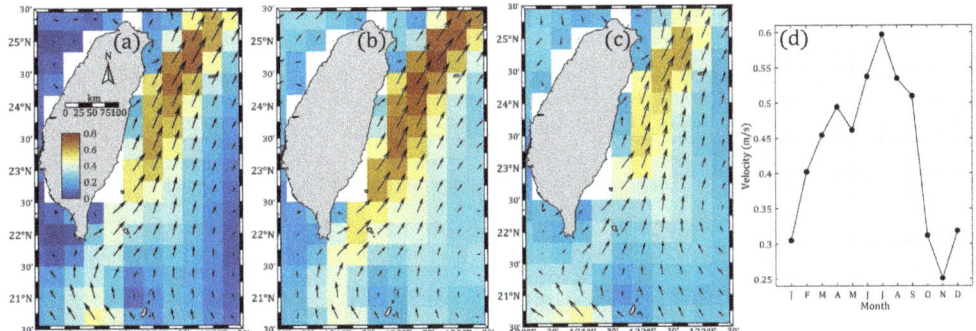

**Figure 12.** The OSCAR sea surface current velocity from 2010 to 2019, (**a**) the annual mean for (**b**) summer and (**c**) winter and (**d**) the average of the incoming current velocity for each month.

**Table 2.** The statistics for incoming speed (m/s) in front of Green Island for the years of 2010 to 2019.

|  | Average | Maximum | Minimum |
|---|---|---|---|
| Spring | 0.47 ± 0.10 | 0.82 | 0.27 |
| Summer | 0.56 ± 0.13 | 0.94 | 0.28 |
| Fall | 0.36 ± 0.18 | 0.84 | 0.02 |
| Winter | 0.34 ± 0.13 | 0.63 | 0.02 |

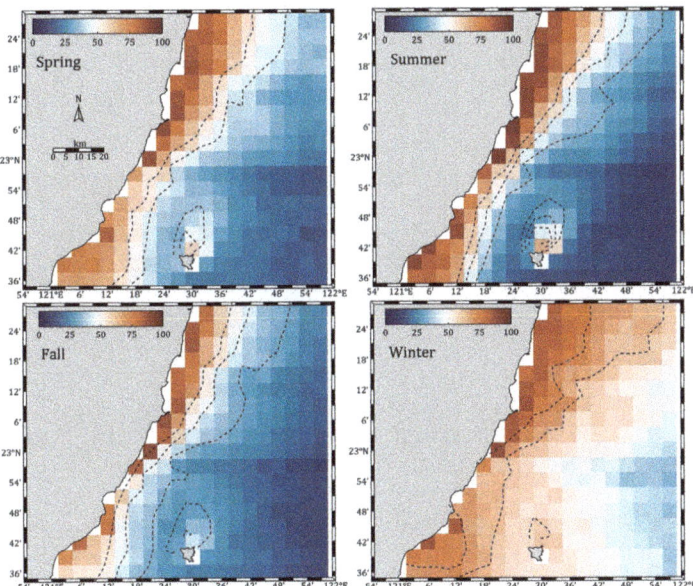

**Figure 13.** The probability distribution of the Chl-a concentration (>0.15 mg/m$^3$) in different seasons.

Of the 101 vortex trajectories in this study, 78 cases appeared in summer (June to August). Velocity and many missing values in winter due to clouds may be the main factors behind this phenomenon. According to theoretical experiments, when $5 < Re < 40$, a fixed pair of symmetric vortices occurred, which is also the most commonly observed cold wake formations in the lee of Green Island, when $40 < Re < 200$, a laminar vortex street occurred, which corresponds to the vortex shedding trains observed in this study. We used the MITgcm model to establish the SST of wake under the same ocean conditions in summer but with different incoming speeds (Figure 14). The three incoming speed conditions correspond to *Re* values of 70, 118, and 156, respectively. When *Re* = 70, no cold wake is generated, when *Re* = 118, the vortex street is generated, when *Re* = 156, the area of a cold wake is significantly expanded, and the SST drop is also significantly increased. The physical parameters obtained in this study can improve the establishment of numerical models of wakes. In addition to changes in the incoming speed, the ocean conditions, which include the values of the sea temperature, seawater stratified structures, Kuroshio meander [29], and local wind effects can affect the evolution of wakes. We will conduct simulation studies of different situations in a future study.

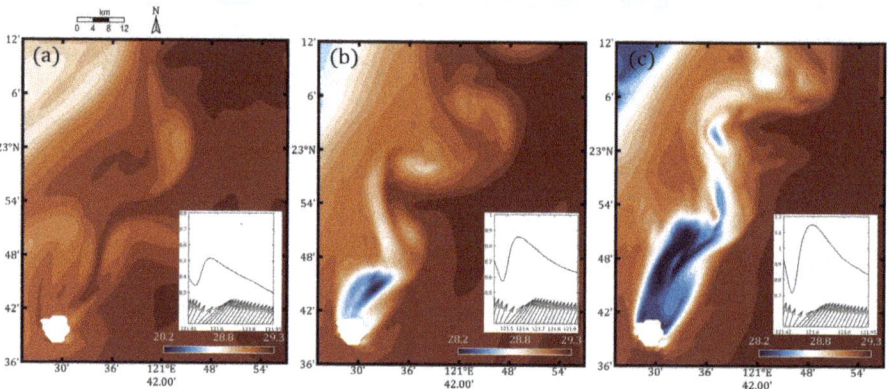

**Figure 14.** The island wake development from the MITgcm simulation for different Reynolds numbers. (**a**) *Re* = 70, (**b**) *Re* = 118, (**c**) *Re* = 156. The sub-image represents the change in speed (m/s) along 22.6°N.

### 4.3. Uncertainties, Errors, and Accuracies

The data of SST and Chl-a used in this study were processed by JAXA. The SST algorithm is based on the method developed for Himawari-8 SST [30]. SGLI SSTs were validated by the comparison with buoy data with 0.8 K difference, the Himawari-8 had an average SST difference of 0.18 K with a tropical atmosphere-ocean array [20]. The Chl-a concentration algorithm is developed based on the empirical algorithms for convenient use with the other sensor products, the estimated errors are −60 to +150% with in-situ data and MODIS data (https://suzaku.eorc.jaxa.jp/GCOM_C/). In this study, GCOM-C data presented significant variations in SST and Chl-a concentrations in the wake region, and the SST drop and changes in Chl-a were similar to previous in-situ measurements [9]. In addition, this study objectively used the lowest SST in the vortex to track its movement, which was hardly affected by the accuracy of the data. All data have been well managed based on cloud masking classification and decision. In this study, we chose cloud-free data for analysis.

### 5. Conclusions

This study used 250 m spatial resolution GCOM-C data and 1-h temporal resolution Himawari-8 imagery to analyze the surface structure and dynamic processes of the Green Island wakes. These two sets of satellite data help us more understand the characteristic of sub-mesoscale eddies in the Kuroshio region. Based on the GCOM-C data, we designed observation lines that were different from the cruise observations because every point on the imaginary lines was certain to occur at the same time. Details of the spatial structure of the wake are revealed in this study, including detailed SST and Chl-a changes, the fronts of SST and Chl-a between the wake and Kuroshio water, the distance between two consecutive vortices, the vorticity transition response to sea surface characteristics, and the different structures of SST and Chl-a in the same vortex. Based on the Himawari-8 imagery, the incoming current speed and the propagation speed of the vortex could be calculated. In total, 101 vortex cases from July 2015 to December 2019 were calculated. About 77% of the cases appeared in summer. The average vortex propagation speed was 0.95 m/s. More than half of the vortices had propagation speeds between 0.8 and 1.2 m/s. The average incoming surface current speed of Green Island was 1.15 m/s, which was calculated by the maximum cross-correlation method. In the 38 cases which have two continuous vortices, the average vortex shedding period was 14.8 h, the corresponding average Strouhal number was 0.114, and the Reynolds number was 64. In this study, the *St-Re* fitting curve relation was used to discuss the calculation results of the Green island wake compared to previous studies, and the results suggest that the size of Green Island is suitably selected from 5 to 6 km. Further, the 100 m$^2$/s of the horizontal eddy viscosity is still suitable. FIn addition, there is a good chance that a Chl-a larger than 0.15 mg/m$^3$ will be produced within 15 km at the lee of the island. This study used new remote sensing data to successfully observe and analyze the dynamic processes of the sub-mesoscale vortices. In the future, Himawari-8, the SGLI of GCOM-C, and the Sentinel-3 data will help establish a wake database to be used for ocean sustainability development and assistance in the protection of fishery resources.

**Author Contributions:** P.-C.H. conceived the project, conducted research, performed initial analyses, visualized data, and wrote the manuscript draft; P.-C.H. and C.-Y.L. processed scientific computing; C.-Y.H. and H.-J.L. Built and processed numerical model; P.-C.H. and C.-R.H. discussed, revised and corrected the manuscript. All authors have read and agreed to the published version of the manuscript.

**Funding:** This research and the APC was funded by the Ministry of Science and Technology of Taiwan through grants MOST 108-2611-M-019-019 and MOST 108-2811-M-019-506.

**Acknowledgments:** The authors appreciate all the data use provided from each open database. The Himawari-8 SST and Chl-a data were supplied by the P-Tree System, Japan Aerospace Exploration Agency (JAXA) (http://www.eorc.jaxa.jp/ptree/); the GCOM-C SGLI data were supplied form the Japan Aerospace Exploration Agency/National Aeronautics and Space Administration; the OSCAR ocean currents data were download from NASA PODAAC (https://doi.org/10.5067/OSCAR-03D01); the Sentinel-3 data provided by EUMETSAT for Copernicus; the open historical CTD data obtained from the Ocean Data Bank of the Ministry of Science and Technology of Taiwan (http://www.odb.ntu.edu.tw/en/); and the open source of MITgcm were provided by http://mitgcm.org/.

**Conflicts of Interest:** The authors declare no conflict of interest.

## References

1. Zeiden, K.L.; Rudnick, D.L.; MacKinnon, J.A. Glider observations of a mesoscale oceanic island wake. *J. Phys. Oceanogr.* **2019**, *49*, 2217–2235. [CrossRef]
2. St. Laurent, L.; Ijichi, T.; Merrifield, S.T.; Shapiro, J.; Simmons, H.L. Turbulence and vorticity in the wake of Palau. *Oceanography* **2019**, *32*, 102–109. [CrossRef]
3. Kodaira, T.; Waseda, T. Tidally generated island wakes and surface water cooling over Izu Ridge. *Ocean Dyn.* **2019**, *69*, 1373–1385. [CrossRef]
4. Tanaka, T.; Hasegawa, D.; Yasuda, I.; Tsuji, H.; Fujio, S.; Goto, Y.; Nishioka, J. Enhanced vertical turbulent nitrate flux in the Kuroshio across the Izu Ridge. *J. Oceanogr.* **2019**, *75*, 195–203. [CrossRef]
5. Chang, M.H.; Tang, T.Y.; Ho, C.R.; Chao, S.Y. Kuroshio-induced wake in the lee of Green Island off Taiwan. *J. Geophys. Res. Ocean.* **2013**, *118*, 1508–1519. [CrossRef]
6. Huang, S.J.; Ho, C.R.; Lin, S.L.; Liang, S.J. Spatial-temporal scales of Green Island wake due to passing of the Kuroshio current. *Int. J. Remote Sens.* **2014**, *35*, 4484–4495. [CrossRef]
7. Zheng, Z.W.; Zheng, Q. Variability of island-induced ocean vortex trains, in the Kuroshio region southeast of Taiwan Island. *Cont. Shelf Res.* **2014**, *81*, 1–6. [CrossRef]
8. Hsu, P.C.; Chang, M.H.; Lin, C.C.; Huang, S.J.; Ho, C.R. Investigation of the island-induced ocean vortex train of the Kuroshio Current using satellite imagery. *Remote Sens. Environ.* **2017**, *193*, 54–64. [CrossRef]
9. Hsu, P.C.; Cheng, K.H.; Jan, S.; Lee, H.J.; Ho, C.R. Vertical structure and surface patterns of Green Island wakes induced by the Kuroshio. *Deep-Sea Res. Part I* **2019**, *143*, 1–16. [CrossRef]
10. Gove, J.M.; McManus, M.A.; Neuheimer, A.B.; Polovina, J.J.; Drazen, J.C.; Smith, C.R.; Merrifield, M.A.; Frienlander, A.M.; Ehses, J.S.; Young, C.W.; et al. Near-island biological hotspots in barren ocean basins. *Nat. Commun.* **2016**, *7*, 1–8. [CrossRef]
11. Chen, T.C.; Ku, K.C.; Ying, T.C. A process-based collaborative model of marine tourism service system–The case of Green Island area, Taiwan. *Ocean Coast. Manag.* **2012**, *64*, 37–46. [CrossRef]

12. Denis, V.; Soto, D.; De Palmas, S.; Lin, Y.T.; Benayahu, Y.; Huang, Y.; Liu, S.L.; Chen, J.W.; Chen, Q.; Sturaro, N.; et al. Mesophotic Coral Ecosystems. In *Coral Reefs of the World*; Loya, Y., Puglise, K., Bridge, T., Eds.; Springer: Cham, Switzerland, 2019; Volume 12, pp. 249–264. [CrossRef]
13. Hsu, T.W.; Doong, D.J.; Hsieh, K.J.; Liang, S.J. Numerical study of monsoon effect on Green Island wake. *J. Coast. Res.* **2015**, *31*, 1141–1150. [CrossRef]
14. Liu, C.L.; Chang, M.H. Numerical studies of submesoscale island wakes in the Kuroshio. *J. Geophys. Res. Ocean.* **2018**, *123*, 5669–5687. [CrossRef]
15. Chang, M.H.; Jan, S.; Liu, C.L.; Cheng, Y.H.; Mensah, V. Observations of island wakes at high Rossby numbers: Evolution of submesoscale vortices and free shear layers. *J. Phys. Oceanogr.* **2019**, *49*, 2997–3016. [CrossRef]
16. Zheng, Q.; Lin, H.; Meng, J.; Hu, X.; Song, Y.T.; Zhang, Y.; Li, C. Sub-mesoscale ocean vortex trains in the Luzon Strait. *J. Geophys. Res. Ocean.* **2008**, *113*. [CrossRef]
17. Taniguchi, N.; Kida, S.; Sakuno, Y.; Mutsuda, H.; Syamsudin, F. Short-Term Variation of the Surface Flow Pattern South of Lombok Strait Observed from the Himawari-8 Sea Surface Temperature. *Remote Sens.-Basel* **2019**, *11*, 1491. [CrossRef]
18. Liu, J.; Emery, W.J.; Wu, X.; Li, M.; Li, C.; Zhang, L. Computing Coastal Ocean Surface Currents from MODIS and VIIRS Satellite Imagery. *Remote Sens.* **2017**, *9*, 1083. [CrossRef]
19. Hu, Z.; Qi, Y.; He, X.; Wang, Y.H.; Wang, D.P.; Cheng, X.; Liu, X.H.; Wang, T. Characterizing surface circulation in the Taiwan Strait during NE monsoon from Geostationary Ocean Color Imager. *Remote Sens. Environ.* **2019**, *221*, 687–694. [CrossRef]
20. Ditri, A.L.; Minnett, P.J.; Liu, Y.; Kilpatrick, K.; Kumar, A. The Accuracies of Himawari-8 and MTSAT-2 sea-surface temperatures in the tropical western Pacific Ocean. *Remote Sens.* **2018**, *10*, 212. [CrossRef]
21. ESR. *OSCAR Third Degree Resolution Ocean Surface Currents*; Ver. 1; PO.DAAC: Pasadena, CA, USA, 2009. [CrossRef]
22. Bonjean, F.; Lagerloef, G.S.E. Diagnostic model and analysis of the surface currents in the tropical Pacific Ocean. *J. Phys. Oceanogr.* **2002**, *32*, 2938–2954. [CrossRef]
23. Johnson, E.S.; Bonjean, F.; Lagerloef, G.S.; Gunn, J.T.; Mitchum, G.T. Validation and error analysis of OSCAR sea surface currents. *J. Atmos. Ocean. Technol.* **2007**, *24*, 688–701. [CrossRef]
24. Marshall, J.; Adcroft, A.; Hill, C.; Perelman, L.; Heisey, C. A finite-volume, incompressible Navier Stokes model for studies of the ocean on parallel computers. *J. Geophys. Res. Ocean.* **1997**, *102*, 5753–5766. [CrossRef]
25. Orlanski, I. A simple boundary condition for unbounded hyperbolic flows. *J. Comput. Phys.* **1976**, *21*, 251–269. [CrossRef]
26. Klymak, J.M.; Legg, S.M. A simple mixing scheme for models that resolve breaking internal waves. *Ocean. Model.* **2010**, *33*, 224–234. [CrossRef]
27. Williamson, C.H.K.; Brown, G.L. A series in $1/\sqrt{Re}$ to represent the Strouhal–Reynolds number relationship of the cylinder wake. *J. Fluids Struct.* **1998**, *12*, 1073–1085. [CrossRef]
28. Apel, J.R. *Principles of Ocean Physics*; Academic Press: London, UK, 1987.
29. Hsu, P.C.; Lin, C.C.; Huang, S.J.; Ho, C.R. Effects of cold eddy on Kuroshio meander and its surface properties, east of Taiwan. *IEEE J.-STARS* **2016**, *9*, 5055–5063. [CrossRef]
30. Kurihara, Y.; Murakami, H.; Kachi, M. Sea surface temperature from the new Japanese geostationary meteorological Himawari-8 satellite. *Geophys. Res. Lett.* **2016**, *43*, 1234–1240. [CrossRef]

© 2020 by the authors. Licensee MDPI, Basel, Switzerland. This article is an open access article distributed under the terms and conditions of the Creative Commons Attribution (CC BY) license (http://creativecommons.org/licenses/by/4.0/).

Article

# Validation of a Primary Production Algorithm of Vertically Generalized Production Model Derived from Multi-Satellite Data around the Waters of Taiwan

Kuo-Wei Lan [1,2], Li-Jhih Lian [3], Chun-Huei Li [4,*], Po-Yuan Hsiao [1] and Sha-Yan Cheng [1]

1. Department of Environmental Biology Fisheries Science, National Taiwan Ocean University, 2 Pei-Ning Rd., Keelung 20224, Taiwan; kwlan@mail.ntou.edu.tw (K.-W.L.); 10831005@mail.ntou.edu.tw (P.-Y.H.); eric@mail.ntou.edu.tw (S.-Y.C.)
2. Center of Excellence for Oceans, National Taiwan Ocean University, 2 Pei-Ning Rd., Keelung 20224, Taiwan
3. Taiwan Cross-Strait Fisheries Cooperation and Development Foundation, 100 Heping W. Rd, Taipei 10070, Taiwan; juno516874@gmail.com
4. Marine Fisheries Division, Fisheries Research Institute, Council of Agriculture, 199 Hou-Ih Rd, Keelung 20246, Taiwan
* Correspondence: chli@mail.tfrin.gov.tw; Tel.: +886-2-24622101 (ext. 2304)

Received: 24 March 2020; Accepted: 17 May 2020; Published: 19 May 2020

**Abstract:** Basin-scale sampling for high frequency oceanic primary production (PP) is available from satellites and must achieve a strong match-up with in situ observations. This study evaluated a regionally high-resolution satellite-derived PP using a vertically generalized production model (VGPM) with in situ PP. The aim was to compare the root mean square difference (RMSD) and relative percent bias (Bias) in different water masses around Taiwan. Determined using light–dark bottle methods, the spatial distribution of VGPM derived from different Chl-a data of MODIS Aqua ($PP_A$), MODIS Terra ($PP_T$), and averaged MODIS Aqua and Terra ($PP_{A\&T}$) exhibited similar seasonal patterns with in situ PP. The three types of satellite-derived PPs were linearly correlated with in situ PPs, the coefficients of which were higher throughout the year in $PP_{A\&T}$ ($r^2 = 0.61$) than in $PP_A$ ($r^2 = 0.42$) and $PP_T$ ($r^2 = 0.38$), respectively. The seasonal RMSR and bias for the satellite-derived PPs were in the range of 0.03 to 0.09 and −0.14 to −0.39, respectively, which suggests the $PP_{A\&T}$ produces slightly more accurate PP measurements than $PP_A$ and $PP_T$. On the basis of environmental conditions, the subareas were further divided into China Coast water, Taiwan Strait water, Northeastern upwelling water, and Kuroshio water. The VPGM PP in the four subareas displayed similar features to Chl-a variations, with the highest PP in the China Coast water and lowest PP in the Kuroshio water. The RMSD was higher in the Kuroshio water with an almost negative bias. The $PP_A$ exhibited significant correlations with in situ PP in the subareas; however, the sampling locations were insufficient to yield significant results in the China Coast water.

**Keywords:** primary productivity; vertically generalized production model; waters around Taiwan; MODIS Aqua and Terra

## 1. Introduction

Primary production (PP) refers to the production of organic carbon during photosynthesis [1]. It sets the upper limit for ocean productivity and is an essential measure of the ocean's capacity to transform carbon dioxide into particulate organic carbon at the base of the food web [2–4]. From a bottom-up perspective [2,5,6], PP is also a good predictor of the potential yield of the world's oceans.

In the marine environment, in situ measurements of PP are taken using materials such as $^{14}$C [7], $^{13}$C [8], chlorophyll a (Chl-a) fluorescence [9], and oxygen isotopes [10]. These shipboard measurements of the snapshot sections vary over short temporal and spatial scales [7–10]. Furthermore, it can be time-consuming to represent minute fractions of ecosystems [6,11]. Scaling these relatively separate in situ measurements of the snapshot sections to a regional scale, let alone basin or global scale projections, therefore remains a significant challenge and needs to rely on remote sensing data and models [11–14].

Basin-scale sampling for high-frequency PP is available from satellites [13]. Ocean color images derived from remote sensing are ideal for assessing PP on a regional to global scale, and provide high-quality spatial and temporal coverage that give daily estimations of the attenuation coefficient, phytoplankton biomass, and photosynthetically available radiation (PAR) [14]. Remote sensing of ocean color cannot provide adequate information on oceanic PP without the support of models and sea truth data [15,16]. Several analytical, empirical, and bio-optical models are currently used to determine ocean PP [2,15,17,18]. Chl-a based models were chosen for this study; a large archive of regional pigment data is available for use [19,20] and only a limited amount of bio-optical data is. The vertically generalized production model (VGPM) formulated by Behrenfeld and Falkowski [19] is among the most commonly used and simplest models for estimating PP from Chl-a data obtained from satellites.

The VGPM is a vertically integrated and light-dependent model that characterizes the environmental factors affecting PP into those that control the optimal efficiency of the productivity profile and influence the relative vertical distribution of PP [19]. The advantage of the VGPM is that it incorporates satellite remote sensing data and employs minimal parameterization of input variables to derive PP [16]. Despite the understanding and knowledge of the ocean optics that determine ocean color signals and the photosynthetic process, PP derived from satellite data often have limited success in reproducing the variability observed in PP data [18,20,21]. Comparisons of PP models have shown that modeled estimates are twice as accurate as that of the carbon-based estimates [22,23]. Their application yields different results; choosing the most realistic one is therefore often a regional issue and the regional dependence of photosynthetic efficiency on hydro-optical and biochemical conditions must be taken into account [21–23]. Consequently, to obtain an appropriate match with in situ observations of PP, satellite-derived models must consider the peculiarities of regional ecosystems.

The Taiwan Strait is an important channel that transports water along the western part of Taiwan and chemical constituents between the South China Sea and East China Sea. Its alternating monsoon-forcing, complex bottom topography, and the conjunction of several current systems means that its ecosystem dynamics and biogeochemical and physical processes vary substantially in space and in time [24–26]. The warm Kuroshio current flows through the eastern part of Taiwan, and a cold dome can often be observed over the edge of the continental shelf's northeast sides [27,28]. In a review of previous studies, four major upwelling regions were identified around Taiwan, namely along the northwestern and southwestern coast of the Taiwan Strait, on the Taiwan Bank, and near the Penghu Islands [27–29]. The seasonal variation and spatial distribution of PP and phytoplankton biomass are largely controlled by the input of nutrients from various water masses [28]. The typhoon and tropical storms led to strong vertical water mixing enhanced nutrients and derived a diatom bloom around the waters of Taiwan in summer [26,30]. In particular, large-scale climatic oscillations, such as the ENSO events, also can cause sea surface temperature (SST) and PP changes on an interannual scale [28,31]. PP is an essential component of both terrestrial and aquatic ecosystems. The total fish and invertebrate production in an ecosystem-based approach to fisheries management is ultimately limited by ecosystem PP [5,6]. The high PP around the waters of Taiwan also sustains commercially important species of fish and cephalopod which develop their life cycle [28,32–34]. The development of site-specific models to estimate PP is therefore extremely desirable. Although in situ measurements of PP have been presented in previous studies (e.g., [24,26,28,35]), high-resolution PP distributions and estimations around Taiwan are rare. The objectives of this study were (1) to evaluate a regionally modified version of the VGPM using high-resolution (1.1 km$^2$) SST and Chl-a derived from satellite remote sensing data with in situ PP, and (2) to compare high-resolution VPGM PP and in situ PP for different water masses around Taiwan to compute the root mean square difference (RMSD) and relative percentage bias (Bias). The high-resolution VPGM PP calculated from satellite data that correspond to 15 cruises provides the first view of PP around Taiwan.

## 2. Data and Methods

### 2.1. In Situ Measurements and Water Sampling

Hydrographic, optical, and biogeochemical properties were investigated in 62 sampling locations around Taiwan in ranges between 21.5°–26°N and 119°–123°E (Figure 1). These covered the period 2009 to 2013 during different seasons on the vessels of Fisheries Research I (Table 1). We defined December, January, and February as winter; March, April, and May as spring; June, July, and August as summer; and September, October, and November as autumn.

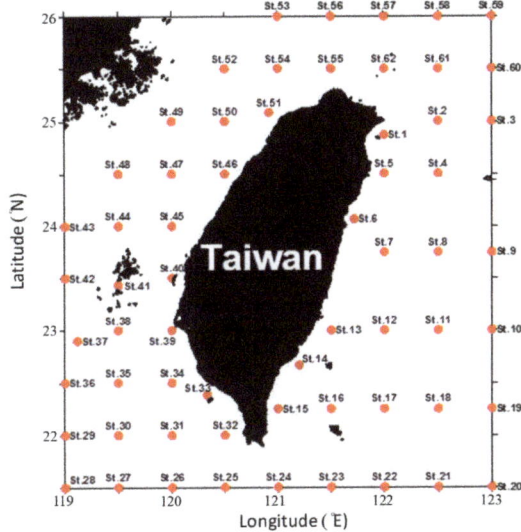

**Figure 1.** Sixty-two sampling locations of Fisheries Research I cruises around the waters of Taiwan between 21.5°–26°N and 119°–123°E.

**Table 1.** Cruise numbers, data, season, and number of PP measurement stations of Fisheries Research I from 2009 to 2013.

| Cruise No. | Date of the Cruise | Season | No. of PP Measurement Stations |
|---|---|---|---|
| FR1-2009-08-25 | 25 August–5 September, 2009 | Summer | 62 |
| FR1-2010-01-07 | 7 January–18 January, 2010 | Winter | 62 |
| FR1-2010-04-08 | 8 April–19 April, 2010 | Spring | 62 |
| FR1-2010-09-27 | 27 September–6 October, 2010 | Autumn | 62 |
| FR1-2011-01-13 | 13 January–24 January 2011 | Winter | 36 |
| FR1-2011-04-21 | 21 April–26 April, 2011 | Spring | 36 |
| FR1-2011-08-09 | 9 August–18 August, 2011 | Summer | 61 |
| FR1-2011-10-17 | 17 October–27 October, 2011 | Autumn | 62 |
| FR1-2011-12-28 | 28–31 December, 2011 1–8 January, 2012 | Winter | 59 |
| FR1-2012-04-18 | 18 April–30 April, 2012 | Spring | 62 |
| FR1-2012-08-19 | 19 August–4 September, 2012 | Summer | 55 |
| FR1-2012-11-02 | 2 November–11 November, 2012 | Autumn | 61 |
| FR1-2013-01-04 | 4 January–15 January, 2013 | Winter | 62 |
| FR1-2013-05-08 | 8 May–18 May, 2013 | Spring | 62 |
| FR1-2013-10-03 | 3 October–14 October, 2013 | Autumn | 61 |

In situ primary production was determined by light–dark bottle methods through incubation in 300 mL dissolved oxygen (DO) bottles. This technique was modified using Winkler's method. We prepared two types of DO bottles, one of which was transparent (light bottle) and the other was wrapped in aluminum foil (dark bottle). Seawater samples were collected at depths of 5, 25, and 50 m using 10 L Niskin bottles in each station, and water samples were filtered using a 300 μm filter cloth. The filtered sample was then packed into two DO bottles: one light bottle and one dark bottle. Each bottle was filled with 300 mL of seawater, following which, measurements were taken of the temperature and dissolved oxygen using a dissolved oxygen meter (YSI Model 52).

One light bottle was placed in the incubator (constant temperature of 25 °C, 4000 lux light). The dark bottle was set in the incubator, which was dark. After one day, the temperature and DO of the three bottles were measured. The PP was the difference in dissolved oxygen between the light and dark bottles, the formula for which was as follows:

In situ PP = ([$O_2$]L−[$O_2$]D) × carbon atom weight/dissolved oxygen atom weight/1 day.

[$O_2$]L: the dissolved oxygen in the light bottle after incubating for 1 day.
[$O_2$]D: the dissolved oxygen in the dark bottle after incubating for 1 day.
The PP (mg C m$^{-2}$ d$^{-1}$) was then integrated in terms of depth (m).
The depth ranged from 0–50 m.

## 2.2. Satellite-Derived PP Estimates

The PP model of the VGPM used in this study was based on Chl-a concentration, and the formulation and parameterization were recommended by Behrenfeld and Falkowski [19]. Maximum photosynthetic efficiency in VGPM is described as an optimal rate of photosynthesis ($P_{opt}^B$) in a water column normalized to Chl-a concentration (mg·C·mg·Chl$^{-1}$·h$^{-1}$). The VGPM estimates the daily integrated PP in a water column of euphotic depth ($PP_{eu}$, mg C m$^{-2}$ day$^{-1}$) as follows:

$$PP_{eu} = 0.66125 \times Chla \times P_{opt}^B \times \frac{PAR}{PAR + 4.1} \times Z_{eu} \times DP \qquad (1)$$

Photosynthetically available radiation (PAR) denotes daily averaged surface photosynthetic active radiation at 400–700 nm (E·m$^{-2}$·day$^{-1}$), $Z_{eu}$ denotes euphotic depth, and DP denotes a day photoperiod. $Z_{eu}$ was calculated from satellite surface chlorophyll-a concentration for lower and higher total chlorophyll conditions following Morel and Berthon [36]. $P_{opt}^B$ is expressed as a seventh-order polynomial function of SST [19], which is formulated as follows:

$$P_{opt}^B = 1.2956 + 2.749 \times 10^{-1} SST + 6.17 \times 10^{-2} SST^2 - 2.05 \times 10^{-2} SST^3 + 2.462 \times 10^{-3} SST^4 - 1.348 \times 10^{-4} SST^5 + 3.4132 \times 10^{-6} SST^6 - 3.27 \times 10^{-8} SST^7 \qquad (2)$$

The satellite data used in VGPM during the study period of 2009–2013 are SST, Chl-a, and PAR. MODIS Aqua/Terra daily Level 1A were downloaded from the NASA Ocean Color website. SeaDAS v6.2 was used to process high-resolution (1.1 km) local area coverage images and Chl-a data (OC3Mv6 algorithm). Daily PAR product data were downloaded from the Ocean Productivity database. Daily SST data were extracted from NOAA AVHRR SST images and had a spatial resolution of 1.1 km. The NOAA HRPT data, including AVHRR scenes, were received at a ground station at National Ocean Taiwan University. Daily DP data were produced by the Central Weather Bureau using a Precision Spectral Pyranometer. The three categories of satellite-derived PP were calculated using different MODIS Chl-a data as follows: (1) MODIS Terra evaluated primary production ($PP_T$), (2) MODIS Aqua evaluated primary production ($PP_A$), and (3) the averaged MODIS Aqua and Terra evaluated primary production ($PP_{A\&T}$).

The divisions of waters around Taiwan also have different oceanographic characteristics. They were divided into 1° gridded areas across 36 sites (Figure 2a). Cluster analysis with normalized

Euclidean distances was used to measure levels of similarity in gridded areas in the waters around Taiwan in 2009–2013, including monthly Chl-a, SST, and PAR. Ward's method was used to illustrate the relationships between them in a dendrogram. The cluster analysis was conducted using STATISTICA 8 statistical software.

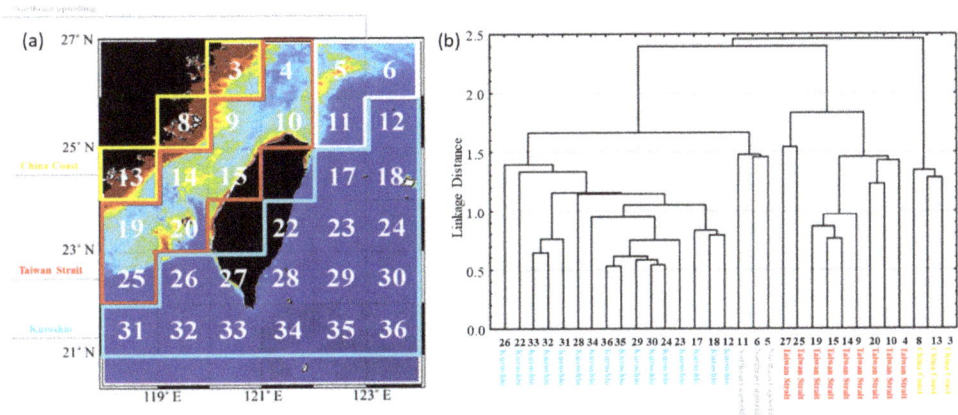

**Figure 2.** (**a**) The waters around Taiwan were divided into 1° gridded areas for 36 sites. The subareas for the China Coast water, Taiwan Strait water, Northeastern upwelling water, and Kuroshio water were divided by the cluster tree diagram results using the monthly Chl-a, SST, and PAR during study period (2009–2013) in (**b**). The monthly mean $PP_{A\&T}$ image in 2009 was used as an example in (**a**).

*2.3. Match-Up Data and the Assessment of Satellite PP Models*

To evaluate the satellite-derived PP with in situ PP, we produced pairs of collocated satellite overpasses and in situ sampling was extracted with a time difference shorter than ± 12 h. The satellite observations have spatial averages of 3 × 3 pixels around each sample site location and were compared with field measurements.

The PP values derived from the VPGM model were regressed against in situ data and a type II linear regression model was applied, as both field and modeled data are subject to errors. The slope, intercept, and the correlation coefficient (r) were then determined. For Chl-a, the regression was performed between log-transformed values. The RMSD statistic assesses model skill such that models with lower values have higher skill, and the model bias assesses whether a model over- or underestimates PP [11]. We calculated the RMSD for n samples of PP:

$$RMSD = \sqrt{\frac{\sum_{i=1}^{n} [\,log(PP_{model,i}) - log(PP_{in\ situ,i})\,]^2}{n}} \qquad (3)$$

where $(PP_{model,i})$ modeled PP and $(PP_{in\ situ,i})$ represents in situ PP estimates at each site. To assess whether a model over- or underestimated PP, we calculated the bias of each model as follows:

$$Bias = \overline{log(PP_{model})} - \overline{log(PP_{in\ situ})} \qquad (4)$$

## 3. Results

*Annual and Seasonal Trends in PP*

The spatial distribution of VPGM-based production derived from AVHRR SST and MODIS Chl-a in the waters around Taiwan showed similar seasonal spatial patterns to in situ PP (Figures 3 and 4). The highest concentration of PPs was observed along the Mainland China Coast and four major upwelling regions (southwestern and northwestern coast, Taiwan Bank, and Penghu Islands) around the waters of Taiwan. Lower PP values were obtained across the whole year within the Kuroshio-influenced region in eastern parts. During the study periods, the total numbers of colocated satellite overpasses and in situ sampling sites were extracted with a time difference shorter than ± 12 h. This contained 102 sets for $PP_{A\&T}$, 151 for $PP_A$, and 150 for $PP_T$, respectively (Table 2).

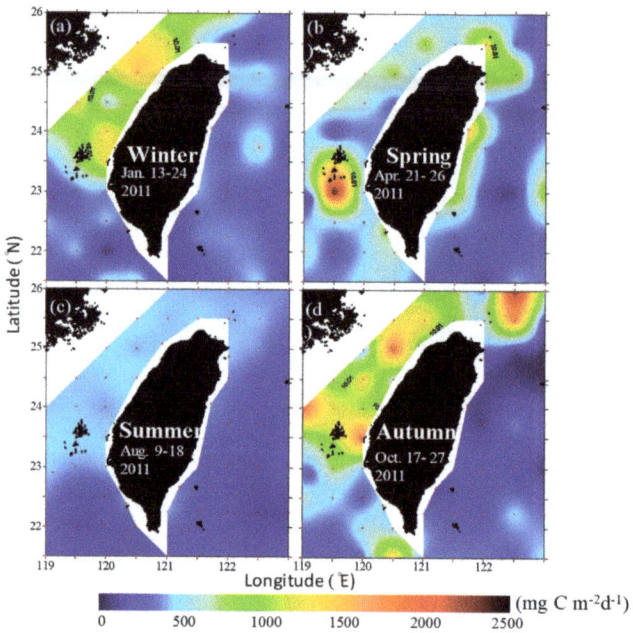

**Figure 3.** Spatial distribution of in situ PP determined using light–dark bottle methods in 2011: (**a**) winter, (**b**) spring, (**c**) summer, and (**d**) autumn.

**Table 2.** Extracted number of satellite-derived PPs with in situ PPs, correlation coefficients ($r^2$), and $p$ values for $PP_{A\&T}$, $PP_A$, and $PP_T$ for whole years and different seasons. The RMSD and bias for $PP_{A\&T}$, $PP_A$, and $PP_T$ for whole years.

|  | Extracted Number | | | Correlation Coefficients | | | $p$ | | |
|---|---|---|---|---|---|---|---|---|---|
|  | PP (A&T) | $PP_A$ | $PP_T$ | PP (A&T) | $PP_A$ | $PP_T$ | PP (A&T) | $PP_A$ | $PP_T$ |
| Years | 102 | 151 | 150 | 0.61 | 0.42 | 0.38 | <0.05 | <0.05 | <0.05 |
| Spring | 18 | 26 | 23 | 0.74 | 0.55 | 0.46 | <0.05 | <0.05 | <0.05 |
| Summer | 25 | 49 | 48 | 0.54 | 0.25 | 0.37 | <0.05 | <0.05 | <0.05 |
| Autumn | 52 | 55 | 59 | 0.51 | 0.46 | 0.42 | <0.05 | <0.05 | <0.05 |
| Winter | 7 | 21 | 20 | 0.33 | 0.14 | 0.07 | 0.31 | 0.22 | 0.41 |

|  | RMSD | | | Bias | | |
|---|---|---|---|---|---|---|
|  | PP (A&T) | $PP_A$ | $PP_T$ | PP (A&T) | $PP_A$ | $PP_T$ |
| Years | 0.37 | 0.34 | 0.34 | −0.24 | −0.197 | −0.174 |

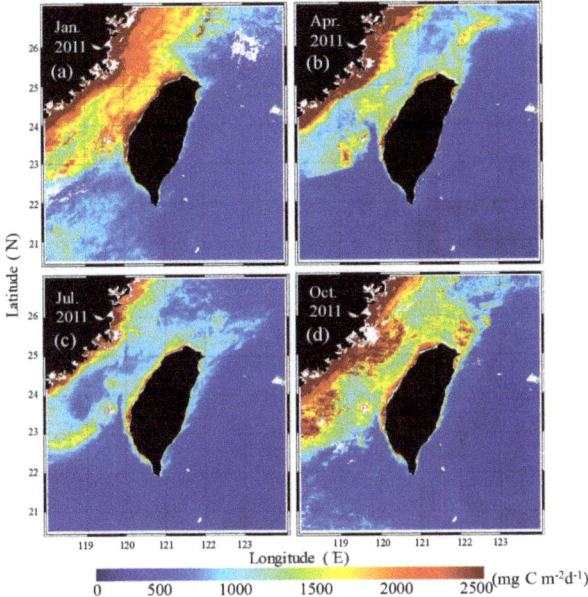

**Figure 4.** Monthly mean spatial distributions of VPGM PP derived from $PP_{A\&T}$ in 2011: (**a**) January, (**b**) April, (**c**) July, and (**d**) October.

## 4. Comparison of Satellite-Derived and in Situ PP

To validate the model results, the three satellite-derived PP values ($PP_{A\&T}$, $PP_A$, $PP_T$) were compared with in situ PP values. They were linearly correlated with the situ PP and the coefficient was higher in $PPA\&T$ ($r^2 = 0.61$) than in $PP_A$ ($r^2 = 0.42$) and $PP_T$ ($r^2 = 0.38$) throughout the year (Table 2). The highest correlations were observed in spring, especially for $PP_{A\&T}$ ($r^2 = 0.74$) (Figure 5).

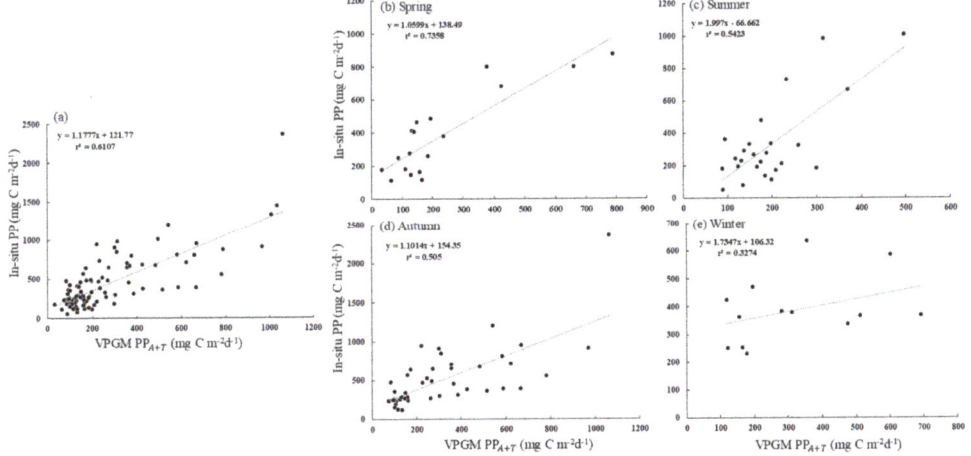

**Figure 5.** Relationship between the VPGM $PP_{A\&T}$ and in situ PP for (**a**) the whole year, (**b**) spring, (**c**) summer, (**d**) autumn, and (**e**) winter during the study period.

The lowest correlations were observed in winter and were nonsignificant. Relative to the in situ PPs, the modeled PPs (PP$_{A\&T}$, PP$_A$, PP$_T$) had a RMSD of 0.37, 0.34, and 0.34, and a bias of −0.24, −0.197 and −0.174, respectively (Table 2). The seasonal RMSR and bias were in the range of 0.03–0.09 and −0.14–0.39, respectively, for the three satellite-derived PPs. This implied that the three PPs had similar results; however, the PP$_{A\&T}$ algorithm produced slightly more accurate PP measurements than the PP$_A$ and PP$_T$ algorithm.

## 5. Cluster Analysis and Characteristics in the Subareas

The cluster analysis showed that the waters around Taiwan can be divided according to environmental conditions into four subareas: China Coast water (CCW), Taiwan Strait water (TSW), Northeastern upwelling water (NUW), and Kuroshio water (KW) (Figure 2b). In terms of the relationship between SST, MODIS Aqua Chl-a, and PP$_A$ in four subareas for each 1 °C gridded, the monthly mean SSTs were in the range of 7–30 °C in CCW, 13–30 °C in TSW, 17–25 °C in NUW, and 22–30 °C in KW, and had no clear relationships with PP$_A$ were revealed (Figure 6a). The strong correlation between PP$_A$ and MODIS Aqua Chl-a in subareas is shown in Figure 6b. The monthly mean PP$_A$ and Chl-a were in the ranges of 600–2500 mg·C·m$^{-2}$·day$^{-1}$ and 2–5 mg·m$^{-3}$ in CCW, 500–1500 mg C m$^{-2}$·day$^{-1}$ and 1–3 mg·m$^{-3}$ in TSW, 500–1000 mg C m$^{-2}$·day$^{-1}$ and 1–2 mg·m$^{-3}$ in NUW, and 0–500 mg C m$^{-2}$·day$^{-1}$ and 0–1.5 mg·m$^{-3}$ in KW.

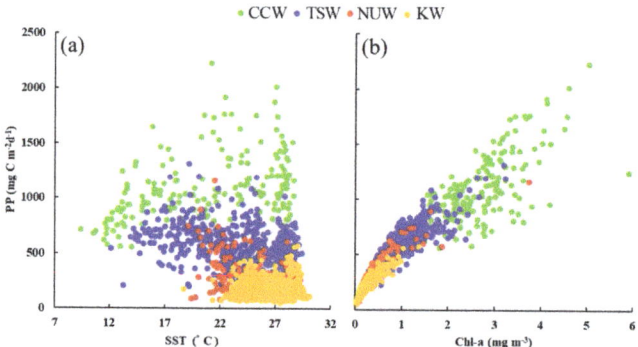

**Figure 6.** Relationship between the monthly mean of (**a**) AVHRR SST and VPGM PP$_A$, and (**b**) MODIS Aqua Chl-a and VPGM PP$_A$ for each 1 °C gridded area in the four subareas of CCW (green circles), TSW (blue circles), NUW (red circles), and KW (yellow circles) from 2009 to 2013.

The comparison of in situ data and model data for each subarea is presented in Table 3. The PP$_A$ had significant correlations with in situ PP in the TSW ($r^2$ = 0.26), NUW ($r^2$ = 0.37) and KW ($r^2$ = 0.14). PP$_{A+T}$ only had significant correlations with in situ PP in the TS, and PP$_T$ had no significant correlations in any of the subareas. The RMSD values were higher in the KW, ranging between 0.06–0.87, with an almost negative bias in the range of −0.74 to 0.38 (Figure 7). The RMSD in the TSW, NUW, and CCW were in the range of 0.07–0.67 with bias in the range of −0.49 to 0.27.

**Table 3.** Extracted number (n) of satellite-derived PPs with in situ PPs, correlation coefficients ($r^2$), and $p$ values for PP$_{A\&T}$, PP$_A$, and PP$_T$ for whole years in four subareas.

| | China Coast | | | Taiwan Strait | | | Northeast Upwelling | | | Kuroshio | | |
|---|---|---|---|---|---|---|---|---|---|---|---|---|
| | n | $r^2$ | p | n | $r^2$ | p | n | $r^2$ | p | n | $r^2$ | p |
| PP$_{A\&T}$ | 3 | 0.16 | 0.51 | 39 | 0.08 | <0.05 | 12 | 0.01 | 0.79 | 48 | 0.02 | 0.07 |
| PP$_A$ | 4 | 0.33 | 0.31 | 52 | 0.26 | <0.05 | 12 | 0.37 | <0.05 | 83 | 0.14 | <0.05 |
| PP$_T$ | 3 | 0.44 | 0.54 | 50 | 0.04 | 0.09 | 14 | 0.01 | 0.7 | 83 | 0.02 | 0.13 |

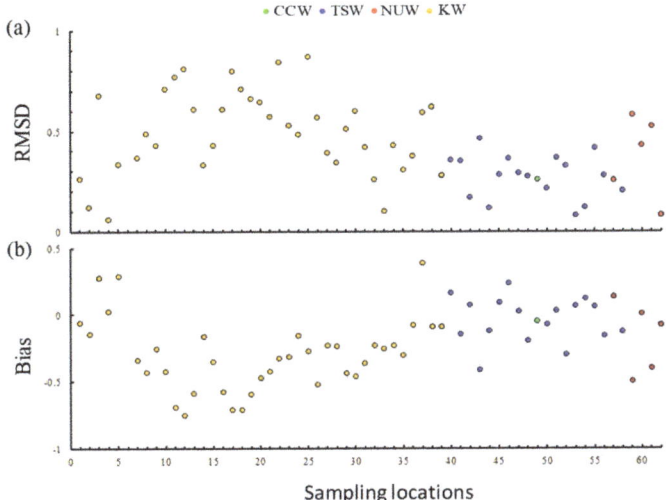

**Figure 7.** (a) RMSD and (b) bias for the 62 sampling locations in the four subareas of the CCW (green circles), TSW (blue circles), NUW (red circles), and KW (yellow circles).

## 6. Discussion

Primary producers reside at the base of food webs, and thus, drive ecosystem dynamics through bottom-up forcing [11]. The global biogeochemical cycles of major elements, particularly the carbon cycle, are greatly influenced by primary producers [2]. The primary producers convert the inorganic to organic carbon by photosynthesis in the light environment, and the carbon production is referred to as gross PP [37]. Net PP is gross PP minus the phytoplankton's respiration, and it supplies to all heterotrophs in the oceans. Net community production is gross PP minus respired by autotrophs and heterotrophs [37]. Both net PP and net community production play an important factor for biological carbon circulation. Therefore, understanding the spatial and temporal dynamics of PP is invaluable in earth and life science research [38,39].

The present study provided the first assessment of a satellite-derived VPGM PP model with in situ PP estimates in the regional waters around Taiwan. In light–dark oxygen methods, not only phytoplankton but also heterotrophic bacteria and zooplankton are in the bottles. The charge in dissolved oxygen of the light bottle is affected by photosynthesis and community respiration, and the gross PP can be estimated by net community production and community respiration [37]. The light–dark oxygen method used in the present study and the $^{14}$C method are often used to measure in situ marine PP. The light–dark oxygen method was the main approach to measure PP before the $^{14}$C method was invented [1], the latter of which is more sensitive and offers good precision, although a radioisotope needs to be added to the water samples. The acquisition, use, and disposal of the radioisotope requires specific procedures and incurs high costs. Nevertheless, it yields reliable data through a careful process using oxygen electrodes [1].

In situ measurements of PP are spatially and temporally limited and require multiple integrated sampling approaches. The use of ocean color data in PP models provides an attractive alternative to field estimations as it offers an estimation at a high spatial and temporal resolution. AVHRR SST and MODIS Chl-a have long been used to study marine characteristics; however, their availability is seriously reduced by cloud coverage. The coverage provided by AVHRR SST and MODIS Chl-a daily images in the current study were in the range 20%–80% (Figure 8 as examples), and notably lower in wintertime (<30%). Although the availability of data from microwave observations was almost 100% [40], the disadvantage of a low spatial resolution meant that it was not appropriate for use in coastal areas. The limited match-up data between the satellite-derived data and in situ measurements were relaxed to within ± 12 h in the present study, and this time difference may have affected the matching accuracy. For example, Lee et al. [41] compared the MODIS SSTs with in situ SSTs and suggested that if the time difference was ± 3 h, this would produce the smallest bias but with a lower match-up of usable data.

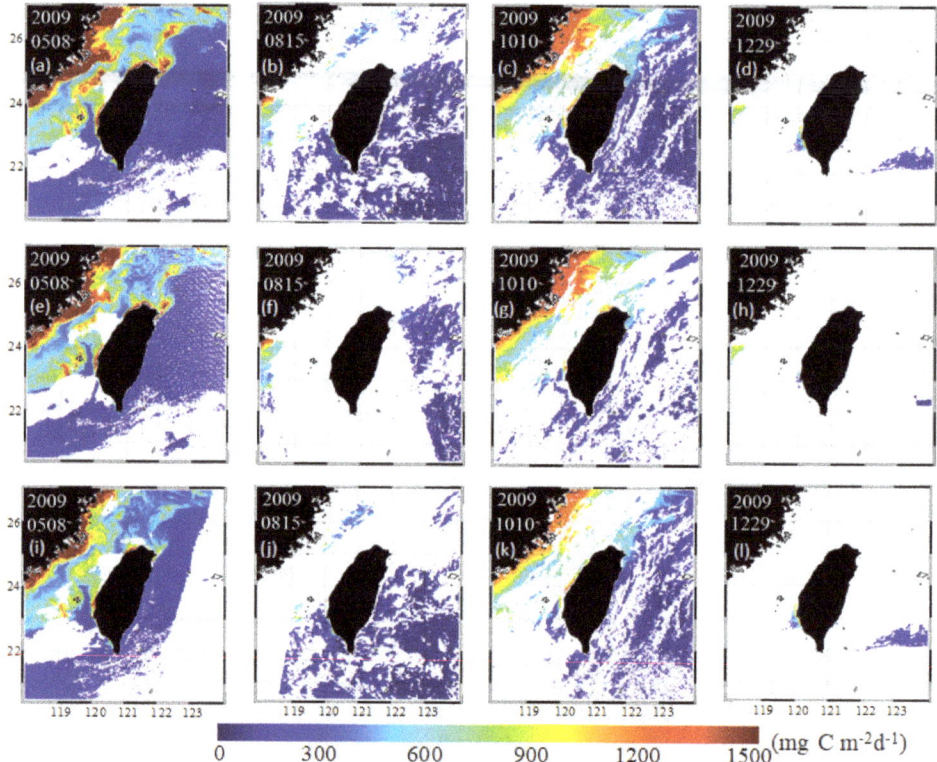

**Figure 8.** Daily spatial distribution maps of (**a**–**d**) $PP_{A\&T}$, (**e**–**h**) $PP_A$, and (**e**–**h**) $PP_T$ in May, August, October, and December 2009.

To make best use of this technique, it is essential to assess the accuracy of the satellite-derived products to determine the uncertainty of the data input into the models [14]. The bias and RMSD between AVHRR SST and in situ data were 0.01 and 0.64 °C and high accuracy than the MODSI SST (bias = 0.03 °C, RMSD = 0.75 °C) in the waters around Taiwan [41,42]. For the MODIS Chl-a data, the previous study suggested that in coastal or upwelled waters, the blue region of the water that left a radiance signal used in standard Chl-a satellite algorithms was affected by colored dissolved organic matter and detrital material in addition to phytoplankton. This resulted in the decreasing accuracy of Chl-a and therefore PP [18,20]. The matching accuracy OC3M algorithm was within 11% of in situ Chl-a in the Arabian Sea and performed better than the Garver Siegel Maritorena Model and Generalized Inherent Optical Property Chla algorithms [43]. The satellite Chl-a estimates tended to be larger than in situ reference values, and also revealed that a nonuniform Chl-a distribution in the water column can be a factor alongside the documented overestimation tendency when larger optical depth match-up stations are considered [44].

When the Chl-a derived from MODIS Aqua and Terra is compared, the MODIS Terra is more accurate in the coastal waters of the Arabian Sea [45], possibly due to differences in sensor design and time differences between the satellites' overpasses. Applying standard products of satellite-derived PP for research in the open ocean is acceptable; however, using such products in regional studies remains questionable. This is especially so when the shelf sea areas are dominated by large rivers containing a large amount of suspended particles and color-dissolved organic matter [26,35]. The estimation errors of satellite-derived PP were often the result of incorrectly applying ocean color chlorophyll algorithms and inaccurate PP data [46]. In our case, the highest correlation coefficients were obtained in the $PP_{A+T}$ around the waters of Taiwan in the spring. However, the highest significant correlation coefficients in the four subareas were observed for $PP_A$. The lowest correlations for $PP_{A+T}$, $PP_T$, and $PP_A$ all occurred in wintertime. Although the RMSD and bias were lowest in $PP_A$, the difference was not significant. We found that VPGM estimated PPs were always underestimated with satellite-derived PPs, which may be due to errors in the $P^B_{opt}$ calculated as a function of SST [47]. However, this photosynthetic parameter also strongly depends on factors such as nutrient supply, irradiance, and dominating phytoplankton species.

Satellite SST represents temperature only in the uppermost ocean layer and $P^B_{opt}$, being a function of SST, differs in the lower layers, providing different photosynthetic efficiency [21]. The VPGM is one of the most widely known and applied depth-integrated/wavelength-integrated models. However, it has rarely been applied to coastal waters. Lobanova et al. [21] compared the accuracy of PP derived from the VPGM, the Platt and Sathyendranth model, and the Absorption-Based Model with in situ data in the North East Atlantic Ocean. The results revealed that the Platt and Sathyendranth model and VPGM had similar accuracy, whereas the Absorption-Based Model was not suitable for the study region. Although using in situ $P^B_{opt}$ and $Z_{eu}$ may have significantly improved the estimation of VPGM PP [16], the scales of in situ measurements were too short to provide adequate coverage of high-quality regional temporal and spatial variations. Improvements in the accuracy of Chl-a from other ocean color sensors, including Medium Spectral Resolution Imaging Spectrometer and Ocean-Colour Climate Change Initiative data, will ultimately lead to an improvement in satellite PP algorithms for further research.

The VPGM PP in the four subareas of CCW, TSW, NUW, and KW also displayed similar features to Chl-a variations with the highest PP in the CCW and lowest in the KW. The RMSD was higher in the KW with an almost negative bias. The $PP_A$ had significant correlations with in situ PP in the TS, NUW, and KW; however, the sampling locations were insufficient to provide significant results in the CCW. The availability of light, the source of energy for photosynthesis; mineral nutrients, (the building blocks for new growth); and temperature, which affects metabolic rates, play crucial roles in regulating PP in the ocean [48,49]. The dominant primary producers in the Taiwan Strait are nano- and pico-phytoplankton [28]. The contribution of the microbial food web to the traditional food web is estimated to be approximately 30%, implying it has fundamental significance in the Taiwan Strait [28]. The high salinity and temperature with low nutrients originate from the TSW source in the South China Sea and KW in summertime; the strong northeastern winds then push the fresh, cold, nutrient-rich CCW southward along the western part of the Taiwan Strait [28,50]. In addition, the KW flowing through the eastern part of Taiwan is relatively deficient in nutrients, and the PP is also lower than in the other currents around Taiwan [51]. However, the nutrients increase from a depth of 200 m under short-term climatic variations such as typhoons [52], and may cause higher RMSD and bias in KW. The spatial distribution and seasonal variation of phytoplankton biomass and primary productivity are largely controlled by the input of nutrients from various water masses.

## 7. Conclusions and Future Research

The present study provided the first assessment of a satellite-derived VPGM PP model with in situ PP estimates in the regional waters around Taiwan. Understanding the PP of waters affected by global warming is critical. In particular, the East China Sea is among the large marine ecosystems that are warming most rapidly. Furthermore, from the 1950s to 2000s, the increased SST has caused 20 °C isotherms in the Taiwan Strait to gradually shift northward in the winter [53]. Climate change caused by global warming along with changes in water temperature also affect the productivity, catchability, and fishing pressure of fish species. It is important to quantify and understand the sources of variation in marine PP and increase confidence in the predictions of future fisheries yielded under uncertainty over future PP and its transfer to higher trophic levels [54]. Our results revealed that the VGPM PP derived from AVHRR SST and three types of MODIS Chl-a were linearly correlated with in situ PPs, as determined by light–dark bottles. The correlation coefficients were highest in the $PP_{A\&T}$ around the waters of Taiwan, especially in springtime. However, the highest significant correlation coefficients in the four subareas were observed in $PP_A$ and wintertime.

Satellite models underestimate in situ PP, probably due to the depth of the phytoplankton in the water column, short-term climatic variations, and optically complex shelf waters. To better understand the PP around Taiwan with complex current systems, substantially more measurements are required across multiple years and seasons. Although initially, these can be used to quantify the productivity of different water masses, they are eventually required to further validate the available biogeochemical models in order to scale up relatively sparse measurements through time and space [18]. Additional studies include making a further comparison of in situ marine PP with the $^{14}C$ method and multi satellite-derived data, such as Ocean-Colour Climate Change Initiative (OC-CCI) data obtained from merged information derived from ocean color sensors. The VPGM is a commonly used model and exhibited significant correlations with in situ data in the present study. However, the simplest model formulation of the Eppley-Square-Root model [15] was tested as the highest skill and lowest bias model in the western boundary of East Australia [18]. The other VPGM-based models, the VGPM-Eppley model [2], and VGPM-Kameda model [17], will provide the crucial next step in conducting more comprehensive investigations by re-parameterizing the original relationships in accordance with in situ data.

**Author Contributions:** K.-W.L. led the study design and wrote the article. L.-J.L. and P.-Y.H. contributed materials of remote sensing data and analysis models. C.-H.L. and S.-Y.C. contributed in situ measurements and water sampling. All authors have read and agreed to the published version of the manuscript.

**Funding:** This research received no external funding.

**Acknowledgments:** This study was financially supported by the Council of Agriculture (108AS-9.2.1-FA-F1, 108AS-25.2.1-AI-A2, 109AS-9.2.2-FA-F1(2) and 109AS-20.2.1-AI-A2) and the National Science Council (MOST 107-2611-M-019-017 and MOST 108-2611-M-019-007).

**Conflicts of Interest:** The authors declare no conflict of interest

## References

1. Cullen, J.J. Primary production methods. *Encycl. Ocean Sci.* **2001**, *4*, 2277–2284.
2. Eppley, R.W.; Peterson, B.J. Particulate organic matter flux and planktonic new production in the deep ocean. *Nature* **1979**, *282*, 677–680. [CrossRef]
3. Marra, J. Approaches to the measurement of plankton production. *Phytoplankton Product. Carbon Assim. Mar. Freshw. Ecosyst.* **2002**, 78–108.
4. Platt, T.; Sathyendranath, S. Fundamental issues in measurement of primary production. *ICES MSS* **1993**, *197*, 3–8.
5. Ryther, J.H. Photosynthesis and fish production in the sea. *Science* **1969**, *166*, 72–76. [CrossRef]
6. Chassot, E.; Bonhommeau, S.; Dulvy, N.K.; Mélin, F.; Watson, R.; Gascuel, D.; Le Pape, O. Global marine primary production constrains fisheries catches. *Ecol. Lett.* **2010**, *13*, 495–505. [CrossRef]
7. Steemann-Nielsen, E. The use of radio-active carbon (C14) for measuring organic production in the sea. *J. Cons.* **1952**, *18*, 117–140. [CrossRef]
8. Hama, T.; Miyazaki, T.; Iwakuma, T.; Takahashi, M.; Ichimura, S. Measurement of photosynthetic production of a marine phytoplankton population using a sTable $^{13}C$ isotope. *Mar. Biol.* **1983**, *73*, 31–36. [CrossRef]
9. Lawrenz, E.; Silsbe, G.; Capuzzo, E.; Ylöstalo, P.; Forster, R.M.; Simis, S.G.; Prášil, O.; Kromkamp, J.C.; Hickman, A.E.; Moore, C.M.; et al. Predicting the electron requirement for carbon fixation in seas and oceans. *PLoS ONE* **2013**, *8*, e58137. [CrossRef]

10. Juranek, L.W.; Quay, P.D. Basin-wide photosynthetic production rates in the subtropical and tropical Pacific Ocean determined from dissolved oxygen isotope ratio measurements. *Glob. Biogeochem. Cycle* **2010**, *24*, GB2006. [CrossRef]
11. Saba, V.S.; Friedrichs, M.A.; Carr, M.E.; Antoine, D.; Armstrong, R.A.; Asanuma, I.; Aumont, O.; Bates, N.R.; Behrenfeld, M.J.; Bennington, V.; et al. Challenges of modeling depth-integrated marine primary productivity over multiple decades: A case study at BATS and HOT. *Glob. Biogeochem. Cycle* **2010**, *24*. [CrossRef]
12. Wu, L.; Cai, W.; Zhang, L.; Nakamura, H.; Timmermann, A.; Joyce, T.; McPhaden, M.J.; Alexander, M.; Qiu, B.; Visbeck, M.; et al. Enhanced warming over the global subtropical western boundary currents. *Nat. Clim. Chang.* **2012**, *2*, 161–166. [CrossRef]
13. Longhurst, A.R. *Ecological Geography of the Sea*; Elsevier: Amsterdam, The Netherlands, 2010.
14. Dogliotti, A.I.; Lutz, V.A.; Segura, V. Estimation of primary production in the southern Argentine continental shelf and shelf-break regions using field and remote sensing data. *Remote Sens. Environ.* **2014**, *140*, 497–508. [CrossRef]
15. Eppley, R.W.; Stewart, E.; Abbott, M.R.; Heyman, U. Estimating ocean primary production from satellite chlorophyll. Introduction to regional differences and statistics for the Southern California Bight. *J. Plankton Res.* **1985**, *7*, 57–70. [CrossRef]
16. Tripathy, S.C.; Ishizaka, J.; Siswanto, E.; Shibata, T.; Mino, Y. Modification of the vertically generalized production model for the turbid waters of Ariake Bay, southwestern Japan. *Estuar. Coast. Shelf Sci.* **2012**, *97*, 66–77. [CrossRef]
17. Kameda, T.; Ishizaka, J. Size-fractionated primary production estimated by a two-phytoplankton community model applicable to ocean color remote sensing. *J. Oeanogr.* **2005**, *61*, 663–672. [CrossRef]
18. Everett, J.D.; Doblin, M.A. Characterising primary productivity measurements across a dynamic western boundary current region. *Deep-Sea Res. Part I-Oceanogr. Res. Pap.* **2015**, *100*, 105–116. [CrossRef]
19. Behrenfeld, M.J.; Falkowski, P.G. A consumer's guide to phytoplankton primary productivity models. *Limnol. Oceanogr.* **1997**, *42*, 1479–1491. [CrossRef]
20. Siegel, D.A.; Maritorena, S.; Nelson, N.B.; Behrenfeld, M.J.; McClain, C.R. Colored dissolved organic matter and its influence on the satellite-based characterization of the ocean biosphere. *Geophys. Res. Lett.* **2005**, *32*, L20605. [CrossRef]
21. Lobanova, P.; Tilstone, G.H.; Bashmachnikov, I.; Brotas, V. Accuracy Assessment of primary production models with and without photoinhibition using Ocean-Colour Climate Change Initiative data in the North East Atlantic Ocean. *Remote Sens.* **2018**, *10*, 1116. [CrossRef]
22. Campbell, J.; Antoine, D.; Armstrong, R.; Arrigo, K.; Balch, W.; Barber, R.; Behrenfeld, M.; Bidigare, R.; Bishop, J.; Carr, M.E.; et al. Comparison of algorithms for estimating ocean primary production from surface chlorophyll, temperature, and irradiance. *Glob. Biogeochem. Cycle* **2002**, *16*. [CrossRef]
23. Carr, M.E.; Friedrichs, M.A.; Schmeltz, M.; Aita, M.N.; Antoine, D.; Arrigo, K.R.; Asanuma, I.; Aumont, O.; Barber, R.; Behrenfeld, M.; et al. A comparison of global estimates of marine primary production from ocean color. *Deep Sea Res. Part II Top. Stud. Oceanogr.* **2006**, *53*, 741–770. [CrossRef]
24. Gong, G.C.; Wen, Y.H.; Wang, B.W.; Liu, G.J. Seasonal variation of chlorophyll a concentration, primary production and environmental conditions in the subtropical East China Sea. *Deep Sea Res. Part II Top. Stud. Oceanogr.* **2003**, *50*, 1219–1236. [CrossRef]
25. Naik, H.; Chen, C.T. Biogeochemical cycling in the Taiwan Strait. *Estuar. Coast. Shelf Sci.* **2008**, *78*, 603–612. [CrossRef]
26. Tseng, H.C.; You, W.L.; Huang, W.; Chung, C.C.; Tsai, A.Y.; Chen, T.Y.; Lan, K.W.; Gong, G.C. Seasonal variations of marine environment and primary production in the Taiwan Strait. *Front. Mar. Sci.* **2020**, *7*, 38. [CrossRef]
27. Wu, C.R.; Lu, H.F.; Chao, S.Y. A numerical study on the formation of upwelling off northeast Taiwan. *J. Geophys. Res.-Oceans* **2008**, *113*, C08025. [CrossRef]
28. Hong, H.; Chai, F.; Zhang, C.; Huang, B.; Jiang, Y.; Hu, J. An overview of physical and biogeochemical processes and ecosystem dynamics in the Taiwan Strait. *Cont. Shelf Res.* **2011**, *31*, 3–12. [CrossRef]
29. Lan, K.W.; Kawamura, H.; Lee, M.A.; Chang, Y.; Chan, J.W.; Liao, C.H. Summertime sea surface temperature fronts associated with upwelling around the Taiwan Bank. *Cont. Shelf Res.* **2009**, *29*, 903–910. [CrossRef]

30. Hung, C.C.; Chung, C.C.; Gong, G.C.; Jan, S.; Tsai, Y.; Chen, K.S.; Chou, W.C.; Lee, M.A.; Chang, Y.; Chen, M.H.; et al. Nutrient supply in the southern East China Sea after typhoon Morakot. *J. Mar. Res.* **2013**, *71*, 133–149. [CrossRef]
31. Tzeng, M.T.; Lan, K.W.; Chan, J.W. Interannual Variability of Wintertime Sea Surface Temperatures in the Eastern Taiwan Strait. *J. Mar. Sci. Technol.-Taiwan* **2012**, *20*, 702–712.
32. Lee, K.T.; Liao, C.H.; Su, W.C.; Hsieh, S.H.; Lu, H.J. The fishing ground formation of sergestid shrimp (Sergia lucens) in the coastal waters of southwestern Taiwan. *J. Mar. Sci. Technol.-Taiwan* **2004**, *12*, 265–272.
33. Lu, H.J.; Lee, H.L. Changes in the fish species composition in the coastal zones of the Kuroshio Current and China Coastal Current during periods of climate change: Observations from the set-net fishery (1993–2011). *Fish Res.* **2014**, *155*, 103–113. [CrossRef]
34. Liao, C.H.; Lan, K.W.; Ho, H.Y.; Wang, K.Y.; Wu, Y.L. Variation in the catch rate and distribution of swordtip squid (Uroteuthis edulis) associated with factors of the oceanic environment in the southern East China. *Mar. Coast. Fish.* **2018**, *10*, 452–464. [CrossRef]
35. Gong, G.C.; Shiah, F.K.; Liu, K.K.; Wen, Y.H.; Liang, M.H. Spatial and temporal variation of chlorophyll a, primary productivity and chemical hydrography in the southern East China Sea. *Cont. Shelf Res.* **2000**, *20*, 411–436. [CrossRef]
36. Morel, A.; Berthon, J.F. Surface pigments, algal biomass profiles, and potential production of the euphotic layer: Relationships reinvestigated in view of remote-sensing applications. *Limnol. Oceanogr.* **1989**, *34*, 1545–1562. [CrossRef]
37. Howarth, R.W.; Michaels, A.F. Light and dark bottle oxygen technique. In *Methods in Ecosystem Science*; Sala, O.E., Jackson, R.B., Moone, H.A., Howarth, R.W., Eds.; Springer: New York, NY, USA, 2000; pp. 74–80.
38. Watson, R.; Zeller, D.; Pauly, D. Primary productivity demands of global fishing fleets. *Fish Fish.* **2014**, *15*, 231–241. [CrossRef]
39. Friedland, K.D.; Stock, C.; Drinkwater, K.F.; Link, J.S.; Leaf, R.T.; Shank, B.V.; Rose, J.M.; Pilskaln, C.H.; Fogarty, M.J. Pathways between primary production and fisheries yields of large marine ecosystems. *PLoS ONE* **2012**, *7*, e28945. [CrossRef]
40. Hosoda, K.; Kawamura, H.; Lan, K.W.; Shimada, T.; Sakaida, F. Temporal scale of sea surface temperature fronts revealed by microwave observations. *IEEE Geosci. Remote Sens. Lett.* **2011**, *9*, 3–7. [CrossRef]
41. Ming-An, L.; Tzeng, M.T.; Hosoda, K.; Sakaida, F.; Kawamura, H.; Shieh, W.J.; Yang, Y.; Chang, Y. Validation of JAXA/MODIS sea surface temperature in water around Taiwan using the Terra and Aqua satellites. *Terr. Atmos. Ocean. Sci.* **2010**, *21*, 7.
42. Ming-An, L.; Chang, Y.I.; Sakaida, F.; Kawamura, H.; Chao-Hsiung, C.; Jui-Wen, C.; Huang, I. Validation of satellite-derived sea surface temperatures for waters around Taiwan. *Terr. Atmos. Ocean. Sci.* **2005**, *16*, 1189–1204.
43. Tilstone, G.H.; Lotliker, A.A.; Miller, P.I.; Ashraf, P.M.; Kumar, T.S.; Suresh, T.; Ragavan, B.R.; Menon, H.B. Assessment of MODIS-Aqua chlorophyll-a algorithms in coastal and shelf waters of the eastern Arabian Sea. *Cont. Shelf Res.* **2013**, *65*, 14–26. [CrossRef]
44. Sá, C.; D'Alimonte, D.; Brito, A.C.; Kajiyama, T.; Mendes, C.R.; Vitorino, J.; Oliveira, P.B.; Da Silva, J.C.; Brotas, V. Validation of standard and alternative satellite ocean-color chlorophyll products off Western Iberia. *Remote Sens. Environ.* **2015**, *168*, 403–419. [CrossRef]
45. Arun Kumar, S.V.V.; Babu, K.N.; Shukla, A.K. Comparative analysis of chlorophyll-a distribution from SEAWIFS, MODIS-AQUA, MODIS-TERRA and MERIS in the Arabian Sea. *Mar. Geod.* **2015**, *38*, 40–57. [CrossRef]
46. Gong, G.C.; Chen, T.Y.; You, W.L. Mixing control on the photosynthesis-irradiance relationship and an estimate of primary production in the winter of the East China Sea. *Terr. Atmos. Ocean. Sci.* **2017**, *28*, 1. [CrossRef]
47. Tilstone, G.; Smyth, T.; Poulton, A.; Hutson, R. Measured and remotely sensed estimates of primary production in the Atlantic Ocean from 1998 to 2005. *Deep-Sea Res. Part II-Top. Stud. Oceanogr.* **2009**, *56*, 918–930. [CrossRef]
48. Dundas, I.; Johannessen, O.M.; Berge, G.; Heimdal, B. Toxic algal bloom in Scandinavian waters, May-June 1988. *Oceanography* **1989**, *2*, 9–14. [CrossRef]
49. Maestrini, S.Y.; Graneli, E. Environmental-conditions and ecophysiological mechanisms which led to the 1988 chrysochromulina-polylepis bloom—An hypothesis. *Oceanol. Acta* **1991**, *14*, 397–413.

50. Jan, S.; Tseng, Y.H.; Dietrich, D.E. Sources of water in the Taiwan Strait. *J. Oceanogr.* **2010**, *66*, 211–221. [CrossRef]
51. Chen, H.Y.; Chen, Y.L.L. Quantity and quality of summer surface net zooplankton in the Kuroshio current-induced upwelling northeast of Taiwan. *Terr. Atmos. Ocean. Sci.* **1992**, *3*, 321–334. [CrossRef]
52. Lin, I.; Liu, W.T.; Wu, C.C.; Wong, G.T.; Hu, C.; Chen, Z.; Liang, W.D.; Yang, Y.; Liu, K.K. New evidence for enhanced ocean primary production triggered by tropical cyclone. *Geophys. Res. Lett.* **2003**, *30*. [CrossRef]
53. Lan, K.W.; Lee, M.A.; Zhang, C.I.; Wang, P.Y.; Wu, L.J.; Lee, K.T. Effects of climate variability and climate change on the fishing conditions for grey mullet (*Mugil cephalus L.*) in the Taiwan Strait. *Clim. Chang.* **2014**, *126*, 189–202. [CrossRef]
54. Brander, K.M. Global fish production and climate change. *Proc. Natl. Acad. Sci. USA* **2007**, *104*, 19709–19714. [CrossRef] [PubMed]

© 2020 by the authors. Licensee MDPI, Basel, Switzerland. This article is an open access article distributed under the terms and conditions of the Creative Commons Attribution (CC BY) license (http://creativecommons.org/licenses/by/4.0/).

*Article*

# Consecutive Dual-Vortex Interactions between Quadruple Typhoons Noru, Kulap, Nesat and Haitang during the 2017 North Pacific Typhoon Season

Yuei-An Liou [1,*], Ji-Chyun Liu [1], Chung-Chih Liu [2], Chun-Hsu Chen [3], Kim-Anh Nguyen [1,4] and James P. Terry [5]

1. Center for Space and Remote Sensing Research, National Central University, No. 300, Jhongda Rd., Jhongli Dist., Taoyuan City 32001, Taiwan
2. Department of Computer Science and Information Engineering, and Natural Sciences Teaching Center, Minghsin University of Science and Technology, No.1, Xinxing Rd., Xinfeng, Hsinchu 30401, Taiwan
3. Computational Intelligence Technology Center, Industrial Technology Research Institute, Hsinchu 31040, Taiwan
4. Institute of Geography, Vietnam Academy of Science and Technology, 18 Hoang Quoc Viet Rd., Cau Giay, Hanoi 10000, Vietnam
5. College of Natural and Health Sciences, Zayed University, P.O. Box 19282, Dubai, UAE
* Correspondence: yueian@csrsr.ncu.edu.tw; Tel.: +886-3-4227151 (ext. 57631)

Received: 16 April 2019; Accepted: 25 July 2019; Published: 7 August 2019

**Abstract:** This study utilizes remote sensing imagery, a differential averaging technique and empirical formulas (the 'Liou–Liu formulas') to investigate three consecutive sets of dual-vortex interactions between four cyclonic events and their neighboring environmental air flows in the Northwest Pacific Ocean during the 2017 typhoon season. The investigation thereby deepens the current understanding of interactions involving multiple simultaneous/sequential cyclone systems. Triple interactions between Noru–Kulap–Nesat and Noru–Nesat–Haitung were analyzed using geosynchronous satellite infrared (IR1) and IR3 water vapor (WV) images. The differential averaging technique based on the normalized difference convection index (NDCI) operator and filter depicted differences and generated a new set of clarified NDCI images. During the first set of dual-vortex interactions, Typhoon Noru experienced an increase in intensity and a U-turn in its direction after being influenced by adjacent cooler air masses and air flows. Noru's track change led to Fujiwhara-type rotation with Tropical Storm Kulap approaching from the opposite direction. Kulap weakened and merged with Noru, which tracked in a counter-clockwise loop. Thereafter, in spite of a distance of 2000–2500 km separating Typhoon Noru and newly-formed Typhoon Nesat, the influence of middle air flows and jet flows caused an 'indirect interaction' between these typhoons. Evidence of this second interaction includes the intensification of both typhoons and changing track directions. The third interaction occurred subsequently between Tropical Storm Haitang and Typhoon Nesat. Due to their relatively close proximity, a typical Fujiwhara effect was observed when the two systems began orbiting cyclonically. The generalized Liou–Liu formulas for calculating threshold distances between typhoons successfully validated and quantified the trilogy of interaction events. Through the unusual and combined effects of the consecutive dual-vortex interactions, Typhoon Noru survived 22 days from 19 July to 9 August 2017 and migrated approximately 6900 km. Typhoon Noru consequently became the third longest-lasting typhoon on record for the Northwest Pacific Ocean. A comparison is made with long-lived Typhoon Rita in 1972, which also experienced similar multiple Fujiwhara interactions with three other concurrent typhoons.

**Keywords:** Typhoons; Fujiwhara effect; cyclone–cyclone interaction; vortex interaction; Liou–Liu formulas; tropical depression (TD)

## 1. Introduction

### 1.1. Tropical Cyclone Hazards

Tropical cyclones (TCs), including typhoons and hurricanes, are considered to be among the most destructive natural hazards in terms of their severity, duration and areas affected. Every year in various parts of the world, they cause loss of human lives, crops and livestock, and extensive damage to infrastructure, transport and communication systems. Information on the distribution and variation of TCs, along with the effects of climate variability and global warming on their occurrence, is therefore crucial for assessing vulnerability and for disaster prevention [1–4]. The Asia region is especially prone to TC occurrence and their negative impacts. Nguyen et al. [4], for example, used 21 indicators to identify vulnerability to typhoons using geospatial techniques by implementing a conceptual framework modified from an eco-environmental vulnerability assessment, suitable for implementation at regional to global scales [5–8]. Accurately predicting the migratory track, intensity and rainfall of TCs is a key research focus for meteorologists and weather forecasters [9]. Factors influencing track orientation, shape, sinuosity, and ultimately points of landfall are of particular interest [10,11]. Satellite-based cloud images are useful for analyzing TC cloud structure and dynamics [12–16].

### 1.2. Dual-Vortex Interactions

When two TCs approach one another, they can influence each other through a cyclone–cyclone vortex interaction. The dual-vortex interaction, known as the 'Fujiwhara effect' (also referred to as the Fujiwhara interaction or a binary interaction), occurs between two TC systems that are close enough (generally less than 1400 km apart) to affect each other significantly and cause a tendency towards mutual rotation. Studying the various possible behavior patterns of a dual-vortex interaction is important as it offers the potential to improve weather forecasting. Several such dual-vortex interactions have been studied in the past [17–20]. For instance, Hart and Evans [21] simulated the interaction of dual vortices in horizontally-sheared environmental flows on a beta plane, and the intensification of Hurricane Sandy in 2012 during the warm seclusion phase of its extratropical transition was investigated by Galarneau et al. [22].

The interaction between TCs and other types of adjacent weaker cyclonic systems such as tropical depressions (TDs) and tropical storms (TSs) has further been identified as an additional type of interaction. For example, Wu et al. [23] proposed that the position of TS Bopha in 2000 between typhoons Saomai and Wukong caused these two systems to interact. Similarly, Liu et al. [24] examined the interaction between typhoons Tembin and Bolaven in 2012. TDs sandwiched between them resulted in an indirect cyclone–cyclone interaction. However, modelling the impacts of a TD positioned between two mature cyclones is problematic and profoundly complicates the numerical weather predictions for such conditions. To characterize multiple dual-vortex interactions, Liou et al. [25] proposed empirical formulas (hereinafter referred to as the 'Liou–Liu formulas') for determining threshold distances between them. The formulas are empirically related to the size factor, height difference, rotation factor, and the current intensity (CI) that takes into account maximum wind speed and intensity. The Liou–Liu formulas successfully predicted and quantified the impacts of intermediate TDs and are therefore adopted in this paper. Nonetheless, because various types of dual-vortex interactions may exist, much further investigation is needed to deepen our understanding of such phenomena and to improve cyclone track predictions in future.

The Northwest Pacific (NWP) is the most active ocean basin for TC (typhoon) formation in the world. Studying the influence of cooler air masses, air flows and outflow jets on typhoons is especially important in the NWP basin because upper cooler air masses exert a temperature gradient to the north, while lower air flows transfer warm and humid air to south of any typhoons that form. Air flow behaves in a fluid manner, meaning air naturally flows from areas of higher pressure to where the pressure is lower. Like any fluid, air flow may exhibit both laminar and turbulent flow patterns. Laminar flow ('air flow' mentioned in this paper) occurs when air can flow smoothly. Turbulent flow ('jet flow' mentioned in this paper) occurs when there is an irregularity which alters the direction of movement. The simultaneous existence of both upper cool air masses and lower warm and humid air flows can greatly affect typhoon intensity. Lee et al. [26], for example, reported that during the winter seasons of 2013 and 2014, typhoons Haiyan and Hagupit both intensified into super-typhoons through their interactions with cold fronts in the NWP Ocean. Cold fronts at the leading edge of cooler air masses were found to exert greater temperature gradients between fronts and the main body of typhoon circulation, assisting typhoon enhancement. During summer over the NWP Ocean, southwest air flows may also play a role in amplifying typhoon intensity [27]. A better visualization of such influences is clearly necessary in studying typhoon intensification processes. Liou et al. [28] built upon earlier work on super-typhoon formation in winter to investigate the seasonal dependence on distribution and profiles. Their findings are potentially helpful in advancing the understanding and predictability of super-typhoons in order to reduce their impacts on human lives and wellbeing.

*1.3. Aims*

The observation and quantification of dual-vortex interactions are important for weather prediction models and forecasts. The present study documents a case of successive dual-vortex interactions from the perspective of satellite observations. It aims to validate cyclonic interactions based on the generalized Liou–Liu formulas, which calculate threshold distances between the centers of two cyclonic systems required for their interaction, as developed by Liou et al. [25] in an earlier study. This is accomplished through investigating an unusual case involving three consecutive interactions in the Pacific region east of Taiwan in 2017. Three sets of dual-vortex interactions occurred between (1) Typhoon Noru and TS Kulap; (2) Typhoon Noru, Typhoon Nesat, and jet flows and air flows separating them; and (3) Typhoon Nesat and TD Haitang. The consecutive phenomena all demonstrate different varieties of dual-vortex interactions. The abovementioned aims are tackled by performing three specific tasks: (1) Examining the main features of multiple TC interactions through a study of quadruple typhoons Noru, Kulap, Nesat and Haitang in 2017; (2) applying useful empirical equations that have the potential to quantify the interactions observed as a function of distance, size, height difference and rotation factors; and (3) highlighting the satellite image analyses techniques that were applied.

## 2. Overview of Typhoons Noru, Kulap, Nesat and Haitang in 2017

When multiple TCs develop in proximity to one another, either simultaneously or in close succession, a triple or quadruple sequence of cyclone–cyclone interactions may occur. The primary focus of our study here is on the life cycle of typhoons Noru, Kalup, Nesat and Haitang in 2017. The goal was to improve our understanding of interactions between simultaneously occurring cyclone systems and the effects of nearby cooler air masses, jet flows and air flows that influenced them. Attention was directed especially on typhoon intensity, corresponding distances between typhoons and influential environmental cooler air masses and air flows, characteristics of a mutual interaction, the induction of outflow jets, and the observed effects on typhoon intensification. Cloud disturbances were used to analyze typhoon development and movement.

Tracks of cyclonic systems Noru, Kulap, Nesat and Haitang over the period 19 July–09 August 2017 are presented in Figure 1. Initially, on 23 July 2017, Typhoon Noru was affected by adjacent environmental cooler air masses and jet flows, and it made a U-turn in its track direction. Over 25–27 July, another system, TS Kulap, then approached Noru in the opposite direction. The interaction weakened Kulap, which resulted in its merger with Noru following another track U-turn. Over 27–28 July, although typhoons Noru and newly-formed Nesat laid some 2000–2500 km apart, it was seen that both typhoons experienced rotation on 28 July 2017 and changed direction after that time. This was therefore regarded as an 'indirect' dual-vortex interaction between Noru and Nesat. Subsequently, from 29 to 31 July, Typhoon Haitang, the last system to form within the study period, tracked close to Typhoon Nesat. A typical effect was observed as the two systems began orbiting cyclonically, creating the third in the observed sequence of dual-vortex interactions. The typhoon positions and timings of the three dual-vortex interactions are noted for 25, 28 and 31 July in Figure 1.

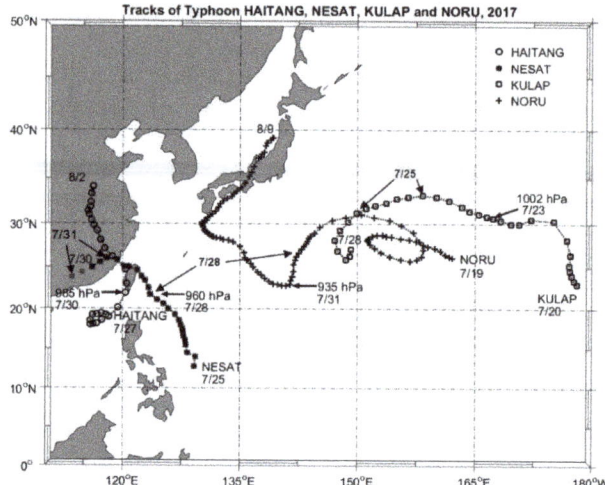

**Figure 1.** Tracks of cyclonic systems Noru, Kulap, Nesat and Haitang over the period 19 July–09 August 2017. Typhoon positions and timings of dual-vortex interactions are indicated on 25, 28 and 31 July.

Owing to this complex sequence of multiple interactions, it is important to locate and clearly recognize the four tropical cyclones of interest over the timeframe of 23 July–01 August. In addition, because a jet flow in the region between typhoons Noru and Nesat exerted a strong influence during this period, the effects of the jet flow on the development and behavior of the two typhoons needed to be examined.

## 3. Data and Methodology

### 3.1. Himawari-8 Geostationary Weather Satellite Images

By utilizing images of the Himawari-8 geostationary weather satellite (longitude 140.7°E), a new empirical technique was recently introduced for the automatic determination of the center of a tropical cyclone system. The data were obtained from the visible (0.55–0.75 μm), water vapor absorption (6.2 μm) and thermal infrared (10.4 μm) channels at thirty (30) minute time intervals. This innovative technique determines the point around which the fluxes of the gradient vectors of brightness temperature (BT) tend to converge [27].

To calculate the variables for measuring an individual typhoon's cloud system, center and intensity, the spectral features of the geostationary satellite IR window and water vapor channel data were used. Hourly data were utilized from two infrared channels: IR1 (10.5 to 11.5 μm) and IR2 (11.5 to 12.5 μm), along with IR3, which is a set of water vapor channels (WV 6.5 to 7.0 μm) [24]. Shortwave infrared (SIR) channels are capable of detecting ice-clouds or ice-covered surfaces within clouds [29–34]. This is achieved by observing cloud effective temperatures and optical depths with detectors at 3.7 [32] or 3.8 μm wavelengths [29,30], thereby facilitating the crude characterization of a cloud vertical structure through a variety of empirical methods [29,30]. Using both IR1 (10.5–11.5 μm) and IR3 (WV 6.5–7.0 μm) channels, which provide cloud images useful for typhoon observation and analysis, an algorithm was proposed to extract cloud systems and spiral TC patterns in the IR images [35].

## 3.2. Differential Averaging Technique

Methods of comparison in microwave technology rely on the comparison of an unknown value of a quantity with a measured quantity using a well-known functional relationship. Among the advantages are a high sensitivity to environmental disturbance. The differential averaging technique is considered as one kind of functional relationship [36]. In applications, the difference values of two quantities are obtained by the differential mode (DM), and the average values of the two quantities are obtained by the common mode (CM). Comparisons are then made using the difference values divided by the average values (DM/CM). Consequently, the desired quantities (as objects) are detected, and the undesired quantities (as background noise) are cancelled through the comparison.

Recently, observations of weather systems began to take measurements at 30-minute intervals within four channels: Visible (0.55–0.75 µm), thermal infrared windows IR1 (10.5–11.5 µm) and IR2 (11.5–12.5 µm), water vapor absorption IR3 (5.6–7.2 µm), and shortwave infrared (SIR) (3.7, 3.8 or 4.0 µm). In many applications requiring the analysis of cloud structure and dynamics, optical cloud images are obtained from the visible channel, cloud-top images from the IR1 and IR2 infrared window channels, cloud water vapor images from the water vapor absorption channel, and ice-cloud images or ice-covered surfaces in clouds from the SIR channels [29–34].

Furthermore, a new set of NDCI (normalized difference convection index) images was generated to provide additional information from satellite images [35]. For analyzing complex weather systems such as typhoons, which are formed from clouds, ice-clouds and water vapor, cloud-image extraction techniques are needed to acquire as much detail as possible on the desired features. One of the available techniques for extraction is the differential averaging technique, which involves the NDCI operator and the filter. The NDCI operator was applied here to geosynchronous satellite infrared (IR1) images ($A_{IR1}$) and IR3 water vapor (WV) images ($A_{WV}$). The NDCI images generated are the output results from differences between the differential mode images and the summary average images in the common mode. The differential mode extracts the water vapor from the IR1 images, whereas the common mode takes the average among IR1 images and WV images. The NDCI operator thus improves the operation by providing clearer images and by depicting the differences better. The differential mode and the common mode are written as:

$$DM = \frac{A_{IR1} - A_{WV}}{2} \quad (1)$$

$$CM = \frac{A_{IR1} + A_{WV}}{2} \quad (2)$$

The equation of the NDCI operator is therefore expressed as:

$$NDCI = \frac{DM}{CM} = \frac{A_{IR1} - A_{WV}}{2} \div \frac{A_{IR1} + A_{WV}}{2} = \frac{A_{IR1} - A_{WV}}{A_{IR1} + A_{WV}} \quad (3)$$

where NDCI values lie between −1 and +1. The NDCI value equals +1 when WV = 0, and it equals −1 when IR1 = 0. A clear sky or atmosphere with thin cloud and dry air will have $0 \leq NDCI \leq +1$. Cloud systems, on the other hand (wet air), will have $-1 \leq NDCI \leq 0$. Generally, a typhoon system exhibits convection with deep vertical development, so it consists of both dry and wet air and thus generates a value of $-1 \leq NDCI \leq +1$.

## 3.3. Application of the Liou–Liu Formulas

The Liou–Liu formulas are helpful because they describe TC interactions in a quantitative way, which should improve the numerical modelling of weather forecasting. Moreover, they can characterize both specific and generalized dual-vortex interactions. For two simultaneous individual cyclones named $TC_1$ and $TC_2$, values $CI_1$ and $CI_2$ are used to represent their current intensities, corresponding to the maximum central wind speed and intensity at the sea surface (Table 1). The pressure–wind relationship for intense TCs was examined with a particular focus on the physical connections between the maximum surface wind and the minimum sea-level pressure [37]. The effects of vortex size, background environmental pressure and the presence of complex vortex features were generally omitted. For a maximum wind speed of 50 km/hr, for example, CI was therefore determined to be 1.346 by using the look-up Table 1, CI = 1 + [(50 − 41) / (54 − 41)] × (1.5 − 1.0).

Table 1. Current Intensity values.

| CI Values | Maximum Wind Speed at Center (km/hr) | Intensity at the Sea Surface (hPa) |
|---|---|---|
| 1.0 | 41 | 1005 |
| 1.5 | 54 | 1002 |
| 2.0 | 67 | 998 |
| 2.5 | 80 | 993 |
| 3.0 | 93 | 987 |
| 3.5 | 106 | 981 |
| 4.0 | 119 | 973 |
| 4.5 | 132 | 965 |
| 5.0 | 145 | 956 |
| 5.5 | 157 | 947 |
| 6.0 | 172 | 937 |
| 6.5 | 185 | 926 |
| 7.0 | 198 | 914 |
| 7.5 | 213 | 901 |
| 8.0 | 226 | 888 |

The application of the Liou–Liu formula [25] for the threshold distance $d_{th}$ (km) indicates whether $TC_1$ and $TC_2$ experience a dual-vortex interaction with each other, as follows:

$$d_{th} = 1000 + 100\left(\frac{CI_1}{4} + \frac{CI_2}{4}\right). \qquad (4)$$

In some situations, a tropical depression (TD) or a tropical storm (TS) (i.e., an area of low pressure) occupies the region lying between two individual cyclone systems. A cyclone's interaction with an intervening TD or TS is a different situation from a regular cyclone-to-cyclone interaction. Upward convection in between may occur because a TD or TS is smaller in size than a cyclone. Such upward convections may strengthen the cyclone and sustain its rotation. Thus, it becomes important to include size ratios, height differences, and rotation in calculating threshold distances. Under such conditions, the Liou–Liu formulas for threshold distances $d_{th1}$ and $d_{th2}$ can determine whether or not two sets of dual-vortex interactions can be identified, between $TC_1$ ($CI_1$) and TD ($CI_d$) and between $TC_2$ ($CI_2$) and TD ($CI_d$), respectively. Following this method, the Liou–Liu formula for the threshold distance D to quantitatively define the dual-vortex interaction can be written as:

$$D = d_{th1} + d_{th2} = 2000 + 100\left(\frac{CI_1}{4} + \frac{CI_d}{4}\right)F_1 + 100\left(\frac{CI_2}{4} + \frac{CI_d}{4}\right)F_2 \qquad (5)$$

Note here that, in a case of two or more cyclonic interactions (Equation (4)), the symbol $d_{th}$ represents the threshold distance between respective cyclone centers, whereas in Equation (5), the same symbols ($d_{th1}$ or $d_{th2}$) represent the threshold distance between TD or TS and the cyclone centers. $F_1$ and $F_2$ are tuning factors dependent on the size factors and are related to height-difference and rotation factors.

$$F_1(q_1, \Delta h_1, \tau_1) = \frac{1}{q_1} \times \frac{1}{\Delta h_1} \times \tau_1 \qquad (6)$$

$$F_2(q_2, \Delta h_2, \tau_2) = \frac{1}{q_2} \times \frac{1}{\Delta h_2} \times \tau_2 \qquad (7)$$

where $q_{1,2}$ the size factor = size (TD or TS)/size($TC_{1,2}$); $\Delta h_{1,2}$ is the height difference between $TC_{1,2}$ and TD or TS, where $\Delta h_{1,2} = (h(TC_{1,2}) - h(TD) \text{ or } h(TS))/(h(TC_{1,2}))$; and $\tau_{1,2}$ is the rotation factor, with $\tau_{1,2} = +1$ for counter-clockwise rotation between $TC_{1,2}$ and TD or TS, or −1 for clockwise rotation [20].

*3.4. Cold Front Detection*

In meteorology, a cold front is defined as the transition zone where a cold air mass is replacing a warmer air mass. It is accompanied by a strong temperature gradient and frontal cloud bands. A meteorological cold front can be identified from distinctive reflectance characteristics in IR imagery owing to the large temperature contrast along the frontal cloud bands. It is noticeable that the cold cloud top is located near the leading edge of the front where the cold air mass interacts with adjacent warm air.

A technique to detect cold air masses and to delimit the position of a cold front in IR cloud imagery is proposed as follows. IR cloud imagery is a type of thermal imagery in which the cloud-top temperature is detected by an IR sensor. For low temperatures (e.g., −70 °C at a cloud top) the image brightness limit is obtained. For high temperatures (e.g., 25 °C at the sea surface) the image darkness limit is obtained. The IR sensor is saturated with a designated lowest saturation temperature (e.g., from −75 °C to −77 °C). The summed values of the saturation temperature and the cold air masses then determine the brightness contours on the cloud imagery. A cold front is the boundary of the low temperature cold air mass, while higher temperature air flow is located in advance of the cold front. Therefore, the position of the cold front is delimited as the abrupt discontinuity in temperature distribution at the leading edge of the cold air mass. The technique described above was used here to mark the cold front position and for calculating its shortest distance from a cyclone center in order to analyze their mutual interaction over time.

## 4. Observations on Dual-Vortex Interactions

*4.1. Advantages of NDCI Images over IR Images*

Before describing the characteristics of the sequence of dual-vortex interactions observed during the study period, it is useful to highlight the advantages of using NDCI images to complement the use of IR1 and IR3 images for analysis. From Figures 2 and 3, the advantage of NDCI images over the individual channel images IR1 and IR3 can be seen. The NDCI images are able to enhance the differences between the individual IR images. The IR1 image exhibits not only the desired objects (typhoons Noru, Nesat, and cooler air masses and jet flows) but also the undesired background noise. In contrast, the NDCI image shows the desired objects more clearly while reducing the undesired background noise.

The best way to demonstrate is to give examples of cloud structural features that are easily visible in the NDCI images (i.e., the third subplot of Figures 2–4), but which are not as clear in the corresponding IR1 or IR3 images. Three examples are provided to serve as illustration. First, in Figure 2, the IR1 image suggests that the cloud associated with the cooler air mass in the north east quadrant appears to be divided into two major sections with a notable gap separating them. However, this apparent gap is much less pronounced in the NDCI image, thus indicating the strong coupling of the two cloud masses that is not obvious from IR1. Note also that the signatures of these cloud masses can hardly be distinguished in IR3. In Figure 3, the second example refers to the southern peripheral spiral cloud arm of Typhoon Noru. The superiority of the NDCI image over the IR1/IR3 images is again clear. In Figure 4, the air flow pattern in the south east quadrant provides a third example which shows the contrast between the relative lack of clarity in the IR3 image compared with the much improved visibility in the NDCI image.

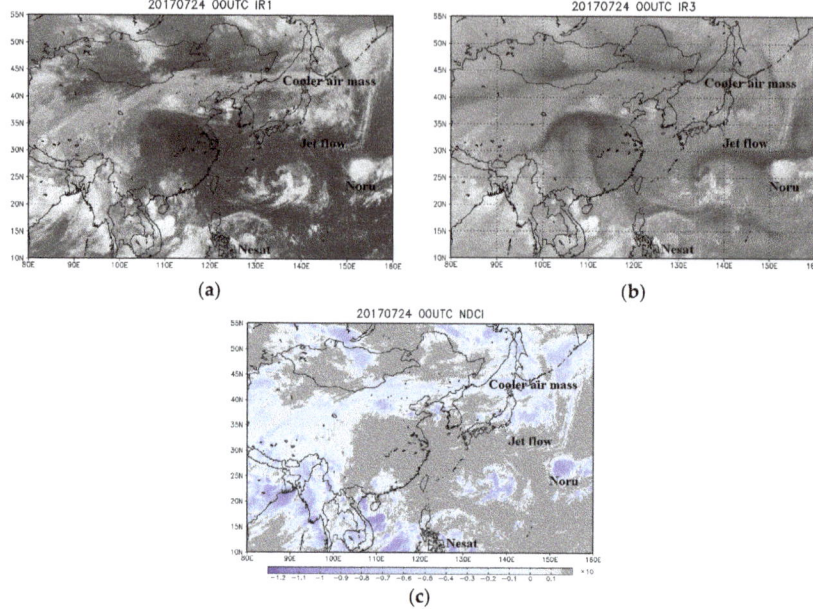

**Figure 2.** Images at 00:00 on 24 July 2017 showing Typhoon Noru, a cooler air mass and jet flow to the north, and Typhoon Nesat to the south. (**a**) IR1 image, (**b**) IR3 (water vapor (WV)) image, and (**c**) normalized difference convection index (NDCI) image.

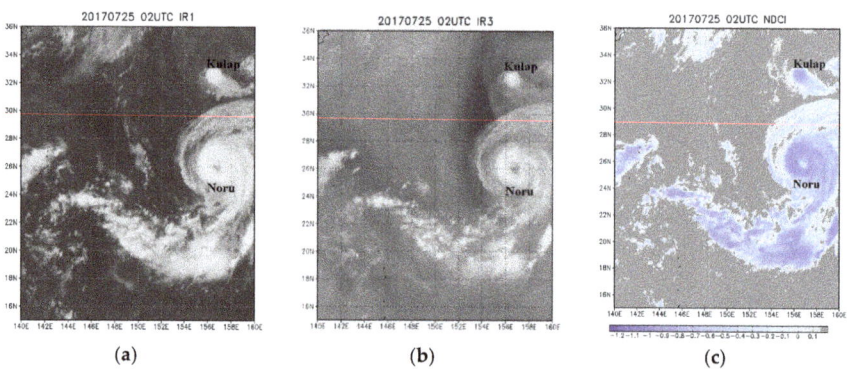

**Figure 3.** Images of tropical storm (TS) Kulap starting to merge with Typhoon Noru at 02:30 on 25 July 2017: (**a**) IR1 image, (**b**) IR3 (WV) image and (**c**) NDCI image.

*4.2. First Dual-Vortex Interaction*

Tropical Storm Noru formed on 21 July 2017, tracking west to northwest on 22 July. Noru intensified significantly over two days until being upgraded to a severe tropical storm on 23 July. Subsequently, Noru slowed down and remained stationary, owing to a dominant steering environment of high pressure ridges on both sides. The steering flows derived from cooler air masses to the west and predominant air flows to the south of the system. Noru rapidly intensified into a typhoon and started to track east-southeastward under the steering influence to the south. Meanwhile, Tropical Storm Kulap moved westward, as shown in Figure 1.

On 24 July, typhoons Noru and Nesat lay some 3100–3200 km apart, while TS Kulap was located to the east of Noru near 33°N 162°E (outside the frame shown). Typhoon Noru and TS Kulap approached one another from opposite directions. A straight-line colder cloud top was observed flowing towards Typhoon Noru in a clear V-shaped pattern. This is the outward jet flow from cooler air masses to the north, as seen in Figure 2. Typhoon Noru was probably influenced by adjacent environmental cooler air masses and jet flows, and it made a U-turn in its track. Noru's directional change facilitated its continuing interaction with these adjacent cooler air masses. The first observed Fujiwhara-type dual-vortex interaction then commenced on 24 July between Noru and Kulap. The interaction caused Kulap to migrate to the north of Noru early on 25 July. Thereafter, Kulap completely merged with Noru. The beginning of the merger of TS Kulap with Typhoon Noru at 02:30 on 25 July is shown in Figure 3.

Air flow to the south became the primary steering influence on Noru, turning the system northeastward and then northward on 26 July. The typhoon then began to track westward along the southern periphery of air flow to the northeast on 27 July. The complex combination of cooler air masses and air flows that surrounded Noru are seen in Figure 4.

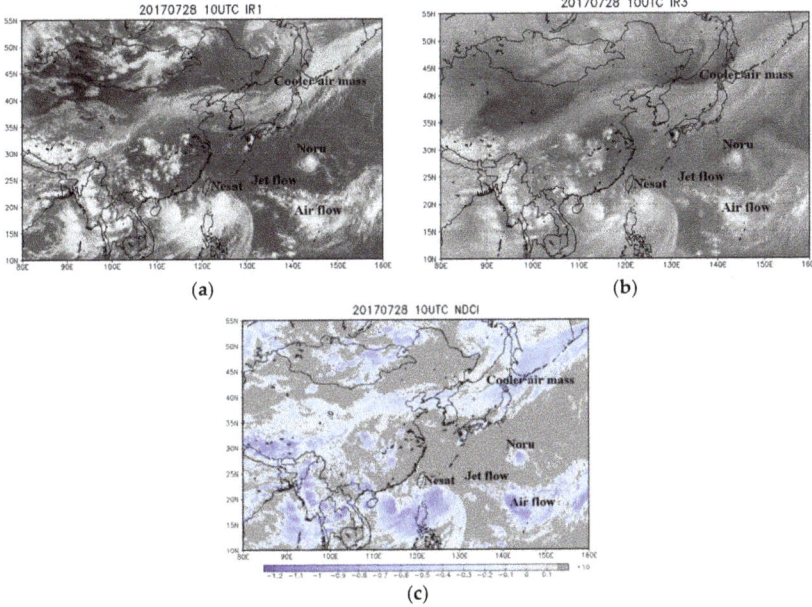

**Figure 4.** Images at 10:00 on 28 July 2017 of Typhoon Noru, Typhoon Nesat, TS Haitang, and a cooler air mass to the north. Note in particular the location of the areas of jet flow and air flow occupying the region between typhoons Noru and Nesat. (**a**) IR1 image, (**b**) IR3 (WV) image and (**c**) NDCI image.

### 4.3. Second Dual-Vortex Interaction

Air flow and jet flow occupying the region (23°N, 133–137°E) between typhoons Noru and Nesat affected both typhoons over 28–29 July. Though Noru and Nesat were separated by a considerable distance of 2000–2500 km, the effects of mid-level air flows and upper jet flows in between the two typhoons can be discerned. The intervening middle air flows and jet flows (900–1100 km) likely had some influence on the intensity and movement of both typhoons. The outward jet flow was the turbulent air flow that occurs between typhoons Noru and Nesat. This is therefore considered to be a special case of an 'indirect' dual-vortex interaction. Evidence suggests that that typhoons Noru and Nesat each intensified to become stronger typhoons, and both experienced changes in their track directions as a result of their mutual interaction (refer to Figure 1). The track behavior incorporating effects of the intervening air flows and jet flows is shown in Figures 5 and 6.

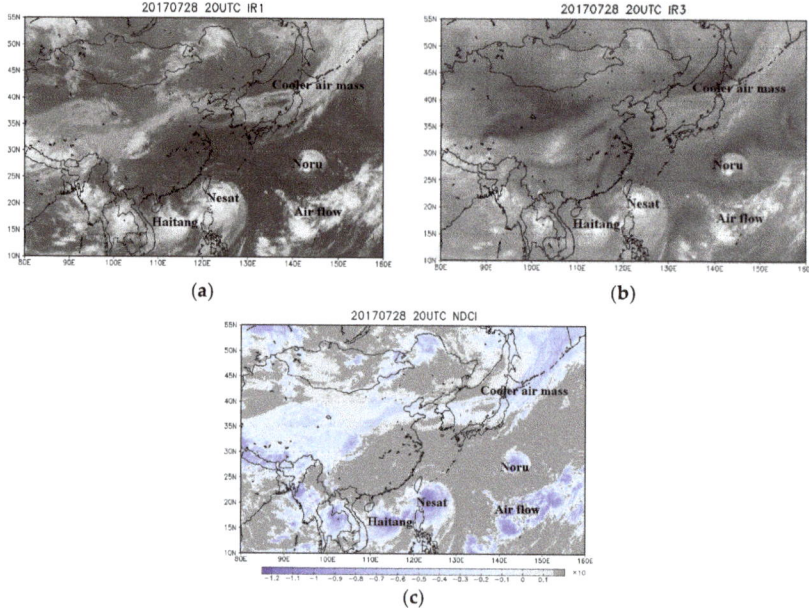

**Figure 5.** Images at 20:00 on 28 July 2017 of Typhoon Noru, Typhoon Nesat, TS Haitang, a cool air mass to the north, and dominant air flow to the south east. (**a**) IR1 image, (**b**) IR3 (WV) image and (**c**) NDCI image.

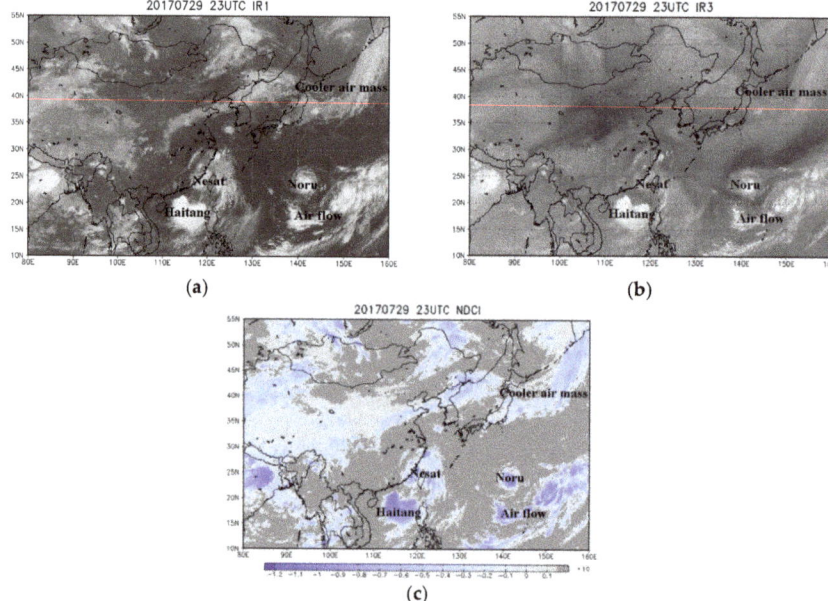

**Figure 6.** Images at 22:50 on 29 July 2017 of Typhoon Noru, Typhoon Nesat, TS Haitang, a cooler air mass to the north, and dominant air flow to the south east. (**a**) IR1 image, (**b**) IR3 (WV) image and (**c**) NDCI image.

## 4.4. Third Dual-Vortex Interaction

In Figure 5, a new cyclonic system, named Tropical Storm Haitang, formed on 27 July. The subsequent interaction between Typhoon Nesat and TS Haitang produced the third consecutive dual-vortex interaction during the period of interest, in Figure 6, specifically the twin circulation of the two typhoons around a common center, shown for 22:50 on 29 July. TS Haitang then crossed south-western Taiwan and made landfall in Fujian Province of mainland China. Simultaneously, Typhoon Noru took on annular characteristics with a symmetrical ring of deep convection surrounding a well-defined 30 km diameter eye, as shown for 09:00 on 31 July in Figure 7. After 2 August, Noru travelled northwards to Japan and made landfall on Kyushu Island. Long-lived Noru hit Japan with heavy rain.

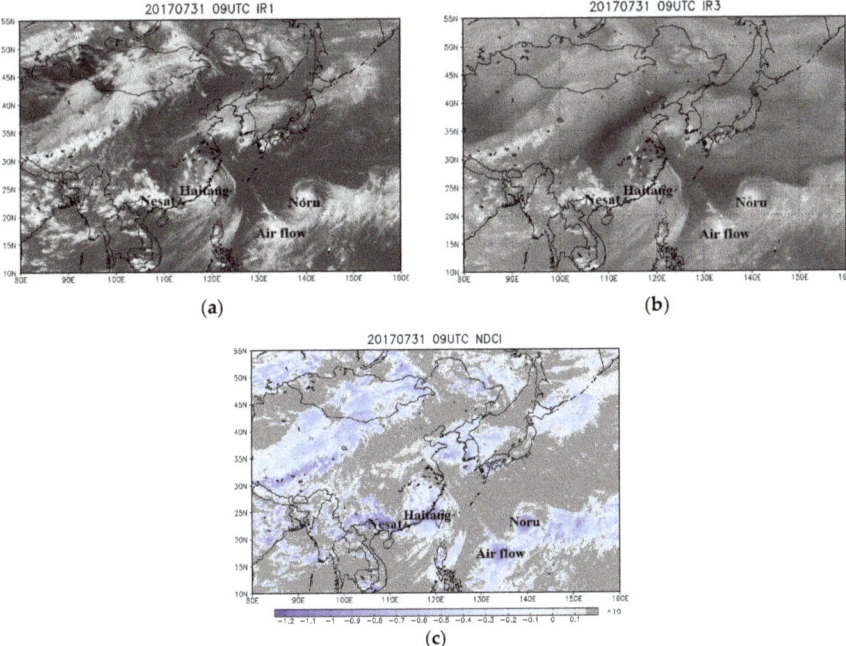

**Figure 7.** Images at 09:00 on 31 July 2017 showing typhoons Noru, Haitang and Nesat, and a dominant area of air flow to the south. (**a**) IR1 image, (**b**) IR3 (WV) image and (**c**) NDCI image.

## 5. Analysis and Discussion

Details of the triple consecutive dual-vortex interaction described above are next tabulated for further analysis. Table 2 shows the results calculated for interactions between Typhoon Noru and TS Kulap over three consecutive three-day intervals (20–23 July, 23–25 July and 25–27 July). Tabulated values for CI (current intensity) and d (measured distance) were calculated using Equation (4). It was found that a dual-vortex interaction was demonstrated during the interval of 25–27 July 2017. During this period, the measured distances (1100–1000 km) between the two systems remained marginally shorter than the calculated threshold distance $d_{th}$ required for an interaction (1133 km). It might be argued that the 25–27 July interactions between Typhoon Noru and TS Kulap cannot be quantitatively confirmed because the threshold distance (1133 km) was only 3% greater than the measured distance, which may have been similar in scale to errors inherent in measurement. Yet, even if it is the case, the threshold distance calculated from the Liou–Liu formula may still be used qualitatively to indicate to possibility of a cyclone interaction.

**Table 2.** Results of the first dual typhoon interaction.

|  |  | 20–23 July | 23–25 July | 25–27 July |
|---|---|---|---|---|
| Typhoon Noru | Pressure (hPa) | 1005–985 | 985–970 | 970 |
|  | CI | 1–3.1 | 3.1–4.1 | 4.1 |
| Tropical Storm Kulap | Pressure (hPa) | 1002 | 1002 | 1002 |
|  | CI | 1.2 | 1.2 | 1.2 |
| Measured distance between Noru and Kulap (d, km) |  | 1700–1900 | 1900–1800 | 1100–1000 |
| Calculated threshold distance ($d_{th}$, km) |  | 1055–1108 | 1108–1133 | 1133 |
| Cyclone–cyclone interaction |  | No | No | Yes |

It was next possible to determine the threshold distances for interaction between typhoons Noru and Nesat over the three intervals: 27–28 July, 28–29 July and 29–31 July. This was accomplished using Equation (5) by substituting $F_1 = F_2 = 1$ for approximation and CI = 1 for the middle jet flows (an unclassified TD at 22°N 136°E). For $F_1 = F_2 = 1$, the variations of size factors (size(TD or TS)–size($TC_{1,2}$)) and related height-difference ($h(TC_{1,2}) \gg (h(TD)$ or $h(TS))$) and rotation factors ($\tau_{1,2} = +1$ counter-clockwise rotation) were considered [20]. Otherwise, applying the wavenumber-one perturbation technique [38] and the vertical maximum values of radar reflectivity with geopotential height [39] were used to obtain the values of $F_1$ and $F_2$. The results are listed in Table 3. The measured distances (2500–2000 km) between two typhoons for the first interval were slightly greater than the calculated threshold distances (2217–2226 km). This means that only a weak interaction was possible between typhoons Noru and Nesat. During the second and third intervals, however, an indirect dual-vortex interaction was indicated, because the measured distances (2000–1600 and 1600–1700 km) between the two typhoons were shorter than the calculated threshold distances (2226–2258 and 2258–2287 km). Thus, the second in the sequence of dual-vortex interactions was clearly demonstrated.

**Table 3.** Results of the second dual typhoon interaction.

|  |  | 27–28 July | 28–29 July | 29–31 July |
|---|---|---|---|---|
| Typhoon Noru | Pressure (hPa) | 970–975 | 975–980 | 980–940 |
|  | CI | 4.1–3.9 | 3.9–3.5 | 3.5–5.8 |
| Typhoon Nesat | Pressure (hPa) | 990–985 | 985–960 | 960–980 |
|  | CI | 2.6–3.1 | 3.1–4.8 | 4.8–3.5 |
| Measured distance between Noru and Nesat (km) |  | 2500–2000 | 2000–1600 | 1600–1700 |
| Measured distance between Noru and intervening jet flows (km) |  | 1000–900 | 900–700 | 700–750 |
| Measured distance between Nesat and intervening jet flows (km) |  | 1500–1100 | 1100–900 | 900–950 |
| Calculated threshold distance (D, km) |  | 2217–2226 | 2226–2258 | 2258–2287 |
| Cyclone–cyclone interaction |  | Partial | Yes | Yes |

Table 4 shows the interactions between Typhoon Nesat and TS Haitang over three consecutive intervals (26–28 July, 28–29 July and 29–31 July). Tabulated CI and d values were again calculated using Equation (4). It was seen that the second and third intervals demonstrated clear dual-vortex interactions, when the measured distances (1000–1200 and 700–900 km) between the two typhoons were shorter than the calculated threshold distances (1155–1198 and 1198–1155 km).

Table 4. Results of the third dual typhoon interaction.

|  |  | 26–28 July | 28–29 July | 29–31 July |
|---|---|---|---|---|
| Typhoon Nesat | Pressure (hPa) | 1005–985 | 985–960 | 960–985 |
|  | CI | 1–3.1 | 3.1–4.8 | 4.8–3.1 |
| Tropical Storm Haitang | Pressure (hPa) | - | 985 | 985 |
|  | CI | - | 3.1 | 3.1 |
| Measured distance between Nesat and Haitang (km) |  | - | 1000–1200 | 700–900 |
| Calculated threshold distance (km) |  | - | 1155–1198 | 1198–1155 |
| Cyclone–cyclone interaction |  | No | Yes | Yes |

As a result of the triple consecutive sets of dual-vortex interactions described, Typhoon Noru survived to become a long-lasting system and travelled approximately 6900 km over 22 days from 19 July to 9 August. Typhoon Noru consequently attained the third longest longevity on record for tropical cyclones in the NWP Ocean, ranked only behind typhoons Rita and Wayne in 1972 and 1986, respectively, as indicated in Table 5.

Table 5. Comparing the three longest-lasting typhoons on record for the Northwest Pacific (NWP) Ocean.

|  | **Typhoon Rita** | **Typhoon Wayne** | **Typhoon Noru** |
|---|---|---|---|
| Year | 1972 | 1986 | 2017 |
| Timing | 5–30 July | 16 August–6 September | 19 July–9 September |
| Duration (days) | 25 | 22 | 22 |
| Minimum pressure (hPa) | 910 | 955 | 935 |
| Distance travelled (km) | 7100 | 7800 | 6900 |
| Dual-vortex interactions | Quadruple system interactions between Rita, Susan, Phyllis and Tess | None | Quadruple system interactions between Noru, Kalup, Nesat and Haitang |

Typhoons Rita, Phyllis, Tess and Susan during the typhoon season of 1972 belonged to a remarkable quadruple-typhoon interaction event, as seen from the track map in Figure 8. Table 6 gives details on three consecutive dual-vortex interactions between typhoons Phyllis and Tess (10 July), typhoons Rita and Susan (12 July), and typhoons Rita and Tess (23 July), as characterized using Equations (4) and (5).

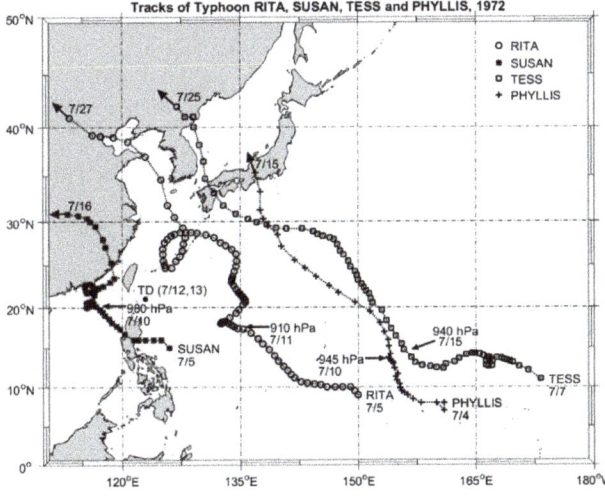

Figure 8. Tracks of typhoons Rita, Phyllis, Tess and Susan over 4–27 July 1972.

**Table 6.** Quantifying interactions between typhoons Rita, Phyllis, Tess, Susan and a tropical depression in July 1972.

|  | 1st Interaction | 2nd Interaction | 3rd Interaction |
|---|---|---|---|
| Date | 10 July | 12 July | 23 July |
| Interacting typhoons | Phyllis / Tess | Rita / Susan | Rita / Tess |
| Positions of typhoons (latitude and longitude) | 14°N 154°E / 14.5°N 164°E | 16°N 133°E / 22°N 118°E | 28°N 128°E / 30°N 132°E |
| Pressure (hPa) | 945 / 970 | 910 / 945 | 960 / 970 |
| CI | 5.4 / 4.2 | 7.2 / 3.5 | 4.8 / 4.1 |
| Measured distance between two typhoons (km) | 1000 | 1400 | 750 |
| Calculated threshold distance for interaction ($d_{th}$, km) | 1240 | 1250 | 1220 |
| Measured distance between Rita and intervening TD / Susan and intervening TD | — | 900 / 500 | — |
| Calculated threshold distance for interaction (D, km) | — | 2273 | — |
| Track response behavior | Tess changed direction | Susan executed a small loop | Rita executed a large loop |

The first dual-vortex interaction occurred between typhoons Phyllis and Tess on 10 July 1972. The interaction was possible because the measured distance separating the typhoons (1000 km) was shorter than the calculated required threshold distance (1240 km). The interaction caused Typhoon Tess to change direction.

Typhoons Rita and Susan were separated by a measured distance of 1400 km on 12 July 1972. According to Equation (4) (Table 6), this exceeded the calculated threshold distance for an interaction of 1250 km. However, the presence of an intervening tropical depression (TD) (i.e., a low pressure area) between the typhoons was influential, because it facilitated an indirect dual-vortex interaction between Rita and Susan in spite of their large separation. Approximating using the tuning factors $F_1 = F_2 = 1$ and a value of CI = 0.1 for the low pressure disturbance (unclassified TD at 27°N 125°E), Equations (4) and (5) are thus able to predict the interaction between Rita and Susan. The measured distances between Typhoon Rita and the TD (900 km) as well as Susan and the TD (400 km) were shorter than the calculated threshold distance (2480 km). This second interaction resulted in Typhoon Susan executing a small loop in its track.

The third interaction occurred between typhoons Rita and Tess on 23 July 1972. The measured distance (750 km) between them was much shorter than the calculated threshold distance (1220 km) according to Equation (4). In consequence, the resulting strong Fujihara-type rotation caused Rita to execute a large loop in its track. Thus, a consecutive triple sequence of dual-vortex interactions can be demonstrated between the four typhoons Rita, Phyllis, Tess and Susan, and their intervening areas of tropical depressions.

From the above discussion, the following points emerge and should be highlighted:

1. Special cases of quadruple typhoons, simultaneously forming in the same region or in quick succession, can give rise to a triple sequence ('trilogy') of dual-vortex interactions.
2. Middle air flows or intervening low pressure areas (tropical depressions) are important environmental influences that should not be ignored in assessment of cyclone interaction behavior.
3. The Liou–Liu formulas (based on current intensity values) can be applied to verify and quantify dual-vortex interactions by comparing measured distances between typhoons with calculated threshold distances required for interaction.
4. Dual-vortex interactions are frequently associated with an increase in typhoon track sinuosity through changes in track direction and/or the execution of complex/looping tracks.
5. The longevity (survival time) of an individual typhoon may be significantly enhanced through a sequence of multiple dual-vortex interactions during its lifespan.

## 6. Conclusions

The integration of remote sensing imagery, a differential averaging technique, and the Liou–Liu formulas was used to identify a 'trilogy' of consecutive dual-vortex interactions between typhoons Noru, Kulap, Nesat, Haitang and environmental air flows in the Northwest Pacific basin during the 2017 typhoon season. Analyzing the satellite imagery of the interaction between typhoons benefits from the application of a differential averaging technique. The NDCI operator and filter applied to geosynchronous satellite IR1 infrared images and IR3 water vapor images were able to depict the differences and generate a processed set of NDCI images with suitable clarity for investigation.

Three distinct dual-vortex interactions among four cyclone systems (and neighboring cooler environmental air masses, air flows and jet flows) occurred in a consecutive sequence. The interactions were validated using the Liou–Liu empirical formulas, which are used for calculating threshold distances for interactions as related to current intensity (CI) values.

The first dual-vortex interaction caused Typhoon Noru to experience Fujiwhara rotation and to merge with the approaching TS Kulap. In the second interaction, typhoons Noru and Nesat were possibly strengthened through the influence of intervening air flows and jet flows, even though a considerable distance separated the typhoons. It has been suggested that this is a special type of 'indirect' cyclone–depression–cyclone interaction, which can nonetheless be influential in mutual typhoon intensification. The third interaction between Haitang and Nesat again resulted in Fujiwhara rotation and changes in typhoon track directions.

Examining the main characteristics of Typhoon Noru in particular, the time series of pressure (intensity) and saturation area shows that a relationship through time exists with distance from other interacting typhoons. Long-lived Typhoon Noru survived from 19 July to 9 August and followed a complex track, with a loop and several U-turns resulting from the sequence of interactions described. On 31 July, Noru became a super-typhoon attaining an intensity of 930 hPa before striking Japan with torrential rain.

Additionally illuminating is the comparison with Typhoon Rita in 1972, which similarly experienced complex multiple interactions with typhoons Phyllis, Tess and Susan. Coincidentally, the 1972 Rita and the 2017 Noru events both involved a sequential trilogy of interactions among four individual typhoon systems, with the interactions leading to significantly enhanced typhoon longevity. It is recommended that further quantitative observations using satellite cloud images of interactions between multiple synchronous/sequential typhoons be undertaken, as this should eventually lead to improved track forecasting and weather prediction in these particularly unusual synoptic circumstances.

**Author Contributions:** Y.-A.L. and J.-C.L. conceived the project, conducted research, performed initial analyses and wrote the first manuscript draft. C.-C.L., C.-H.C., and K.-A.N. provided helpful discussions during conception of the project. J.-C.L. edited the first manuscript. Y.-A.L and J.P.T. finalized the manuscript for the communication with the journal.

**Funding:** This work was supported by the Ministry of Science and Technology under Grant MOST 105-2221-E-008-056-MY3, 107-2111-M-008-036 and Grant 105-2221-E-008-056-MY3. J.P. Terry acknowledges funding support from Zayed University.

**Acknowledgments:** Constructive comments from anonymous reviewers helped the authors make significant improvements to the original manuscript.

**Conflicts of Interest:** The authors declare no conflict of interest.

## References

1. Gierach, M.M.; Subrahmanyam, B. Satellite data analysis of the upper ocean response to hurricanes Katrina and Rita (2005) in the Gulf of Mexico. *IEEE Geosci. Remote Sens. Lett.* **2007**, *4*, 132–136. [CrossRef]
2. Terry, J.P. *Tropical Cyclones: Climatology and Impacts in the South Pacific*; Springer: New York, NY, USA, 2007; 210p.

3. Acker, J.; Lyon, P.; Hoge, F.; Shen, S.; Roffer, M.; Gawlikowski, G. Interaction of hurricane Katrina with optically complex water in the Gulf of Mexico: Interpretation using satellite-derived inherent optical properties and chlorophyll concentration. *IEEE Geosci. Remote Sens. Lett.* **2009**, *6*, 209–213. [CrossRef]
4. Nguyen, A.K.; Liou, Y.A.; Terry, J.P. Vulnerability and adaptive capacity maps of Vietnam in response to typhoons. *Sci. Total Environ.* **2019**, *682*, 31–46. [CrossRef] [PubMed]
5. Nguyen, A.K.; Liou, Y.A.; Li, M.H.; Tran, T.A. Zoning eco-environmental vulnerability for environmental management and protection. *Ecol. Indic.* **2016**, *69*, 100–117. [CrossRef]
6. Liou, Y.-A.; Nguyen, A.K.; Li, M.H. Assessing spatiotemporal eco-environmental vulnerability by Landsat data. *Ecol. Indic.* **2017**, *80*, 52–65. [CrossRef]
7. Nguyen, A.K.; Liou, Y.A. Global mapping of eco-environmental vulnerability from human and nature disturbances. *Sci. Total Environ.* **2019**, *664*, 995–1004. [CrossRef]
8. Nguyen, A.K.; Liou, Y.A. Mapping global eco-environment vulnerability due to human and nature disturbances. *MethodsX* **2019**, *6*, 862–875. [CrossRef]
9. Lin, C.Y.; Hsu, H.M.; Sheng, Y.F.; Kuo, C.H.; Liou, Y.A. Mesoscale processes for super heavy rainfall of Typhoon Morakot (2009) over Southern Taiwan. *Atmos. Chem. Phys.* **2011**, *11*, 345–361. [CrossRef]
10. Terry, J.P.; Kim, I.-H.; Jolivet, S. Sinuosity of tropical cyclone tracks in the South West Indian Ocean: Spatio-temporal patterns and relationships with fundamental storm attributes. *Appl. Geogr.* **2013**, *45*, 29–40. [CrossRef]
11. Terry, J.P.; Kim, I.-H. Morphometric analysis of tropical storm and hurricane tracks in the North Atlantic basin using a sinuosity-based approach. *Int. J. Climatol.* **2015**, *35*, 923–934. [CrossRef]
12. Piñeros, M.F.; Ritchie, E.A.; Tyo, J.S. Objective measures of tropical cyclone structure and intensity change from remotely sensed infrared image data. *IEEE Trans. Geosci. Remote Sens.* **2008**, *46*, 3574–3580. [CrossRef]
13. Chang, P.L.; Jou, B.J.D.; Zhang, J. An algorithm for tracking eyes of tropical cyclones. *Weather Forecast.* **2009**, *24*, 245–261. [CrossRef]
14. Wimmers, A.J.; Velden, C.S. Objectively determining the rotational center of tropical cyclones in passive microwave satellite imagery. *J. Appl. Meteorol. Climatol.* **2010**, *49*, 2013–2034. [CrossRef]
15. Wimmers, A.J.; Velden, C.S. Advancements in objective multisatellite tropical cyclone center fixing. *J. Appl. Meteorol. Climatol.* **2016**, *55*, 197–212. [CrossRef]
16. Liou, Y.A.; Liu, J.C.; Chane-Ming, F.; Hong, J.S.; Huang, C.Y.; Chiang, P.K.; Jolivet, S. *Remote Sensing for Improved Forecast of Typhoons*; Barale, V., Gade, M., Eds.; Springer: Cham, Switzerland, 2019; ISBN 978-3-319-94065-6. [CrossRef]
17. Fujiwhara, S. On the growth and decay of vortical systems. *Q. J. R. Meteorol. Soc.* **1923**, *49*, 75–104. [CrossRef]
18. Prieto, R.; McNoldy, B.D.; Fulton, S.R.; Schubert, W.A. A classification of binary tropical cyclone-like vortex interactions. *Mon. Weather Rev.* **2003**, *131*, 2656–2666. [CrossRef]
19. Liu, C.C.; Shyu, T.Y.; Chao, C.C.; Lin, Y.F. Analysis on typhoon Long Wang intensity changes over the ocean via satellite data. *J. Mar. Sci. Technol.* **2009**, *17*, 23–28.
20. Zhang, C.J.; Wang, X.D. Typhoon cloud image enhancement and reducing speckle with genetic algorithm in stationary wavelet domain. *IET Image Process.* **2009**, *3*, 200–216. [CrossRef]
21. Hart, R.; Evans, J. Simulations of dual-vortex interaction within environmental shear. *J. Atmos. Sci.* **1999**, *56*, 3605–3621. [CrossRef]
22. Galarneau, J.T.; Davis, C.A.; Shapiro, M.A. Intensification of hurricane sandy (2012) through extratropical warm core seclusion. *Mon. Weather Rev.* **2013**, *141*, 4296–4321. [CrossRef]
23. Wu, C.C.; Huang, T.S.; Huang, W.P.; Chou, K.H. A new look at the binary interaction: Potential vorticity diagnosis of the unusual southward movement of Typhoon Bopha (2000) and its interaction with Typhoon Saomai (2000). *Mon. Weather Rev.* **2003**, *131*, 1289–1300. [CrossRef]
24. Liu, J.C.; Liou, Y.A.; Wu, M.X.; Lee, Y.J.; Cheng, C.H.; Kuei, C.P.; Hong, R.M. Interactions among two tropical depressions and typhoons Tembin and Bolaven (2012) in Pacific Ocean: Analysis of the depression-cyclone interactions with 3-D reconstruction of satellite cloud images. *IEEE Trans. Geosci. Remote Sens.* **2015**, *53*, 1394–1402. [CrossRef]
25. Liou, Y.A.; Liu, J.C.; Wu, M.X.; Lee, Y.J.; Cheng, C.H.; Kuei, C.P.; Hong, R.M. Generalized empirical formulas of threshold distance to characterize cyclone-cyclone. *IEEE Trans. Geosci. Remote Sens.* **2016**, *54*, 3502–3512. [CrossRef]

26. Lee, Y.S.; Liou, Y.A.; Liu, J.C.; Chiang, C.T.; Yeh, K.D. Formation of winter super-typhoons Haiyan (2013) and Hagupit (2014) through interactions with cold fronts as observed by multifunctional transport satellite. *IEEE Trans. Geosci. Remote Sens.* **2017**, *55*, 3800–3809. [CrossRef]
27. Jaiswal, N.; Kishtawal, C.M. Objective detection of center of tropical cyclone in remotely sensed infrared images. *IEEE J. Sel. Top. Appl. Earth Obs. Remote Sens.* **2013**, *6*, 1031–1035. [CrossRef]
28. Liou, Y.A.; Liu, J.C.; Liu, C.P.; Liu, C.C. Season-dependent distributions and profiles of seven super-typhoons (2014) in the Northwestern Pacific Ocean from satellite cloud images. *IEEE Trans. Geosci. Remote Sens.* **2018**, *56*, 2949–2957. [CrossRef]
29. Minnis, P.; Sun-Mack, S.; Young, D.F.; Heck, P.W.; Garber, D.P.; Chen, Y.; Spangenberg, D.A.; Arduini, R.F.; Trepte, Q.Z.; Smith, W.L.; et al. CERES Edition 2 cloud property retrievals using TRMM VIRS and Terra and Aqua MODIS data: Part I: Algorithms. *IEEE Trans. Geosci. Remote Sens.* **2011**, *49*, 4374–4400. [CrossRef]
30. Minnis, P.; Sun-Mack, S.; Chen, Y.; Khaiyer, M.M.; Yi, Y.; Ayers, J.K.; Brown, R.R.; Dong, X.; Gibson, S.C.; Heck, P.W.; et al. CERES Edition-2 cloud property retrievals using TRMM VIRS and Terra and Aqua MODIS Data—Part II: Examples of average results and comparisons with other data. *IEEE Trans. Geosci. Remote Sens.* **2011**, *49*, 4401–4430. [CrossRef]
31. King, M.D.; Menzel, W.P.; Kaufman, Y.J.; Tanre, D.; Gao, B.C.; Platnick, S.; Ackerman, S.A.; Remer, L.A.; Pincus, R.; Hubanks, P.A. Cloud and aerosol properties, precipitable water, and profiles of temperature and humidity from MODIS. *IEEE Trans. Geosci. Remote Sens.* **2003**, *41*, 442–458. [CrossRef]
32. Platnick, S.; King, M.D.; Ackerman, S.A.; Menzel, W.P.; Baum, B.A.; Riédi, J.C.; Frey, R.A. The MODIS cloud products: Algorithms and examples from Terra. *IEEE Trans. Geosci. Remote Sens.* **2003**, *41*, 459–473. [CrossRef]
33. Huang, H.L.; Yang, P.; Wei, H.; Baum, B.A.; Hu, Y.; Antonelli, P.; Ackerman, S.A. Inference of ice cloud properties from high spectral resolution infrared observations. *IEEE Trans. Geosci. Remote Sens.* **2004**, *42*, 842–853. [CrossRef]
34. Hong, G.; Yang, P.; Huang, H.L.; Baum, B.A.; Hu, Y.; Platnick, S. The sensitivity of ice cloud optical and microphysical passive satellite retrievals to cloud geometrical thickness. *IEEE Trans. Geosci. Remote Sens.* **2007**, *45*, 1315–1323. [CrossRef]
35. Liu, C.C.; Shyu, T.Y.; Lin, T.H.; Liu, C.Y. Satellite-derived normalized difference convection index for typhoon observations. *J. Appl. Remote Sens.* **2015**, *9*, 096074. [CrossRef]
36. Liu, J.C.; Chaung, J.C.; Chou, H.C. Improved differential averaging technique for comparison decrement method. *IEE Proc. H* **1997**, *145*, 377–381. [CrossRef]
37. Kieu, C.Q.; Chen, K.H.; Zhang, D. An examination of the pressure-wind relationship for intense tropical cyclones. *Weather Forecast.* **2010**, *25*, 895–907. [CrossRef]
38. Nolan, D.S.; Montgomery, M.T.; Grasso, L.D. The wavenumber-one instability and trochoidal motion of hurricane-like vortices. *J. Atmos. Sci.* **2001**, *58*, 3243–3270. [CrossRef]
39. Jian, G.; Wu, C. A numerical study of the track deflection of super typhoon Haitang (2005) prior to its landfall in Taiwan. *Mon. Weather Rev.* **2008**, *136*, 598–615. [CrossRef]

© 2019 by the authors. Licensee MDPI, Basel, Switzerland. This article is an open access article distributed under the terms and conditions of the Creative Commons Attribution (CC BY) license (http://creativecommons.org/licenses/by/4.0/).

*Article*

# Habitat Suitability Estimation Using a Two-Stage Ensemble Approach

**Jehyeok Rew †, Yongjang Cho †, Jihoon Moon and Eenjun Hwang \***

School of Electrical Engineering, Korea University, 145, Anam-ro, Seongbuk-gu, Seoul 02841, Korea; rjh1026@korea.ac.kr (J.R.); dydwkd486@korea.ac.kr (Y.C.); johnny89@korea.ac.kr (J.M.)
\* Correspondence: ehwang04@korea.ac.kr; Tel.: +82-2-3290-3256
† These authors contributed equally to this work.

Received: 20 March 2020; Accepted: 4 May 2020; Published: 6 May 2020

**Abstract:** Biodiversity conservation is important for the protection of ecosystems. One key task for sustainable biodiversity conservation is to effectively preserve species' habitats. However, for various reasons, many of these habitats have been reduced or destroyed in recent decades. To deal with this problem, it is necessary to effectively identify potential habitats based on habitat suitability analysis and preserve them. Various techniques for habitat suitability estimation have been proposed to date, but they have had limited success due to limitations in the data and models used. In this paper, we propose a novel scheme for assessing habitat suitability based on a two-stage ensemble approach. In the first stage, we construct a deep neural network (DNN) model to predict habitat suitability based on observations and environmental data. In the second stage, we develop an ensemble model using various habitat suitability estimation methods based on observations, environmental data, and the results of the DNN from the first stage. For reliable estimation of habitat suitability, we utilize various crowdsourced databases. Using observational and environmental data for four amphibian species and seven bird species in South Korea, we demonstrate that our scheme provides a more accurate estimation of habitat suitability compared to previous other approaches. For instance, our scheme achieves a true skill statistic (TSS) score of 0.886, which is higher than other approaches (TSS = 0.725 ± 0.010).

**Keywords:** habitat suitability estimation; deep neural network; two-stage modeling; ensemble approach

## 1. Introduction

For decades, the importance of biodiversity conservation has been emphasized globally because high biodiversity offers a variety of natural services that support sustainable human living [1]. Despite this importance, ecosystem services have rapidly declined for a variety of reasons, such as indiscriminate resource development, rapid urban expansion, and global climate change. The loss of biodiversity can have adverse consequences on the ecosystem because of the complex interactions that exist among species [2,3]. To maintain biodiversity levels, ecologists have devised and applied various methods to protect habitats by analyzing the characteristics of target species and their habitats [4,5].

Habitat suitability models, also known as species distribution models (SDMs), environmental niche models (ENMs), and predictive habitat distribution models, have been used to predict the habitat of target species based on various environmental factors, such as temperature, precipitation, seasonality, and terrain [2,3,6]. Habitat suitability models can be used to assess not only the relationship among various environmental factors, such as global climatic conditions, landscape information, and species habitats, but also landscape management and the conservation of endangered species [6–9]. With the development of remote sensing technology, the performance of habitat suitability models has improved significantly. Until a few decades ago, the prediction of habitat suitability for particular species over a wide range of areas with reasonable accuracy was very challenging. This is because remote sensing technology at that time had several limitations, including high costs, poor spatial resolution, complicated digital maps at scales larger than remote sensing images, and human error during interpretation analysis. Recently, state-of-the-art remote sensing technologies have overcome previous data-processing issues, making it possible to obtain reliable temporal and spatial data for factors such as land cover and the climate and to subsequently construct inference models for habitat suitability that can cover very small to large areas.

Based on remote sensing data, several ecological researchers have attempted to construct effective habitat suitability models using a profile, statistical, and machine learning methods (Table 1). The surface range envelope (SRE) model, a profile approach, has been used to estimate habitat suitability [7–9]. Araújo et al. [7] introduced a series of surface envelop models based on the associations between climatic variables and the distributions of species to determine suitable conditions for the maintenance of a viable population. Heikkinen et al. [8] presented several critical methodological issues that may lead to uncertainty in predictions based on bioclimatic modeling. They concluded that bioclimatic envelop models have several advantages, one of which is that the modeling results are simple and easy to understand.

Statistical methods, such as flexible discriminant analysis (FDA), the multivariate adaptive regression spline (MARS), and the generalized linear model (GLM), investigate multiple linear relationships between species distributions and environmental layers. Statistical methods all have advantages and disadvantages, so various statistical methods are often employed together to improve habitat suitability estimation [10–14]. To determine the potential habitat and distribution of species, Elith et al. [10] presented a practical guide that included how to efficiently use statistical methods. They compared 16 modeling methods, including the MARS and GLM, using 226 species from six regions of the world. They found that the GLM, and BIOCLIM outperformed the other profile and statistical modeling methods. Leathwick et al. [11] utilized two statistical modeling methods, the generalized additive model (GAM) and MARS, to analyze the relationships between the distributions of 15 freshwater fish species and the corresponding environment. They reported that the MARS model performed strongly with low-prevalence species and that it could be used to analyze a large dataset.

Recently, habitat suitability modeling has been conducted using machine-learning methods, such as the generalized boosting model (GBM), maximum entropy (MAXENT), and random forest (RF). These machine-learning methods have been reported to produce more accurate predictions than profile and statistical methods [15–27]. For instance, Phillips et al. [15] utilized various machine-learning methods for the habitat modeling of ocean sunfish species. They used observations of ocean sunfishes and a number of environmental variables to conduct species distribution modeling using MARS, the SRE model, classification tree analysis (CTA), FDA, RF, and GLM. Reiss et al. [16] predicted the distribution of benthic species in the North Sea. They compared nine different methods: the support vector machine (SVM), GLM, GAM, GBM, MAXENT, FDA, BIOCLIM, and MARS. In their experiments, the machine-learning methods MAXENT, GBM, and RF produced a better predictive performance than the profile and statistical methods. Guisan et al. [17] employed various machine learning-based prediction methods to determine a suitable model for species habitats using remote sensing data. They examined multiple steps in predictive modeling by considering the conceptual model and its statistical formulation and calibration. Phillips et al. [18] employed maximum entropy-based modeling to predict the habitat of the *Bradypus variegatus*. They used two remote-sensed datasets, climate, and elevation that were derived from the Intergovernmental Panel on Climate Change (IPCC) and the United States Geological Survey (USGS), respectively. They evaluated the effectiveness of the model by comparing it to a rule-set-based genetic algorithm. Heikkinen et al. [19] conducted species habitat modeling for endangered butterfly species and predicted the distribution of Apollo butterflies using various machine-learning methods such as GLM, GAM, CTA, a shallow neural network (SNN), MARS, and boosted regression tree (BRT). They concluded that statistical analysis and machine-learning methods were useful for conservation planning and protecting endangered species.

**Table 1.** Overview of habitat suitability modeling techniques.

| Models | Descriptions | References | Category |
|---|---|---|---|
| SRE | Profiling technique that uses the environmental conditions of locations of occurrence data to profile the environments where a species can be found. | Araújo et al. [7], Heikkinen et al. [8], Thuiller et al. [9] | Profile |
| FDA | Classification technique based on a mixture of linear regression models. | Hastie et al. [14] | Statistical regression |
| GLM | Parametric regression technique based on a random component, a systematic component, and a link function describing a relation between the former the random and systematic component. | Elith et al. [10], Zuur et al. [12] | Statistical regression |
| MARS | Non-parametric regression technique that builds multiple linear regression models across the range of predictor values. | Elith et al. [10], Leathwick et al. [11], Friedman et al. [13] | Statistical regression |
| GBM | Machine learning technique based on the combinations of decision tree algorithms and boosting methods. | Thomaes et al. [20], De'Ath et al. [21] | Machine learning |
| CTA | Machine learning technique that is a supervised non-parametric statistical classification approach based on binary recursive partitioning techniques | Breiman et al. [22], De'ath G et al. [23] | Machine learning |
| SNN | Machine learning technique based on non-linear mapping structures inspired on the biological system of the brain. | D'heygere et al. [24], Fukuda et al. [25], Phillips et al. [17] | Machine learning |
| RF | Machine learning technique using bootstrap aggregation to create a set of decision trees. | Breiman L et al. [26], Cutler DR et al. [27] | Machine learning |
| MAXENT | Machine learning technique using the principle of maximum entropy to make a prediction from incomplete knowledge. | Phillips et al. [15], Reiss et al. [16] | Machine learning |

More recently, habitat modeling based on deep neural networks (DNNs) has been investigated in ecological research [28–31]. In general, DNNs provide more accurate predictions in terms of identifying potential habitats for target species than conventional models such as GLM, GAM, MARS, and BRT. This is because DNNs can automatically extract features and learn complex non-linearities from extracted features [32]. For instance, Rademaker et al. [28] determined the niches of wild and domesticated ungulate species using modeling schemes based on a DNN. They focused on the applicability of the DNN and employed it in habitat estimation modeling. In their experiment, they showed that DNN could effectively identify potential habitats using sufficient observational data. Botella et al. [29] proposed a deep-learning approach for SDM. They applied a convolution neural network (CNN) and DNN to overcome the shortcomings of the traditional SDM. To evaluate its performance, they used part of the GBIF dataset and 46 environmental layers, including climate, digital elevation, and land cover. They subsequently found that both models performed better than traditional models such as the GAM and MAXENT.

However, despite the versatility of DNNs [30–35], a DNN-based habitat model trained on a small observational dataset for a species has been shown to produce inferior estimations to traditional machine-learning models, such as MAXENT and RF. Thus, constructing an accurate habitat suitability model is challenging because obtaining observational data is difficult [28]. Collecting sufficient observational data is particularly crucial when constructing habitat suitability models for endangered species. To overcome these problems, in this paper, we propose a novel two-stage based ensemble model called TSEM for the development of an effective habitat suitability model using an ensemble of various habitat suitability estimation techniques and DNN. Our TSEM was trained and tested on the crowdsourced datasets composed of volunteers' observation data. Strictly speaking, the observation data may indicate where the species actually lives or where the observations were made. As a result, the estimation results of our model could have similar characteristics. To improve the performance of habitat suitability models, we focus on three major issues, which are the main contributions of this paper.

- Using crowdsourcing databases [36–39] and a diverse range of environmental data, we employ data pre-processing to generate framed data, which consist of observation data for the target species and related environmental data.
- We propose a two-stage modeling scheme. In the first stage, we construct a DNN model using framed data. Then, we build an ensemble model using a diverse range of habitat suitability estimation methods and the results of the DNN model in the first stage to improve estimation performance in the second stage.
- We compare our ensemble model with other estimation models based on a variety of evaluation metrics and statistical analysis and verify the superiority of our model.

The rest of this paper is organized as follows. We first introduce the steps required to construct the TSEM in Section 2. Then, we present several experiments conducted to evaluate the performance of our proposed model and visualize the results using a map-overlay function in Section 3. Finally, we summarize the major findings and provide directions in Section 4.

## 2. Materials and Methods

### 2.1. Overview of Two-Stage Habitat Suitability Estimation Model

We describe in detail our two-stage habitat suitability estimation model with the overall structure (Figure 1). Observational, global climatic [40], and Korean land cover data [41] were initially collected to configure the independent variables for our model. In the first stage of the model, we constructed a DNN model as a sub-model. Next, a stacking ensemble-based estimation model was constructed using the results of the DNN sub-model as input to improve estimation performance. Finally, the habitat suitability results and other widely used evaluation metrics, including area under the curve (AUC), sensitivity, specificity, the kappa statistic, and the TSS were visualized for a performance comparison between the TSEM and previously reported models [42–47].

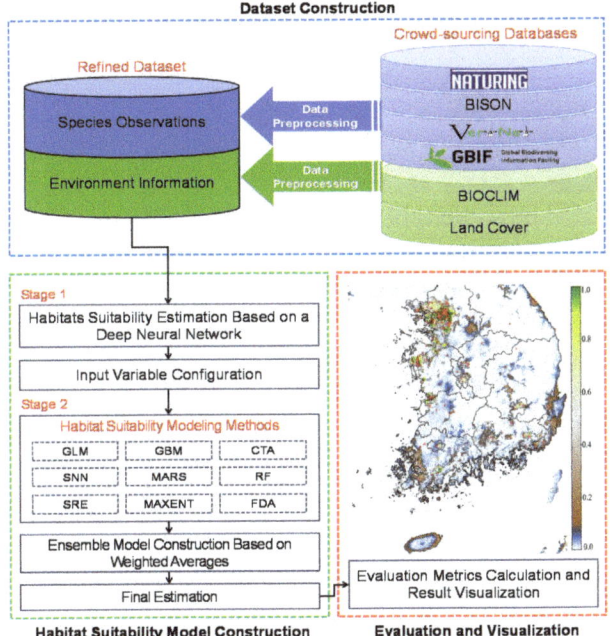

Figure 1. Overall process of the TSEM.

## 2.2. Dataset Construction

In general, the performance of a DNN-based estimation model depends on the quality and size of the dataset used for training. To construct our dataset, we first reviewed several crowdsourced databases that contain global observations of various species, selected 11 target species that were primarily found in South Korea, and then collected observational data for these species. These species are all considered conservation targets in Korean wildlife conservation projects. We listed the target species and their number of observations, as reported in the global biodiversity information facility (GBIF), VertNet, biodiversity information serving our nation (BISON), and Naturing databases (Table 2). Species habitats are closely related to the climate and land conditions [42–47]. Therefore, to construct a valid habitat model, we collected various layers of environment information from Worldclim Bioclimatic [40] and a land cover dataset for South Korea [41], which are both widely used for ecological modeling. The land cover dataset for South Korea was generated using Korea multi-purpose satellite No. 2, also known as KOMPSAT No. 2 or Arirang 2, and satellite pour l'observation de la terre 5 (SPOT 5) remote sensing images from 2009. KOMPSAT No. 2, which is equipped with a 1-m high-resolution multi-spectral camera (MSC), has orbited Earth approximately 46,800 times in nine years, capturing approximately 75,400 high-quality satellite images of Korea, while SPOT 5 can capture satellite images with a coverage of 60 km × 60 km and a resolution of 5 m. Compared to the GlobCover, the land cover dataset for South Korea provides more detailed information due to its higher resolution. Because this land cover dataset consists of categorical variables, we converted them into proximity distance layers, which have continuous values. Proximity distance for land cover layers is regarded as crucial for modeling because the unique survival traits of species and their habitat characteristics are closely related. Indeed, several studies have improved the performance of habitat suitability estimation models by considering the distance between the environmental layers and species observations. For this reason, we also employed proximity distance as an input variable. In total, we used 41 environmental layers as input variables for our habitat suitability modeling (Table 3).

Table 2. Target species and their observations.

| Target Species (Scientific Name) | Image | IUCN Red List | Number of Observations in South Korea | Suitable Habitats |
|---|---|---|---|---|
| Streptopelia orientalis | | Least Concern | 1523 | Shrubland, Terrestrial, Forest |
| Passer montanus | | Least Concern | 1498 | Shrubland, Terrestrial, Forest |
| Ardea cinerea | | Least Concern | 1116 | Marine neritic, Forest, Wetlands, Grassland |
| Hypsipetes amaurotis | | Least Concern | 1162 | Terrestrial, Forest |
| Hynobius leechii | | Least Concern | 1336 | Wetlands, Forest |
| Anas zonorhyncha | | Least Concern | 856 | Wetlands, Artificial/Aquatic and Marine, Terrestrial, Marine coastal |
| Rana huanrenensis | | Least Concern | 906 | Wetlands, Forest, Grassland |
| Anas platyrhynchos | | Least Concern | 714 | Wetlands, Artificial/Aquatic and Marine |
| Cyanopica cyanus | | Least Concern | 679 | Forest, Terrestrial |
| Rana dybowskii | | Least Concern | 511 | Wetlands, Aquatic and Marine, Terrestrial, Vegetation, Shrubland, Forest |
| Hyla japonica | | Least Concern | 1261 | Wetlands, Aquatic and Marine, Terrestrial, Vegetation, Shrubland, Forest |

Table 3. List of input variables.

| Variable Name | Description | Type |
|---|---|---|
| BIO_1 | Annual mean temperature | Continuous |
| BIO_2 | Mean diurnal range | Continuous |
| BIO_3 | Isothermality | Continuous |
| BIO_4 | Temperature seasonality | Continuous |
| BIO_5 | Max. temperature of the warmest month | Continuous |
| BIO_6 | Min. temperature of the coldest month | Continuous |
| BIO_7 | Temperature annual range | Continuous |
| BIO_8 | Mean temperature of the wettest quarter | Continuous |
| BIO_9 | Mean temperature of the driest quarter | Continuous |
| BIO_10 | Mean temperature of the warmest quarter | Continuous |
| BIO_11 | Mean temperature of the coldest quarter | Continuous |
| BIO_12 | Annual precipitation | Continuous |
| BIO_13 | Precipitation of the wettest month | Continuous |
| BIO_14 | Precipitation of the driest month | Continuous |
| BIO_15 | Precipitation seasonality | Continuous |
| BIO_16 | Precipitation of the wettest quarter | Continuous |
| BIO_17 | Precipitation of the driest quarter | Continuous |
| BIO_18 | Precipitation of the warmest quarter | Continuous |
| BIO_19 | Precipitation of the coldest quarter | Continuous |
| Distance_1 | Proximity distance from each cell to a residential area (detached residential and common residential areas) | Continuous |
| Distance_2 | Proximity distance from each cell to an industrial area | Continuous |
| Distance_3 | Proximity distance from each cell to a commercial area (commercial/business and mixed residential/business areas) | Continuous |
| Distance_4 | Proximity distance from each cell to a leisure facility area | Continuous |
| Distance_5 | Proximity distance from each cell to a transportation area (airport, harbor, railway, road, and other transportation and communication facilities) | Continuous |
| Distance_6 | Proximity distance from each cell to a public facility area (basic environmental, education/administrative, and other public facilities) | Continuous |
| Distance_7 | Proximity distance from each cell to paddy fields (land consolidation success and undergoing land consolidation in paddy fields) | Continuous |
| Distance_8 | Proximity distance from each cell to dry fields (land consolidation success and undergoing land consolidation in dry fields) | Continuous |
| Distance_9 | Proximity distance from each cell to a greenhouse | Continuous |
| Distance_10 | Proximity distance from each cell to an orchard | Continuous |
| Distance_11 | Proximity distance from each cell to other plantations (pastureland and other plantations) | Continuous |
| Distance_12 | Proximity distance from each cell to broadleaf forest | Continuous |
| Distance_13 | Proximity distance from each cell to coniferous forest | Continuous |
| Distance_14 | Proximity distance from each cell to mixed forest | Continuous |
| Distance_15 | Proximity distance from each cell to natural pasture | Continuous |
| Distance_16 | Proximity distance from each cell to artificial pasture (golf course, cemetery, and other pastures) | Continuous |
| Distance_17 | Proximity distance from each cell to coastal wetland (tidal mudflat and saltern) | Continuous |
| Distance_18 | Proximity distance from each cell to inland wetland | Continuous |
| Distance_19 | Proximity distance from each cell to naturally barren areas (beaches, riverbanks, and rocks) | Continuous |
| Distance_20 | Proximity distance from each cell to artificially barren areas (mining area, playground, and other barrens) | Continuous |
| Distance_21 | Proximity distance from each cell to inland water (rivers and lakes) | Continuous |
| Distance_22 | Proximity distance from each cell to ocean water | Continuous |

## 2.3. Data Preprocessing

To conduct habitat suitability estimation, preprocessing of the collected observation data and environmental variables was required, consisting of a number of steps. We carried out this data preprocessing using Quantum Geographic Information System (QGIS) 3.8.1. We present the steps used to prepare the training and testing datasets in Figure 2. First, we set the resolution to 3000 × 3000 pixels and cropped the collected layers based on the study area, which corresponded to a rectangle on the map. For this, we used the World Geodetic System 1984 (WGS84), which is an Earth-centered, Earth-fixed terrestrial reference system and a geodetic datum. In this system, the entire South Korea region is represented by the latitude and longitude coordinates (125.000, 38.083), (129.583, 38.083), (125.000, 33.166), and (129.583, 33.166). Because the bioclimatic and land cover layers (i.e., the classified land cover in South Korea) are all in a shape (.shp) file format, we converted them into a gridded data (.grd) file format, which consists of scalar values on a regular rectangular grid, either in longitude or latitude space.

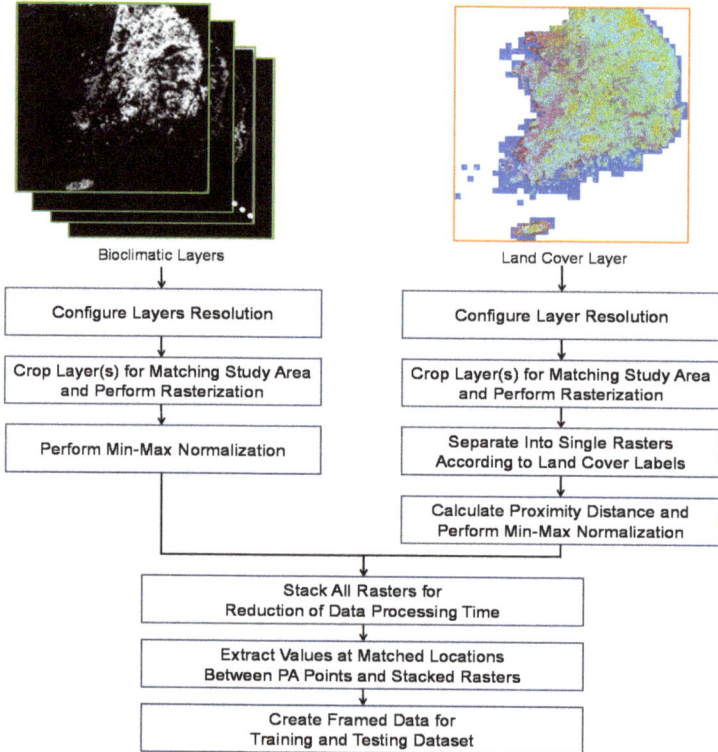

**Figure 2.** The overall preprocessing process for the present study.

We configured each pixel of the raster to have a coverage of approximately 135 m, which produces 900 M grids if we convert the entire South Korean region into a grid space. We then applied min-max normalization to all variables in the cropped bioclimatic layers. For the preprocessing of the land cover layer, we first divided it into multiple layers according to the land cover labels and conducted rasterization. We then calculated the proximity distance based on the separate single rasters and applied min-max normalization to each distance raster. Finally, we stacked all of the rasters to generate a data frame that included labeled presence and absence, and that matched the values of the environmental rasters in the given presence and absence locations.

## 2.4. Stage 1: Habitat Suitability Estimation Based on a DNN

Recently, several DNN-based habitat suitability models have been proposed and have performed well when compared with previous methods [28–30]. Hence, we constructed a DNN model in the first stage to determine the probability of a species' presence or absence, with a higher probability of presence for a species indicating higher habitat suitability. In general, DNNs consist of three types of layers: an input layer, one or more hidden layers, and an output layer. The input layer receives input variables, while the hidden layers are involved in hidden feature processing. The output layer then produces the final prediction. The prediction performance of a DNN model is determined by the configuration of each layer and the model design. For example, it has been shown that the learning rate, the optimizer, regularization, and the activation function significantly affect prediction performance [33]. To obtain the best performance from the DNN model, we carefully determined the optimal hyperparameters using grid search and considering related research [28]. We used the GridSearchCV function of the scikit-learn library [48]. The number of repetitions of grid search was set to infinite, and the number of cross-validations was set to five times. Consequently, when we constructed our DNN model, we used four hidden layers containing 250, 200, 150, and 100 neurons (Figure 3). We set the batch size to 75 and the number of epochs to 5000 with early stopping to optimize model training. To decide the batch size in the training stage, we carefully considered the results of the grid search. For training optimization, we tested three optimizers, including the stochastic gradient descent (SGD), root mean square propagation (RMSprop), adaptive moment estimation (ADAM), and then selected the ADAM as the best optimizer. In addition, we utilized the he-normal (HE) initialization to sort initial weights for individual inputs in a neuron model. The activation function controls the non-linearity of individual neurons. We tested five popular activation functions: linear, soft-max, rectified linear unit (ReLU), tangent, and sigmoid. Through the grid search, we selected ReLU as the activation function of our training model. The learning rate is a hyper-parameter that controls how much we are adjusting the weights of our network with respect to the loss gradient. If the learning rate is set too small, it might take a long time to converge on the performance goal. On the contrary, if the learning rate is set too large, the average loss will increase. To obtain the optimal learning rate, we performed a grid search with ADAM as the optimizer and ReLU as the activation function. Our training model was able to achieve optimal learning efficiency when the learning rate was 0.001.

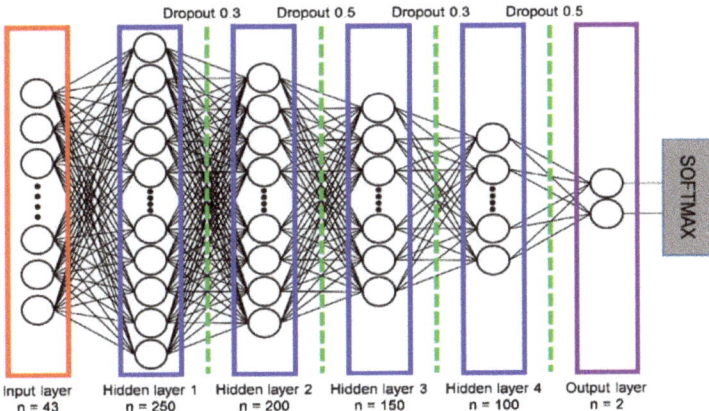

**Figure 3.** Construction of the DNN model used in Stage 1 of the present study.

The finished DNN model was to generate the probability of both presence and absence for a species. The more suitable an area was as habitat for a particular species, the closer the probability of that species' presence was to 1, while the probability of absence followed the opposite trend.

## 2.5. Stage 2: Ensemble-Based Habitat Suitability Estimation

Ensembles of machine-learning techniques have been widely adopted to solve various prediction problems in past research [43–47]. Compared to one machine-learning model, ensembles can improve prediction performance by combining several models. In the field of ecological modeling, ensemble models are widely known as a useful approach for the construction of potential habitat estimation models [43–47]. Therefore, in the second stage, we developed an ensemble-based habitat suitability estimation model using the BIOMOD2 package [49] for R programming. We present the overall construction process for our ensemble model in Stage 2 in Figure 4.

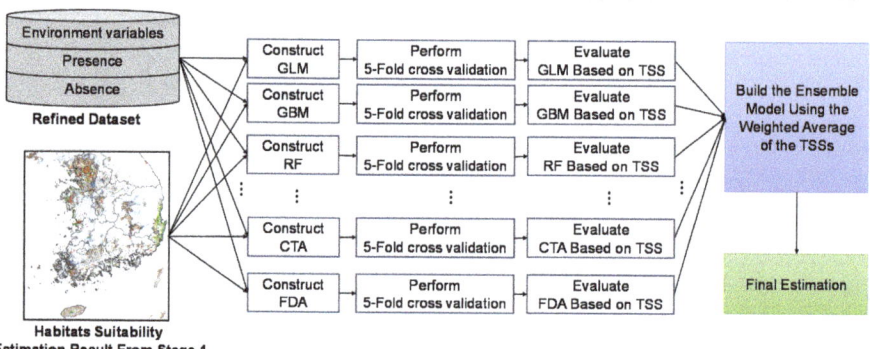

**Figure 4.** Construction of the ensemble model employed in Stage 2 of the present study.

According to the authors of [28,50], a low number of observations (i.e., n < 100) can degrade the estimation performance of a habitat model because using very few observations in model construction leads to overfitting and bias [8–10]. To solve this issue, we used 41 environmental layers and the results of habitat suitability from the DNN in Stage 1 as input variables for the ensemble model in Stage 2. This modeling method is known as stacking and can effectively avoid the possibility of overfitting and bias [51]. We built our ensemble model by combining GLM, GBM, CTA, SNN, FDA, MARS, RF, the SRE model, and MAXENT and used a weighted-average algorithm, which returns a weighted value for each model based on selected evaluation scores. Therefore, an accurate estimation model will have a relatively high weighted value when it combines all of the models. Because the TSS has been proven to be a reliable evaluation metric when measuring and assessing the performance of habitat models [47], we used the TSS to calculate the weighted value. Equation (1) was used to calculate the final estimation using the weighted average value for each model, in which $i$ and $j$ represent the class label for presence and absence and the number of models, respectively, $\hat{y}$ indicates the estimated class label, and $p_{ij}$ is the calculated probability of the $j$th model. In addition, $w_j$ is the weighted value of the $j$th model, which was calculated using Equation (2). We evaluated each model using five-fold cross-validation and obtained the final TSS value as the average of the TSSs generated by the individual models.

$$\hat{y} = \underset{i}{argmax} \sum_{j=1}^{m} w_j * p_{ij} \qquad (1)$$

$$w_j = \frac{TSS_j - Min(TSS)}{Max(TSS) - Min(TSS)} \qquad (2)$$

## 3. Results and Discussion

We evaluated our proposed model and compared its performance with other approaches to habitat suitability modeling. We first explained the evaluation metrics used to assess the quality of habitat suitability estimation and then evaluated the performance of our proposed model and other commonly used models using these metrics. We also visualized the results for habitat suitability analysis using map overlays.

### 3.1. Evaluation Metrics

To evaluate the performance of our model, we used five metrics: sensitivity, specificity, the kappa statistic, AUC, and TSS. These have all been regularly used to assess habitat modeling performance in ecology [47]. The percentage of correctly predicted sites was excluded as a measure of prediction accuracy for the proposed model because, even though it is simple to calculate, its usefulness is severely limited for rare species [19]. To evaluate the estimation results, we used a confusion matrix in which $a$, $b$, $c$, and $d$ indicate true positive, false positive, false negative, and true negative, respectively (Table 4). For instance, when the ground truth is the presence and the prediction result from the proposed model is also the presence, then we counted it as a true positive. Sensitivity, specificity, the kappa statistic, and TSS were calculated using Equations (3)–(6), respectively, based on this confusion matrix.

$$Sensitivity = \frac{a}{a+c} \quad (3)$$

$$Specificity = \frac{d}{b+d} \quad (4)$$

$$Kappa\ statistic = \frac{\left(\frac{a+d}{n}\right) - \frac{(a+c)(a+b) + (b+d)(c+d)}{n^2}}{1 - \frac{(a+c)(a+b) + (b+d)(c+d)}{n^2}} \quad (5)$$

$$TSS = sensitivity + specificity - 1 \quad (6)$$

**Table 4.** Confusion matrix for the evaluation of our presence–absence model.

| Predicted | Observed | |
|---|---|---|
| | Presence | Absence |
| Presence | a | b |
| Absence | b | d |

Sensitivity represents the probability that a model will correctly predict the presence of a species, while specificity measures the probability of a model accurately predicting the absence of a species. TSS normalizes overall accuracy [47,50]. AUC is widely used to assess the accuracy of habitat suitability models because it is easy to interpret, thus allowing comparison between models. ROC curves are frequently used as a single threshold-independent measure for model performance. In previous studies designed to predict habitat suitability [52–54], models with an AUC greater than 0.8 were considered valid as predictive models. The kappa statistic is also a common evaluation metric used for habitat suitability estimation models, but it has been criticized for being heavily dependent on prevalence. TSS, on the other hand, avoids this problem while offering the advantages of the kappa statistic. In general, most ecological modeling research uses sensitivity, specificity, the kappa statistic, and TSS together to analyze the performance of habitat suitability models [51–55]. Thus, we used these five metrics together to compare their weaknesses, strengths, and commonalities.

### 3.2. Performance Evaluation

We described the comparison results for the habitat suitability models using the five metrics discussed in Section 3.1. We compared as many estimation models as possible, including GLM, GBM, CTA, SNN, FDA, MARS, RF, SRE, DNN, ensemble models not including DNN (EMED), and our proposed approach. As mentioned above, EMED has demonstrated satisfactory performance in the previous studies. We constructed the EMED model in the present study using GLM, GBM, CTA, SNN, FDA, MARS, RF, and SRE. All of these models were trained and tested using the BIOMOD2 package in

R and were verified using five-fold cross-validation. We used 80% of the species observation as the training set and 20% as the test set. We present the selected parameters and training strategies for each model in the following (Table 5).

Table 5. Selected parameters and training strategies for the estimation models.

| Estimation Model | Selected Parameters and Training Strategies |
| --- | --- |
| GLM | Quadratic-type regression |
|  | Akaike information criterion (AIC) for environmental layer selection |
| GBM | Bernoulli distribution, 2500 trees, 7 depths, 5 terminal nodes, 0.001 learning rate |
| CTA | Categorical classification, Default tree parameter (auto-optimized by BIOMOD2) |
| SNN | Single hidden layer, Auto-optimized neuron size, 200 iterations |
| FDA | Mars method |
| MARS | Simple pricewise linear, 0.001 threshold, Backward pruning method |
| RF | Maximum of 500 trees, Default number of variables at each split (auto-optimized by BIOMOD2), 5 nodes |
| SRE | 0.025 quantile for environmental variable selection |
| MAXENT | Maximum of 200 iterations, Linear and quadratic variables |
|  | Default parameters for threshold and hinge (auto-optimized by BIOMOD2) |
| EMED | Assigning weights using TSS evaluation, Weighted average-based model assembly, 0.7 for the ensemble threshold, Committee averaging |

We calculated the estimation performance of various models for target species using five metrics and present their averages in Table 6. Detailed experimental results, including sensitivity, specificity, AUC, kappa statistic, and TSS, can be found in the Supplementary Materials (Tables S1–S5). To objectively assess the estimation results, we show the evaluation criteria of AUC, kappa statistic, and TSS in Table 7. We can observe that our proposed model showed the best performance, while the DNN exhibited weak performance because model training was insufficient due to the lack of training data. Likewise, the SRE model showed a poor performance for the prediction (AUC < 0.6). Even though the SRE model is intuitive and fast, it does not fully reflect the interactions between environmental conditions and species distributions in modeling. All other models except the DNN and SRE models performed reasonably well in terms of predicting the presence of a species. In contrast, in terms of specificity, DNN was the best performing model. The AUC has long been regarded as the standard metric for assessing the performance of habitat suitability models. In most cases, TSEM demonstrated the best estimation performance, while EMED also generated high AUC values, with an average of 0.972. This demonstrates that the two-stage based ensemble approach can improve estimation performance. While TSEM performed best for kappa statistic and TSS, SRE was the worst-performing model. For *Hyla japonica*, EMED yielded a higher TSS (EMED = 0.786 and TSEM = 0.783) than TSEM because the DNN model in the first stage produced very poor estimation results. However, for all other species, our model outperformed the other models. In summary, based on these comparisons, clearly TSEM is more suitable for deriving ecological insights related to habitat suitability estimation.

To confirm whether the estimation results of our model are valid, we selected *Rana huanrenensis* as a visualization case. The visualizations of the results for other target species can be found in the Supplementary Materials (Figures S1–S10). *Rana huanrenensis*, also known as the Korean stream brown frog, lives mainly in Korea and Japan, and its habitats are identified as wetland, forest, and grassland (Table 2). Due to the low number of confirmed populations of this species, it could be listed as vulnerable (VN) under the IUCN Red List criterion, but is listed as least concern (LC) based on the assumption of widespread occurrence, especially in Korea. The *Rana huanrenensis* lives in valleys in high montane regions, above 500 m in elevation [56]. This species is mainly observed from March to April, which is very closely related to the breeding season of this species. *Rana huanrenensis* breeds in slow-moving montane streams and rivers, and their eggs are laid in moderately small masses that are attached to submerged rocks [57]. Indeed, the distribution of observation data for *Rana huanrenensis* fits well with their habitat characteristics. We marked the blue points as a training set and yellow points as a test set. The areas marked in black represent the entire coniferous forest, mixed forest, and broad-leaved forest (Figure 5). In Figure 6, green indicates suitable species habitats and blue indicates uninhabitable areas. In evaluating the habitat suitability estimation of *Rana huanrenensis*,

TSEM showed a better estimation performance (TSS = 0.949) than other estimation models. Although EMED and RF performed slightly worse than TSEM (TSS of EMED = 0.819 and TSS of RF = 0.822), these models showed excellent estimation performance based on evaluation criteria. The distribution of *Rana huanrenensis* habitat estimated by TSEM is very similar to the mountainous terrain of South Korea. The SNN, MARS, GBM, EMED, and RF also showed the distribution of habitat estimations similar to mountainous terrain. However, we confirmed that the MARS, GBM, EMED, and RF estimated that the *Rana huanrenensis* was suitable for habitation in some regions, including residential, industrial, and commercial areas. The TSEM estimated that the areas of mixed forests, coniferous forests, and broad-leaved forests were relatively more suitable as the *Rana huanrenensis* habitat. Even though the TSEM was trained and tested on crowdsourced datasets of the *Rana huanrenensis*, the estimation results matched well with their actual habitats because most of the observation data for the *Rana huanrenensis* were near the main habitats such as wetland, montane streams, and forest. We demonstrated that the habitat suitability results estimated by the TSEM are well-matched when compared with the existing studies of the *Rana huanrenensis* habitats [56,57].

Table 6. Performance comparison of estimation models.

| Estimation Model | Evaluation Metrics (Avg.) | | | | |
|---|---|---|---|---|---|
| | Sensitivity | Specificity | AUC | Kappa Statistic | TSS |
| GLM | 0.855 | 0.833 | 0.888 | 0.662 | 0.689 |
| GBM | 0.879 | 0.906 | 0.950 | 0.780 | 0.785 |
| CTA | 0.854 | 0.884 | 0.901 | 0.724 | 0.738 |
| SNN | 0.857 | 0.880 | 0.920 | 0.723 | 0.738 |
| FDA | 0.865 | 0.892 | 0.938 | 0.750 | 0.758 |
| MARS | 0.859 | 0.871 | 0.927 | 0.711 | 0.731 |
| RF | 0.885 | 0.947 | 0.967 | 0.838 | 0.832 |
| SRE | 0.567 | 0.904 | 0.735 | 0.499 | 0.470 |
| MAXENT | 0.781 | 0.816 | 0.862 | 0.665 | 0.658 |
| DNN (Stage 1) | 0.757 | **0.957** | 0.886 | 0.753 | 0.759 |
| EMED | 0.905 | 0.911 | 0.972 | 0.862 | 0.816 |
| TSEM (Stage 2) | **0.966** | 0.920 | **0.983** | **0.887** | **0.886** |

The highest values are in bold.

Table 7. Evaluation criteria of AUC, kappa statistic and TSS.

| | AUC | Kappa Statistic | TSS |
|---|---|---|---|
| Excellent | ≥ 0.9 | ≥ 0.9 | ≥ 0.8 |
| Good | 0.8 – 0.9 | 0.8 – 0.9 | 0.6 – 0.8 |
| Fair | 0.6 – 0.8 | 0.7 – 0.8 | 0.4 – 0.6 |
| Poor or no predictive ability | ≤ 0.6 | ≤ 0.6 | ≤ 0.4 |

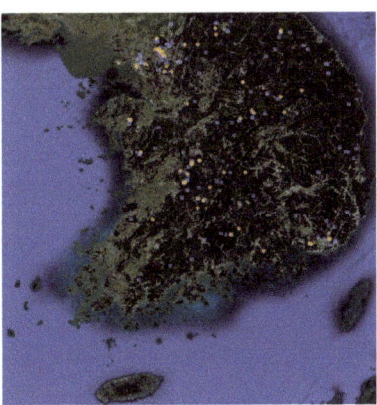

Figure 5. Observations of *Rana huanrenensis* in South Korea.

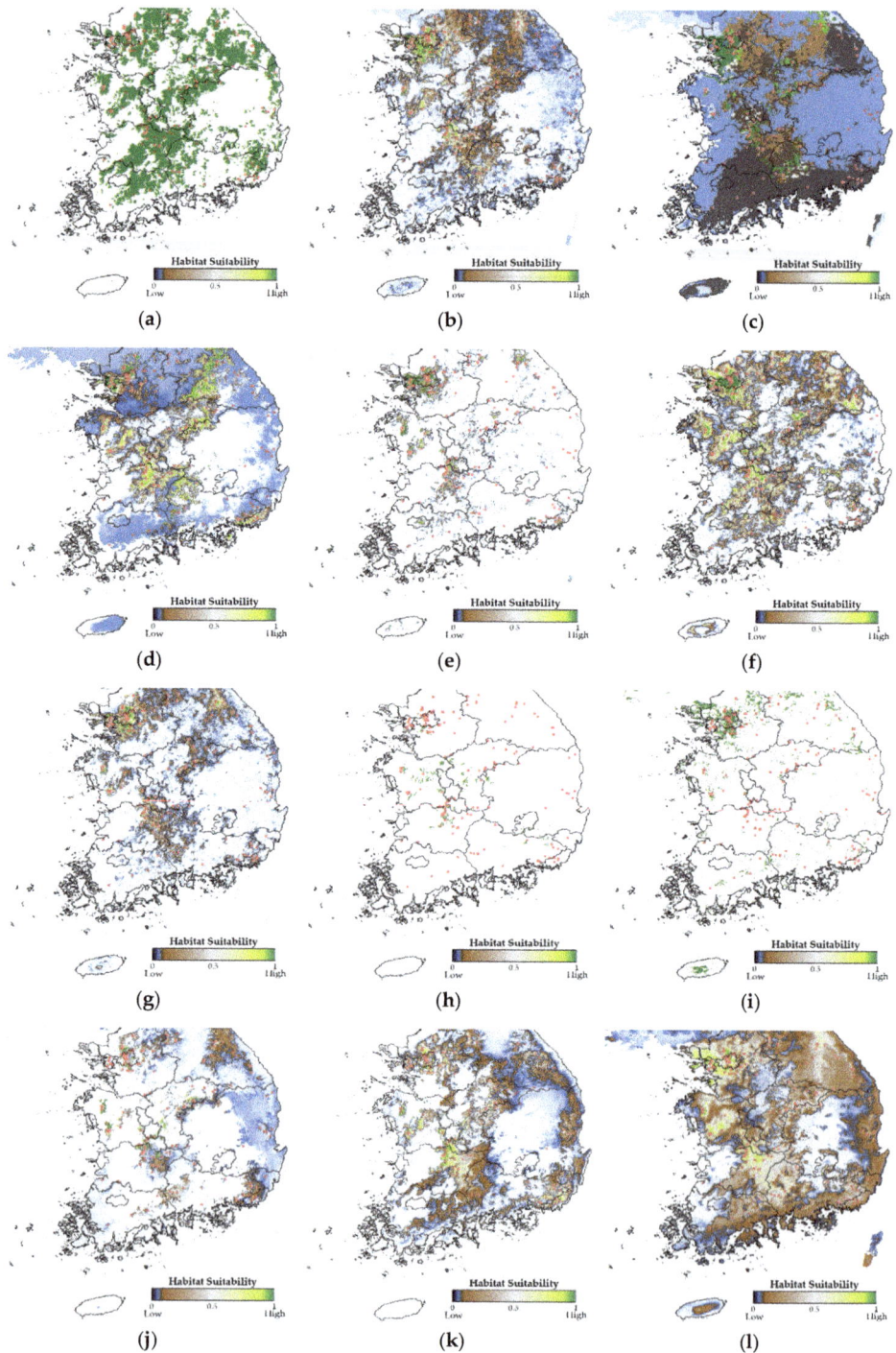

**Figure 6.** Habitat suitability visualization of Rana huanrenensis–(**a**) GLM; (**b**) GBM; (**c**) CTA; (**d**) SNN; (**e**) FDA; (**f**) MARS; (**g**) RF; (**h**) SRE; (**i**) MAXENT; (**j**) DNN; (**k**) EMED; (**l**) TSEM.

## 3.3. Statistical Evaluation

To demonstrate the superiority of our proposed method, we performed Wilcoxon signed-rank and Friedman tests [58,59]. The Wilcoxon signed-rank test is a non-parametric statistical hypothesis test used to compare two related samples [58]. It can be used as an alternative to the *t*-test when one or more of the samples are not normally distributed. It establishes a null hypothesis to determine whether there is a significant difference between the two samples. If the *p*-value is less than a certain significance level, the null hypothesis is rejected, and the two samples are assumed to be significantly different. In the statistical hypothesis testing, the *p*-value is the probability of obtaining test results at least as extreme as the results actually observed during the test, assuming that the null hypothesis is correct. The Friedman test is a multiple comparison test that aims to identify significant differences between three or more samples [59]. It first ranks each row (block) together, and then considers the values of the ranks by column. The data are organized into a matrix with B rows (blocks) and T columns (treatments) with a single operation in each cell of the matrix. To verify the results of these tests, we used AUC, the kappa statistic, and TSS for each machine-learning method. The results of the Wilcoxon signed-rank test, for which the significance level was set to 0.05, and the Friedman test are listed in Table 8. Based on these tests, our proposed method exhibited significantly better prediction performance than the other machine-learning methods because the *p*-value was below the significance level in most cases.

Table 8. The results of the Wilcoxon signed-rank and Friedman tests for our proposed method.

| Estimation Models | Wilcoxon Signed-Rank Test (*p*-Value < 0.05) | | | Friedman Test | | |
|---|---|---|---|---|---|---|
| | AUC | Kappa | TSS | AUC | Kappa | TSS |
| GLM | 0.00098 | 0.00098 | 0.00098 | | | |
| GBM | 0.00379 | 0.00098 | 0.00384 | | | |
| CTA | 0.00384 | 0.00098 | 0.00384 | | | |
| SNN | 0.00382 | 0.00098 | 0.00098 | | | |
| FDA | 0.00098 | 0.00098 | 0.00382 | | | |
| MARS | 0.00381 | 0.00382 | 0.00382 | $2.236 \times 10^{-17}$ | $1.446 \times 10^{-18}$ | $2.236 \times 10^{-17}$ |
| RF | 0.00379 | 0.00384 | 0.00384 | | | |
| SRE | 0.00379 | 0.00098 | 0.89390 | | | |
| MAXENT | 0.00098 | 0.00098 | 0.00098 | | | |
| DNN (Stage 1) | 0.00098 | 0.00098 | 0.00098 | | | |
| EMED | 0.00368 | 0.00382 | 0.00195 | | | |

AUC, area under the curve; Kappa, kappa statistic; TSS, true skill statistic.

## 4. Conclusions

In this study, we focused on a two-stage modeling scheme that can be applied to habitat suitability estimation for various species. First, we investigated and selected 11 species that are present in South Korea and regarded as targets for species conservation research. To obtain a sufficient number of observations for the target species, we extracted observational data for these species from several crowdsourced databases and added them to our database. Since spatial bias is a well-known problem in crowdsourced data, we tried to alleviate this bias by using three global datasets and one domestic dataset. In particular, the domestic dataset, Naturing database, contains data of target species observed quite evenly across South Korea. We also employed 41 environmental layers that included information on the global climate and the land cover of South Korea as input variables. To effectively estimate habitat suitability, we used a DNN model and an ensemble of habitat suitability estimation models in the first and second stages, respectively. To evaluate the effectiveness of the proposed model, we compared it with previously employed models and visualized these results using a suitability map overlay. The experimental results demonstrate that the proposed model has significant potential for use in estimating habitat suitability.

For model training and testing, we used crowdsourced datasets. This implies that there could be some bias in the observation data and the estimation results, as mentioned above. Hence, even though our model showed better performance than other models, estimation results might indicate where the observation was made, in other words, the species can be observed. To the best of our knowledge, it is an inevitable limitation of prediction models based on crowdsourced data. In future work, to ensure the reliability of our habitat suitability model, we plan to develop a method that can alleviate the potential biases of crowdsourcing datasets.

**Supplementary Materials:** The following are available online at http://www.mdpi.com/2072-4292/12/9/1475/s1, Table S1: Sensitivity comparison; Table S2: Specificity comparison; Table S3: AUC comparison; Table S4: Kappa statistic comparison; Table S5: TSS comparison; Figure S1: Habitat suitability visualization of *Anas platyrhynchos*; Figure S2: Habitat suitability visualization of *Anas zonorhyncha*; Figure S3: Habitat suitability visualization of *Ardea cinerea*; Figure S4: Habitat suitability visualization of *Cyanopica cyanus*; Figure S5: Habitat suitability visualization of *Hyla japonica*; Figure S6: Habitat suitability visualization of *Hynobius leechii*; Figure S7: Habitat suitability visualization of *Hypsipetes amaurotis*; Figure S8: Habitat suitability visualization of *Passer montanus*; Figure S9: Habitat suitability visualization of *Rana dybowskii*; Figure S10: Habitat suitability visualization of *Streptopelia orientalis*.

**Author Contributions:** Conceptualization, J.R.; methodology, J.R., Y.C., and J.M.; software, J.R. and Y.C.; validation, J.R., Y.C., and J.M.; investigation, Y.C. and J.M.; data curation, J.R. and Y.C.; visualization, Y.C.; writing—original draft preparation, J.R., and J.M.; writing—review and editing, E.H.; supervision, E.H.; and project administration, E.H. All authors have read and agreed to the published version of the manuscript.

**Funding:** This work was supported by Korea Environment Industry & Technology Institute (KEITI) through Public Technology Program based on Environmental Policy, funded by Korea Ministry of Environment (MOE) (2017000210001).

**Conflicts of Interest:** The authors declare no conflicts of interest.

## References

1. Pimm, S.L.; Russell, G.J.; Gittleman, J.L.; Brooks, T.M. The future of biodiversity. *Science* **1995**, *269*, 347–350. [CrossRef] [PubMed]
2. Dirzo, R.; Raven, P.H. Global State of Biodiversity and Loss. *Annu. Rev. Environ. Resour.* **2003**, *28*, 137–167. [CrossRef]
3. Jenkins, M. Prospects for Biodiversity. *Science* **2003**, *302*, 1175–1177. [CrossRef] [PubMed]
4. Corsi, F.; Duprè, E.; Boitani, L. A large-scale model of wolf distribution in Italy for conservation planning. *Conserv. Biol.* **1999**, *13*, 150–159. [CrossRef]
5. Peterson, A.T.; Soberón, J.; Sánchez-Cordero, V. Conservatism of ecological niches in evolutionary time. *Science* **1999**, *285*, 1265–1267. [CrossRef]
6. Franklin, J. Species distribution models in conservation biogeography: Developments and challenges. *Divers. Distrib.* **2013**, *19*, 1217–1223. [CrossRef]
7. Araújo, M.B.; Peterson, A.T. Uses and misuses of bioclimatic envelope modeling. *Ecology* **2012**, *93*, 1527–1539. [CrossRef]
8. Heikkinen, R.K.; Luoto, M.; Araújo, M.B.; Virkkala, R.; Thuiller, W.; Sykes, M.T. Methods and uncertainties in bioclimatic envelope modelling under climate change. *Prog. Phys. Geogr.* **2006**, *30*, 751–777. [CrossRef]
9. Thuiller, W.; Lafourcade, B.; Araujo, M. Presentation manual for BIOMOD. *Ecography* **2010**, *32*, 369–373. [CrossRef]
10. Elith, J.; Graham, C.H.; Anderson, R.P.; Dudík, M.; Ferrier, S.; Guisan, A.; Hijmans, R.J.; Huettmann, F.; Leathwick, J.R.; Lehmann, A.; et al. Novel methods improve prediction of species' distributions from occurrence data. *Ecography* **2006**, *29*, 129–151. [CrossRef]
11. Leathwick, J.R.; Elith, J.; Hastie, T. Comparative performance of generalized additive models and multivariate adaptive regression splines for statistical modelling of species distributions. *Ecol. Modell.* **2006**, *199*, 188–196. [CrossRef]
12. Zuur, A.F.; Ieno, E.N.; Elphick, C.S. A protocol for data exploration to avoid common statistical problems. *Methods Ecol. Evol.* **2010**, *1*, 3–14. [CrossRef]
13. Friedman, J.H. Multivariate adaptive regression splines. *Ann. Stat.* **1991**, 1–67. [CrossRef]
14. Hastie, T.; Tibshirani, R.; Buja, A. Flexible discriminant analysis by optimal scoring. *J. Am. Stat. Assoc.* **1994**, *89*, 1255–1270. [CrossRef]

15. Phillips, N.D.; Reid, N.; Thys, T.; Harrod, C.; Payne, N.L.; Morgan, C.A.; White, H.J.; Porter, S.; Houghton, J.D.R. Applying species distribution modelling to a data poor, pelagic fish complex: The ocean sunfishes. *J. Biogeogr.* **2017**, *44*, 2176–2187. [CrossRef]
16. Reiss, H.; Cunze, H.; König, K.; Neumann, K.; Kröncke, I. Species distribution modelling of marine benthos: A North Sea case study. *Mar. Ecol. Prog. Ser.* **2011**, *442*, 71–86. [CrossRef]
17. Guisan, A.; Thuiller, W. Predicting species distribution: Offering more than simple habitat models. *Ecol. Lett.* **2005**, *8*, 993–1009. [CrossRef]
18. Phillips, S.J.; Anderson, R.P.; Schapire, R.E. Maximum entropy modeling of species geographic distributions. *Ecol. Model.* **2006**, *190*, 231–259. [CrossRef]
19. Heikkinen, R.K.; Luoto, M.; Kuussaari, M.; Toivonen, T. Modelling the spatial distribution of a threatened butterfly: Impacts of scale and statistical technique. *Landsc. Urban Plan.* **2007**, *79*, 347–357. [CrossRef]
20. Thomaes, A.; Kervyn, T.; Maes, D. Applying species distribution modelling for the conservation of the threatened saproxylic Stag Beetle (*Lucanus cervus*). *Biol. Conserv.* **2008**, *141*, 1400–1410. [CrossRef]
21. De'ath, G.; Fabricius, K.E. Classification and regression trees: A powerful yet simple technique for ecological data analysis. *Ecol. Lett.* **2000**, *81*, 3178–3192. [CrossRef]
22. Breiman, L.; Friedman, J.; Stone, C.J.; Olshen, R.A. *Classification and regression trees*; CRC press: Boca Raton, FL, USA, 1984. [CrossRef]
23. De'Ath, G. Boosted trees for ecological modeling and prediction. *Ecol. Lett.* **2007**, *88*, 243–251. [CrossRef]
24. D'heygere, T.; Goethals, P.L.; De Pauw, N. Genetic algorithms for optimisation of predictive ecosystems models based on decision trees and neural networks. *Ecol. Model.* **2006**, *195*, 20–29. [CrossRef]
25. Fukuda, S.; De Baets, B.; Waegeman, W.; Verwaeren, J.; Mouton, A.M. Habitat prediction and knowledge extraction for spawning European grayling (Thymallus thymallus L.) using a broad range of species distribution models. *Environ. Model. Softw.* **2013**, *47*, 1–6. [CrossRef]
26. Breiman, L. Random forests. *Mach. Learn.* **2001**, *45*, 5–32. [CrossRef]
27. Cutler, D.R.; Edwards, T.C., Jr.; Beard, K.H.; Cutler, A.; Hess, K.T.; Gibson, J.; Lawler, J.J. Random forests for classification in ecology. *Ecology* **2007**, *88*, 2783–2792. [CrossRef]
28. Rademaker, M.; Hogeweg, L.; Vos, R. Modelling the niches of wild and domesticated Ungulate species using deep learning. *BioRxiv* **2019**, 744441. [CrossRef]
29. Botella, C.; Joly, A.; Bonnet, P.; Monestiez, P.; Munoz, F. A deep learning approach to species distribution modelling. In *Multimedia Tools and Applications for Environmental & Biodiversity Informatics*; Springer: Cham, Switzerland, 2018; pp. 169–199. [CrossRef]
30. Hulleman, W.; Vos, R.A. Modeling abiotic niches of crops and wild ancestors using deep learning: A generalized approach. *BioRxiv* **2019**, 826347. [CrossRef]
31. Rew, J.; Park, S.; Cho, Y.; Jung, S.; Hwang, E. Animal movement prediction based on predictive recurrent neural network. *Sensors* **2019**, *19*, 4411. [CrossRef]
32. Moon, J.; Park, S.; Rho, S.; Hwang, E. A comparative analysis of artificial neural network architectures for building energy consumption forecasting. *Int. J. Distrib. Sens. Netw.* **2019**, *15*. [CrossRef]
33. Moon, J.; Kim, Y.; Son, M.; Hwang, E. Hybrid Short-Term Load Forecasting Scheme Using Random Forest and Multilayer Perceptron. *Energies* **2018**, *11*, 3283. [CrossRef]
34. Kim, H.; Kim, H.; Hwang, E. Real-time facial feature extraction scheme using cascaded networks. In Proceedings of the IEEE International Conference on Big Data and Smart Computing (BigComp), Kyoto, Japan, 27 February–2 March 2019; pp. 1–7. [CrossRef]
35. Kim, J.; Moon, J.; Hwang, E.; Kang, P. Recurrent inception convolution neural network for multi short-term load forecasting. *Energy Build.* **2019**, *194*, 328–341. [CrossRef]
36. GBIF Homepage. Available online: https://www.gbif.org (accessed on 5 March 2020).
37. VertNet Homepage. Available online: http://vertnet.org (accessed on 5 March 2020).
38. BISON Homepage. Available online: https://bison.usgs.gov (accessed on 5 March 2020).
39. Naturing Homepage. Available online: https://www.naturing.net (accessed on 5 March 2020).
40. Worldclim Homepage. Available online: https://www.worldclim.org (accessed on 5 March 2020).
41. Land Cover of South Korea Homepage. Available online: http://www.neins.go.kr/gis/mnu01/doc03a.asp (accessed on 5 March 2020).
42. Ferraz, K.M.P.M.d.B.; Ferraz, S.F.d.B.; Paula, R.C.d.; Beisiegel, B.; Breitenmoser, C. Species distribution modeling for conservation purposes. *Nat. Conserv.* **2012**, *10*, 214–220. [CrossRef]

43. Wan, J.; Wang, C.; Han, S.; Yu, J. Planning the priority protected areas of endangered orchid species in northeastern China. *Biodivers. Conserv.* **2014**, *23*, 1395–1409. [CrossRef]
44. Buisson, L.; Thuiller, W.; Casajus, N.; Lek, S.; Grenouillet, G. Uncertainty in ensemble forecasting of species distribution. *Glob. Change Biol.* **2010**, *16*, 1145–1157. [CrossRef]
45. Forester, B.R.; DeChaine, E.G.; Bunn, A.G. Integrating ensemble species distribution modelling and statistical phylogeography to inform projections of climate change impacts on species distributions. *Divers. Distrib.* **2013**, *19*, 1480–1495. [CrossRef]
46. Ranjitkar, S.; Xu, J.; Shrestha, K.K.; Kindt, R. Ensemble forecast of climate suitability for the Trans-Himalayan Nyctaginaceae species. *Ecol. Model.* **2014**, *282*, 18–24. [CrossRef]
47. Allouche, O.; Tsoar, A.; Kadmon, R. Assessing the accuracy of species distribution models: Prevalence, kappa and the true skill statistic (TSS). *J. Appl. Ecol.* **2006**, *43*, 1223–1232. [CrossRef]
48. Sckit-learn Homepage. Available online: https://https://scikit-learn.org/stable (accessed on 5 March 2020).
49. Thuiller, W.; Lafourcade, B.; Engler, R.; Araújo, M.B. BIOMOD–a platform for ensemble forecasting of species distributions. *Ecography* **2009**, *32*, 369–373. [CrossRef]
50. Miller, J. Species distribution modeling. *Geogr. Compass* **2010**, *4*, 490–509. [CrossRef]
51. Moon, J.; Jung, S.; Rew, J.; Rho, S.; Hwang, E. Combination of short-term load forecasting models based on a stacking ensemble approach. *Energy Build.* **2020**, *216*, 109921. [CrossRef]
52. Thuiller, W.; Araújo, M.B.; Lavorel, S. Generalized models vs. classification tree analysis: Predicting spatial distributions of plant species at different scales. *J. Veg. Sci.* **2003**, *14*, 669–680. [CrossRef]
53. Giannini, T.C.; Chapman, D.S.; Saraiva, A.M.; Alves-dos-Santos, I.; Biesmeijer, J.C. Improving species distribution models using biotic interactions: A case study of parasites, pollinators and plants. *Ecography* **2013**, *36*, 649–656. [CrossRef]
54. Barbet-Massin, M.; Jetz, W. A 40-year, continent-wide, multispecies assessment of relevant climate predictors for species distribution modelling. *Divers. Distrib.* **2014**, *20*, 1285–1295. [CrossRef]
55. Duque-Lazo, J.; van Gils, H.; Groen, T.A.; Navarro-Cerrillo, R.M. Transferability of species distribution models: The case of Phytophthora cinnamomi in Southwest Spain and Southwest Australia. *Ecol. Modell.* **2016**, *320*, 62–70. [CrossRef]
56. Liu, M.; Liu, Y.; Li, J. Reproductive habits of Rana huanrenensis. *Sichuan J. Zool.* **2004**, *23*, 183–184.
57. Yang, S.Y.; Kim, J.B.; Min, M.S.; Suh, J.H.; Kang, Y.J.; Matsui, M.; Fei, L. First record of a brown frog Rana huanrenensis (Family Ranidae) from Korea. *Korean J. Biol. Sci.* **2000**, *4*, 45–50. [CrossRef]
58. Park, S.; Moon, J.; Jung, S.; Rho, S.; Baik, S.W.; Hwang, E. A two-stage industrial load forecasting scheme for day-ahead combined cooling, heating and power scheduling. *Energies* **2020**, *13*, 443. [CrossRef]
59. Moon, J.; Kim, J.; Kang, P.; Hwang, E. Solving the Cold-Start Problem in Short-Term Load Forecasting Using Tree-Based Methods. *Energies* **2020**, *13*, 886. [CrossRef]

© 2020 by the authors. Licensee MDPI, Basel, Switzerland. This article is an open access article distributed under the terms and conditions of the Creative Commons Attribution (CC BY) license (http://creativecommons.org/licenses/by/4.0/).

MDPI  
St. Alban-Anlage 66  
4052 Basel  
Switzerland  
Tel. +41 61 683 77 34  
Fax +41 61 302 89 18  
www.mdpi.com

*Remote Sensing* Editorial Office  
E-mail: remotesensing@mdpi.com  
www.mdpi.com/journal/remotesensing